Springer Water

The book series Springer Water comprises a broad portfolio of multi- and interdisciplinary scientific books, aiming at researchers, students, and everyone interested in water-related science. The series includes peer-reviewed monographs, edited volumes, textbooks, and conference proceedings. Its volumes combine all kinds of water-related research areas, such as: the movement, distribution and quality of freshwater; water resources; the quality and pollution of water and its influence on health; the water industry including drinking water, wastewater, and desalination services and technologies; water history; as well as water management and the governmental, political, developmental, and ethical aspects of water.

Naef A. A. Qasem · Muhammad M. Generous ·
Bilal A. Qureshi · Syed M. Zubair

Thermodynamic and Thermophysical Properties of Saline Water

Models, Correlations and Data for Desalination and Relevant Applications

Springer

Naef A. A. Qasem
Department of Aerospace Engineering
King Fahd University of Petroleum
and Minerals (KFUPM)
Dhahran, Saudi Arabia

Muhammad M. Generous
Department of Mechanical Engineering
King Fahd University of Petroleum
and Minerals (KFUPM)
Dhahran, Saudi Arabia

Bilal A. Qureshi
Department of Mechanical Engineering
King Fahd University of Petroleum
and Minerals (KFUPM)
Dhahran, Saudi Arabia

Syed M. Zubair
Department of Mechanical Engineering
King Fahd University of Petroleum
and Minerals (KFUPM)
Dhahran, Saudi Arabia

ISSN 2364-6934 ISSN 2364-8198 (electronic)
Springer Water
ISBN 978-3-031-35192-1 ISBN 978-3-031-35193-8 (eBook)
https://doi.org/10.1007/978-3-031-35193-8

This Springer imprint is published by the registered company Springer Nature Switzerland AG
The registered company address is: Gewerbestrasse 11, 6330 Cham, Switzerland

Preface

All the analytical analysis, thermodynamic analysis, and design models of the desalination technologies depend on saline water properties. As scientists and researchers working on different desalination technologies, we found it difficult to find all saline water properties in one source, including multicomponent and binary salty solutions, under different conditions (salinity, temperature, and pressure). We have published the thermodynamic and thermophysical properties as two-part articles in *Arabian Journal for Science and Engineering* (https://doi.org/10.1007/s13369-020-05019-y and https://doi.org/10.1007/s13369-020-05020-5). However, the journal articles are limited and cannot show all details. Furthermore, these article reviewers advised us to expand this subject to a book to provide optimal benefit to the literature (faculty, researchers, engineers, and designers). Therefore, we introduce this book for this purpose.

This book is based on published articles and includes new additional details about the thermodynamic and thermophysical properties of saline water, involving thermodynamic approaches, multicomponent models, and simple correlations and data, comparison between the correlations of properties in figures, recommendation of the most accurate correlations and methods, and the used EES codes. We aim for this book to be a principal source for all interests in desalination and water treatment subjects.

Dhahran, Saudi Arabia

Naef A. A. Qasem
Muhammad M. Generous
Bilal A. Qureshi
Syed M. Zubair

The original version of this book has been revised: An Acknowledgements page has been added to the Front Matter.

Acknowledgements

The authors express their gratitude for the assistance provided by King Fahd University of Petroleum and Minerals (KFUPM) in Dhahran, Saudi Arabia, during the book's writing phase. Specifically, Syed Zubair wishes to extend appreciation for the support received through the DUP20101 project.

Contents

About the Authors

Naef A. A. Qasem is currently (2023) Assistant Professor of thermofluid sciences in the Department of Aerospace Engineering at King Fahd University of Petroleum and Minerals (KFUPM), Dhahran, Saudi Arabia. Before that (2017–2019), he was Postdoctoral Fellow in the Department of Mechanical Engineering at KFUPM. He earned his Ph.D. in Mechanical Engineering from KFUPM in 2017. His current research focuses on desalination, renewable energy, clean propulsion, and other fields related to thermofluid sciences. He published over 80 journal articles (the majority are in Q1 journals with h-index > 24), seven issued patents, and some filed. He served as Reviewer for many prestigious journals. He currently serves as "Young Editor" in the prestigious journal "*Applied Energy*, Elsevier, Q1, Top 10%". Recently, he has been listed in the Top 2% Scientists Worldwide for the year 2021, published by Elsevier and compiled by Stanford University Researchers.

Muhammad M. Generous is currently (2023) a Ph.D. candidate in the Mechanical Engineering Department at Khalifa University, UAE. He earned an M.S. degree in Mechanical Engineering from King Fahd University of Petroleum and Minerals (KFUPM), Dhahran, Saudi Arabia. He published about ten journal articles (the majority are in Q1 journals with h-index > 7) related to desalination and heat transfer.

Bilal A. Qureshi is currently (2023) Assistant Professor in the Department of Mechanical Engineering at King Fahd University of Petroleum and Minerals (KFUPM), Dhahran, Saudi Arabia. Before that (2014–2017), he was Postdoctoral Fellow in the Department of Mechanical Engineering at KFUPM. He earned his Ph.D. in Mechanical Engineering from KFUPM in 2014. His current research focuses on desalination and renewable energy. He published over 40 journal articles (the majority are in Q1 journals with h-index > 22) related to desalination, two-phase flow, and heat exchanger topics.

Syed M. Zubair is Distinguished Professor in the Department of Mechanical Engineering at KFUPM, Dhahran, Saudi Arabia. He earned his Ph.D. in Mechanical Engineering from the Georgia Institute of Technology in Atlanta. His research interest involves applied and fundamental areas of energy and desalination systems, refrigeration and air-conditioning systems, and fouling of heat exchangers. He has participated in several research projects, including projects supported by MIT-KFUPM Center for Clean Water and Clean Energy. The heating and dehumidification (HDH) technology he developed with MIT collaborators ended up in a commercial application for treating produced water from oil and gas wells. He has published over 300 research papers (the majority are in Q1 journals with h-index = 60) and was awarded fourteen US patents. Currently, he serves on the editorial board of the *International Journal of Refrigeration* and as Associate Editor of *Nature*—npj Clean Water. He has been listed in the Top 2% Scientists Worldwide for three years, 2019, 2020, and 2021, published by Elsevier and compiled by Stanford University Researchers.

Nomenclature

Abbreviations

ATM	Atmospheric pressure and temperature
BPE	Boiling point elevation
ED	Electrodialysis
EDR	Electrodialysis reversal
FO	Forward osmosis
HDH	Humidification-dehumidification
MD	Membrane distillation
MEE	Multi-effect evaporation
MSE	Mean square error
MSF	Multi-stage flash desalination
MVC	Mechanical vapor compression
ppb	Parts per billion
ppm	Parts per million
ppt	Parts per thousand
RO	Reverse osmosis
STP	Standard temperature and pressure

Symbols

a	Activity
A, B	Constants used in McCleskey multicomponent model
A^{φ}	Modified Debye–Hückel parameter, $kg^{1/2}/mol^{1/2}$
b, s, n	Parameters related to the Wang model for BPE calculations
B_{ij}, $B_{ij}{}^{\phi}$	Pitzer parameter, second virial coefficient, kg/mol
$B_{ij}{}'$	Pitzer binary interaction parameter, kg^2/mol
C	Concentration, mol/m^3
C_{ij}, $C_{ij}{}^{\phi}$	Pitzer parameter, unlike-charged interactions, kg^2/mol^2
Cl	Chlorinity, g/kg
c_o, c_1, c_2, c_3, c_4	Constants
C_P	Specific heat, J/kg·K
C_{Pw}	Partial heat capacity of water in solution, J/kg·K or cal/g·K
$^{\phi}C_P$	Apparent heat capacity, J/kg·K
D	Diffusion coefficient, m^2/s
d_s	Density of salt, kg/m^3
e	Unit electrical charge, 1.602×10^{-19} C
F	Faraday's constant, 96485.3329 A·s/mol
F_1, F_2, F_3	Constants for Pawlowicz method
g	Specific Gibb's energy, kJ/kg
G^{ex}	Excess Gibbs energy, kJ
$\bar{h}$	Partial enthalpy, kJ/kg
h	Enthalpy of saline water, kJ/kg
$^{\phi}h$,$^{\phi}L$	Apparent enthalpy, kJ/kg
I	Ionic strength, mol/kg
k	Thermal conductivity, W/m·K
K_{ATM}	Compressibility factor at atmospheric pressure
K_b	Boltzmann constant, $1.3805 \times 10^{-23} J/K$
K_e	Electrical conductivity, mS/m
K_f	Compressibility factor
L	Latent heat of vaporization, kJ/kg
m	Molality, mol/kg

M	Molarity, mol/L
m_i	Molality of ith ion, mol/kg
M_i	Molar mass of ionic species, g/mol
m_s	Mass of solute, g or kg
M_w	Molar mass of water, g/mol
n	Number of ions
N_A	Avogadro's number, 6.022×10^{23}
P	Pressure, MPa or bar
p	Vapor pressure, MPa or bar
p_w	Vapor pressure of pure water, MPa or bar
R	Universal gas constant, kJ/mol·K
S	Salinity, g/kg
s	Specific entropy, J/kg·K
$\bar{S}$	Partial specific entropy, J/kg·K
$^{\phi}S$	Apparent specific entropy, J/kg·K
T	Temperature, °C or K
u	Ionic mobility, m^2/V·s
v	Molar volume, m^3/mol
V	Volume, m^3
x	Mole fraction
Z	Ionic charge, C

Greek Symbols

Δ	Change
μ	Viscosity of saline water, kg/m·s
α	Thermal expansivity, 1/K or Pitzer parameter, $kg^{1/2}/mol^{1/2}$
β	Thermal compressibility, 1/Pa or Pitzer parameter, kg/mol
γ	Activity coefficient
θ	Pitzer parameter, kg/mol
λ	Equivalent conductivity, S·cm^2/eq
λ_{ij}	Pitzer parameter, uncharged interactions, kg/mol
μ	Chemical potential, kJ/mol
ν	Stoichiometric coefficient
ρ	Density of saline water, kg/m^3
σ	Surface tension, N/m
ϕ	Osmotic coefficient
Φ_{ij}, Φ_{ij}^{ϕ}	Pitzer parameter, like-charged interactions, kJ/mol
Φ_{ij}'	Pitzer parameter, like-charged interactions, kg^2/mol^2

Ψ_{ijk}	Pitzer parameter, ternary interactions, kg^2/mol^2
K	Adiabatic compressibility
ε	Haline contraction coefficient

Subscripts

a, X	Anion
c, M	Cation
e	Electrical
F	Freezing
n, N	Neutral species
p	Isobaric
R	Relative
S	Salt
s	Solute
SW	Seawater
W	Water

Superscripts

o	Infinite dilution

Chapter 1
Introduction

Abstract Saline water properties are essential for desalination and water treatment technologies. These properties could be classified into thermodynamic and thermophysical properties. Thermodynamic properties are saltwater concentration, density, specific heat capacity, latent heat of vaporization, boiling point elevation, vapor pressure, specific enthalpy, specific entropy, and freezing point. Thermophysical properties are viscosity, electrical conductivity, thermal conductivity, osmotic coefficient, activity coefficient, surface tension, and thermal expansivity. This chapter introduces these properties and explains the difference between pure water and saline water. In addition, the estimation of saline water concentration, salinity, molality, and molarity are presented. A roadmap of the book content is exposed.

Keywords Saline water properties · Thermodynamic · Thermophysical · Concentration

1.1 Saline Water and Properties

It is not hidden that water is one of the essential elements for life on Earth and its atmosphere. However, the presence of freshwater in the total available water is less than 3%, with only 0.76% available in the liquid phase (Shiklomanov et al. 2013). There is a significantly increasing demand for freshwater owing to the rapid increase in the human population and industrialization. It was reported that the majority (around 80%) of the population is endangered by water scarcity (Vörösmarty et al. 2010). Thus, all efforts have focused on securing the needed amount of freshwater using industrial processes such as desalination and water treatment.

N. A. A. Qasem et al., *Thermodynamic and Thermophysical Properties of Saline Water*, Springer Water, https://doi.org/10.1007/978-3-031-35193-8_1

Desalination is categorized into thermal- and membrane-based technologies. The thermal-based methods involve multistage flash (MSF) desalination (Lv et al. 2019), multi-effect (ME) evaporation (Park et al. 2005), humidification-dehumidification (HDH) (Narayan et al. 2013; Qasem and Zubair 2019). The membrane-based methods involve reverse osmosis (RO) (Mitra et al. 2017; Qasim et al. 2019), electrodialysis (ED) (Fidaleo and Moresi 2011; Qasem et al. 2018), and membrane distillation (MD) (Khalifa et al. 2017). Each method has certain operating conditions. For instance, the salinity should not go > 40 g/kg for RO desalination (Tsiakis and Papageorgiou 2005), while it is recommended to be < 10 g/kg (brackish water) or > 50 g/kg for ED to minimize the salinity to be feasible for RO (Chehayeb et al. 2017; Kwon et al. 2015). The technologies that depend on thermal energy are valid for all salinities (Lv et al. 2019; Mitra et al. 2013; Narayan et al. 2013).

The saline water properties are important for designing and operating desalination processes. These properties could be classified into thermodynamic and thermophysical. Thermodynamic properties are used to estimate the energy and entropy (1st and 2nd Laws of Thermodynamics) of the process involving saltwater concentration, density, specific heat capacity, latent heat of vaporization, boiling point elevation, vapor pressure, specific enthalpy, specific entropy, and freezing point. Thermophysical properties are also important to describe the physical description of saline water and help calculate fluid flow and electrical fluxes, etc. Such properties include saltwater viscosity, electrical conductivity, thermal conductivity, osmotic coefficient, surface tension, activity coefficient, and thermal expansivity.

Some literature focused on reporting seawater under limited conditions. Sharqawy et al. (2010) reviewed the thermophysical properties of seawater in terms of existing correlations and data. Later, Nayar et al. (2016) extended the review of Sharqawy et al. (2010) by accommodating the pressure influence on seawater properties. The majority of properties are suggested for sodium chloride (NaCl) as it is the major component in seawater (Nayar et al. 2016; Sharqawy et al. 2010). Multicomponent consideration for saline water is unignorable for brackish water with low salinity (Xu et al. 2008) and oil and gas hydraulic fracturing with high salinity (Thiel and Lienhard 2014).

The investigated studies to examine multicomponent and high saline water properties are comparatively very low. A good book reference for the thermodynamics of solutions was by Zemaitis et al. (1986). It introduced density, activity coefficient, heat capacity, and osmotic coefficient for single electrolyte and multiple components (ions) electrolytes. This reference expressed some methods for binary and multicomponent saline water, such as Pitzer (1987), Bromley (Bromley 1973; Bromley et al. 1967), Meissner (Kusik and Meissner 1978), and Chen (1980).

1.2 Difference Between Pure Water and Saline Water

Pure water is composed of two hydrogen atoms and one oxygen atom, while saline water contains a variety of dissolved salts, such as sodium chloride. Pure water has a neutral pH of 7, whereas saline water has a higher pH due to the presence of dissolved salts. The thermodynamic and thermophysical properties of pure water were established well, as many human and industrial applications depend on them. The mass and energy transfer of pure water depends on the temperature and pressure conditions. Adding salt makes hydrothermal fluids' mass and energy-transporting capacities less sensitive to changing pressure and temperature conditions as the salinity increases (Klyukin et al. 2016).

The properties of the saline water are relatively closer to the salty water for low salinities. The increases of dissolved salt in water vary the solution properties. For example, with salinity increases, the density of saltwater increases because the mass increases. Also, for the rise in salinity, the specific heat capacity, specific enthalpy, specific entropy, and latent heat decrease since their values of pure water are higher than the salt at the same conditions. Moreover, the increase in salinity leads to a decrease in vapor pressure as the dissolved salt ions have lower free energy than the water molecules, an increase in the viscosity due to addition of the salt ions, an increase in the thermal conductivity due to the increased dissolved salt (Poling et al. 2001), a change (decrease and increase, showing minima values) in the activity and osmotic coefficients, an increase in the surface tension due to the dipole nature of water molecules and their interacts with salt ions (Boniewicz-Szmyt and Pogorzelski 2016; Lima et al. 2013; Nayar 2014; Nayar et al. 2014; Okur et al. 2017; Salameh et al. 2016; Wang et al. 2018), an increase in the electric conductivity due to the higher conductivity of salt ions, an increase in the thermal expansivity due to that the water molecules become tightly less bonded to each other with the presence of salt ions, and a decrease in the freezing point because the increased salt molecules try block water molecules from packing together.

1.3 Saline Water Concentration

Saltwater consists of water and salt components, called a solution; the solutes are the salt components (impurities). The solution concentration could be expressed in different ways, such as parts per million (ppm), salinity, weight percent, molarity (mol/L), molality (mol/kg), and normal concentration (mol/m^3). Different concentration scales are listed in Table 1.1. More explanation is also discussed in Sects. 1.4 and 1.5.

Table 1.1 Methods to estimate the saline water concentration (Overview 1930)

No.	Unit	Abbreviation	Factor time wt. fraction	Formula
1	Salinity	S (g/kg)	–	$\frac{\sum \text{mass of solute}}{\text{mass of solution}}$
2	Molar concentration (molarity)	M (molar)	–	$\frac{\text{Moles of solute}}{\text{Volume of solution}}$
3	Molality	m (molal)	–	$\frac{\text{Moles of solute}}{\text{mass of solvent}}$
4	% mass	%	10^2	Salinity × 10^2
5	Parts per thousand	‰ (ppt)	10^3	Salinity × 10^3
6	Parts per ten-thousand	‱	10^4	Salinity × 10^4
7	Parts per million	ppm	10^6	Salinity × 10^6
8	Parts per billion	ppb	10^9	Salinity × 10^9

The multicomponent models usually use the ionic strength (I) to estimate the participation of all impurities and their influences on the saltwater properties. The ionic strength is a function of molality (m) and ionic charge (Z) and can be written as:

$$I = \frac{1}{2}\sum_{i}^{n} m_i Z_i^2 \tag{1.1}$$

For NaCl, it is equal to molality.

1.4 Salinity

Salinity could be used instead of concentration of saltwater. However, it mainly expresses the masses (in a gram) of all solutes per unit kg of the solution.

Table 1.2 illustrates the methods used to calculate seawater salinity. The typical seawater salinity is 35.17 g/kg, as calculated in Table 1.2. The method used in Table 1.2 is valid for any solution.

1.5 Molarity and Molality

As described in Table 1.1, molarity (M) is the ratio of solute moles per liter of solution. For multicomponent saltwater, M of each component is calculated separately, and M of all components is given by:

$$M = \frac{1}{2}\sum_{i=1}^{n} M_i Z_i^2 \tag{1.2}$$

Table 1.2 Estimation of the typical seawater salinity

#	Species	No. of moles per kg of seawater (mol/kg)	Molar mass (g/mol)	Salinity (g/kg)
1	Cl^-	0.546	35.453	19.35734
2	Na^+	0.469	22.9897	10.78217
3	Mg^{2+}	0.0528	24.305	1.283304
4	$SO_4{}^{2-}$	0.0282	96.060	2.708892
5	Ca^{2+}	0.0103	40.078	0.412803
6	K^+	0.0102	39.098	0.398800
7	$CO_3{}^{2-}$	0.00206	60.008	0.123616
8	Br^-	0.000844	79.904	0.067439
9	$H_3BO_3{}^-$	0.000416	61.830	0.025721
10	Sr^{2+}	0.000091	87.620	0.007973
11	F^-	0.000068	18.9984	0.001292
				$\sum S_i = 35.169$ g/kg

On the other hand, molality (m) is the ratio of solute moles per kg of only solvent (i.e., water).

The following is an example to estimate the molality and molarity of seawater according to Table 1.2 data. For the molecular weight of seawater = 58.44 g/mol, m = 35.169/58.44 = 0.6 mol/kg (molal). Also, for the density of seawater = 1.025 kg/L, M = 0.6 × 1.025 = 0.615 mol/L.

1.6 Importance and Coverage

As most of the literature introduced the properties of seawater under some typical conditions, this book aims to compile and estimate the existing methods, data, and correlations of all important thermodynamic and thermophysical properties used for all saltwater types (multicomponent and binary). Various conditions will be applied for salinity (up to > 200 g/kg), temperature, and pressure. Every method to estimate the required property is plotted against the available experimental data to judge its accuracy under the suggested operating conditions. An assessment to recommend some correlations is also conducted. New correlations for some properties are also proposed. The brief content of this book is shown in Fig. 1.1. This book is an extension of the authors' published review papers (Generous et al. 2020; Qasem et al. 2021) with more elaboration in addition to some properties. It also involves the engineering equations solver (EES) codes for the tested properties.

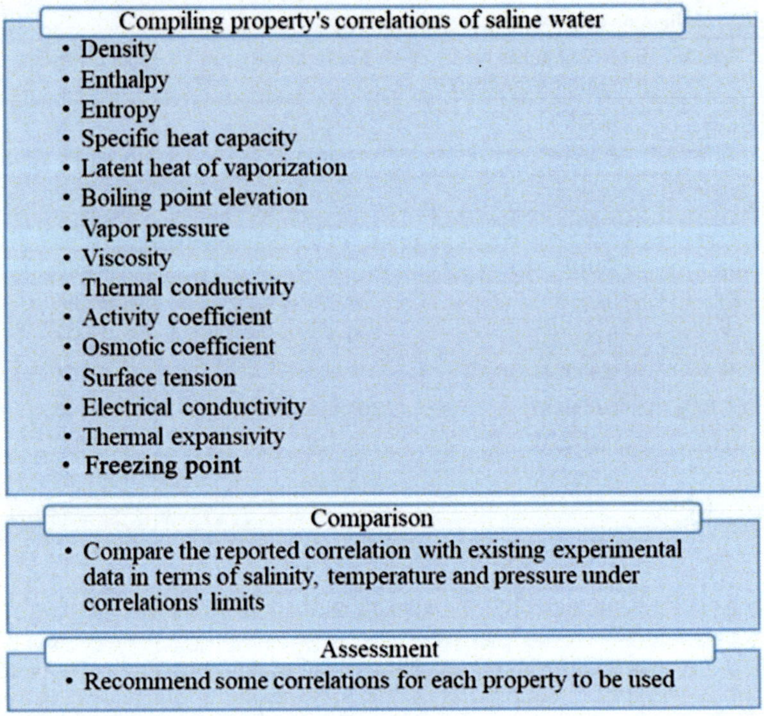

Fig. 1.1 The saline water properties discussed in this book

References

Boniewicz-Szmyt K, Pogorzelski SJ (2016) Thermoelastic surface properties of seawater in coastal areas of the Baltic Sea. Oceanologia 58(1):25–38. https://doi.org/10.1016/j.oceano.2015.08.003

Bromley LA (1973) Thermodynamic properties of strong electrolytes in aqueous solutions. AIChE J 19(2):313–320. https://doi.org/10.1002/aic.690190216

Bromley LA, Desaussure VA, Cupp JC, Wright JS (1967) Heat capacities of sea water solutions at salinities of 1 to 12% and temperatures of 2° to 80°. J Chem Eng Data 12(2):202–206. https://doi.org/10.1021/je60033a013

Chehayeb KM, Farhat DM, Nayar KG, Lienhard JH (2017) Optimal design and operation of electrodialysis for brackish-water desalination and for high-salinity brine concentration. Desalination 420:167–182. https://doi.org/10.1016/j.desal.2017.07.003

Chen C-C (1980) Computer simulation of chemical processes with electrolytes. http://dspace.mit.edu/handle/1721.1/15864

Fidaleo M, Moresi M (2011) Electrodialytic desalting of model concentrated NaCl brines as such or enriched with a non-electrolyte osmotic component. J Membr Sci 367(1–2):220–232. https://doi.org/10.1016/j.memsci.2010.10.069

Generous MM, Qasem NAA, Qureshi BA, Zubair SM (2020) A comprehensive review of saline water correlations and data-part I: thermodynamic properties. Arab J Sci Eng 45(11):8817–8876. https://doi.org/10.1007/s13369-020-05019-y

Khalifa AE, Alawad SM, Antar MA (2017) Parallel and series multistage air gap membrane distillation. Desalination 417:69–76. https://doi.org/10.1016/j.desal.2017.05.003

Klyukin YI, Driesner T, Steele-MacInnis M, Lowell RP, Bodnar RJ (2016) Effect of salinity on mass and energy transport by hydrothermal fluids based on the physical and thermodynamic properties of H_2O-NaCl. Geofluids 16(3):585–603. https://doi.org/10.1111/gfl.12181

Kusik CL, Meissner HP (1978) Electrolyte activity coefficients in inorganic processing. Am Inst Chem Eng Symp Ser 74(173):14–20

Kwon K, Han J, Park BH, Shin Y, Kim D (2015) Brine recovery using reverse electrodialysis in membrane-based desalination processes. Desalination 362:1–10. https://doi.org/10.1016/j.desal.2015.01.047

Lima ERA, De Melo BM, Baptista LT, Paredes MLL (2013) Specific ion effects on the interfacial tension of water/hydrocarbon systems. Braz J Chem Eng 30(1):55–62. https://doi.org/10.1590/S0104-66322013000100007

Lv H, Wang Y, Wu L, Hu Y (2019) Numerical simulation and optimization of the flash chamber for multistage flash seawater desalination. Desalination 465:69–78. https://doi.org/10.1016/j.desal.2019.04.032

Mitra S, Srinivasan K, Kumar P, Murthy SS, Dutta P (2013) Solar driven adsorption desalination system. Energy Procedia 49:2261–2269. https://doi.org/10.1016/j.egypro.2014.03.239

Mitra S, Thu K, Saha BB, Dutta P (2017) Performance evaluation and determination of minimum desorption temperature of a two-stage air cooled silica gel/water adsorption system. Appl Energy 206:507–518. https://doi.org/10.1016/j.apenergy.2017.08.198

Narayan GP, St. John MG, Zubair SM, Lienhard JH (2013) Thermal design of the humidification dehumidification desalination system: an experimental investigation. Int J Heat Mass Transfer 58(1–2):740–748. https://doi.org/10.1016/j.ijheatmasstransfer.2012.11.035

Nayar KG (2014) An experimental investigation of the surface tension of seawater (issue December 2014). MIT

Nayar KG, Panchanathan D, McKinley GH, Lienhard JH (2014) Surface tension of seawater. J Phys Chem Ref Data 43(4). https://doi.org/10.1063/1.4899037

Nayar KG, Sharqawy MH, Banchik LD, Lienhard JH (2016) Thermophysical properties of seawater: a review and new correlations that include pressure dependence. Desalination 390:1–24. https://doi.org/10.1016/j.desal.2016.02.024

Okur HI, Chen Y, Wilkins DM, Roke S (2017) The Jones–Ray effect reinterpreted: surface tension minima of low ionic strength electrolyte solutions are caused by electric field induced water-water correlations. Chem Phys Lett 684:433–442. https://doi.org/10.1016/j.cplett.2017.06.018

Overview S (1930) Water quality measurement, pp 105–138. https://doi.org/10.1016/B978-0-12-088796-5.50007-7

Park IS, Park SM, Ha JS (2005) Design and application of thermal vapor compressor for multi-effect desalination plant. Desalination 182(1–3):199–208. https://doi.org/10.1016/j.desal.2005.02.027

Pitzer KS (1987) A thermodynamic model for aqueous solutions of liquid-like density. Rev Mineral 17:97–142

Poling BE, Prausnitz JM, O'Connell JP (2001) Properties of gases and liquids. In: Principles of environmental physics, 5th edn. McGraw-Hill Education

Qasem NAA, Zubair SM (2019) Performance evaluation of a novel hybrid humidification-dehumidification (air-heated) system with an adsorption desalination system. Desalination 461:37–54. https://doi.org/10.1016/j.desal.2019.03.011

Qasem NAA, Qureshi BA, Zubair SM (2018) Improvement in design of electrodialysis desalination plants by considering the Donnan potential. Desalination 441. https://doi.org/10.1016/j.desal.2018.04.023

Qasem NAA, Generous MM, Qureshi BA, Zubair SM (2021) A comprehensive review of saline water correlations and data: part II—thermophysical properties. Arab J Sci Eng 46(3):1941–1979. https://doi.org/10.1007/s13369-020-05020-5

Qasim M, Badrelzaman M, Darwish NN, Darwish NA, Hilal N (2019) Reverse osmosis desalination: a state-of-the-art review. Desalination 459:59–104. https://doi.org/10.1016/j.desal.2019.02.008

Salameh A, Vorka F, Daskalakis V (2016) Correlation between surface tension and the bulk dynamics in salty atmospheric aquatic droplets. J Phys Chem C 120(21):11508–11518. https://doi.org/10.1021/acs.jpcc.6b01842

Sharqawy MH, Lienhard JH, Zubair SM (2010) Thermophysical properties of seawater: a review of existing correlations and data. Desal Water Treat 16(1):354–380. https://doi.org/10.5004/dwt.2010.1079

Shiklomanov IA, Rodda JC, Babkin VI, Penkova NV, Georgievsry VY, Zaretskaya IP, Izmailova AV, Balonishnikova JA, Grigorkina TE, Grube TV, Skoryatina EL, Tsytsenko VK, Yunitsyna VP (2013) World water resources at the beginning of the twenty-first century. Choice Rev Online 41(07). https://doi.org/10.5860/choice.41-4063

Thiel GP, Lienhard JH (2014) Treating produced water from hydraulic fracturing: composition effects on scale formation and desalination system selection. Desalination 346:54–69. https://doi.org/10.1016/j.desal.2014.05.001

Tsiakis P, Papageorgiou LG (2005) Optimal design of an electrodialysis brackish water desalination plant. Desalination 173:173–186. https://doi.org/10.1016/j.desal.2004.08.031

Vörösmarty CJ, Sullivan CA, Davies PM, Bunn SE, McIntyre PB, Prusevich A, Liermann CR, Gessner MO, Dudgeon D, Green P, Glidden S (2010) Global threats to human water security and river biodiversity. Nature 467(7315):555–561. https://doi.org/10.1038/nature09440

Wang X, Chen C, Binder K, Kuhn U, Pöschl U, Su H, Cheng Y (2018) Molecular dynamics simulation of the surface tension of aqueous sodium chloride: from dilute to highly supersaturated solutions and molten salt. Atmos Chem Phys 18(23):17077–17086. https://doi.org/10.5194/acp-18-17077-2018

Xu P, Drewes JE, Heil D, Wang G (2008) Treatment of brackish produced water using carbon aerogel-based capacitive deionization technology. Water Res 42(10):2605–2617. https://doi.org/10.1016/j.watres.2008.01.011

Zemaitis JF, Clark DM, Rafal M, Scrivner NC (1986) Handbook of aqueous electrolyte thermodynamics. Wiley. https://doi.org/10.1002/9780470938416

Chapter 2
Density

Abstract The density of saline water is a critical thermodynamic property for designing and rating all the relevant processes, including desalination. It is a reciprocate of the specific volume and important for thermal–hydraulic performance. This chapter introduces the thermodynamic approaches, multicomponent models, correlation, and data that are used to estimate the saline water density at different salinity, temperature, and pressure values. Experimental data of seawater density are used to compare the estimated density values from the correlations and methods and assess their accuracy. Some correlations and methods are recommended based on their accuracy and operating conditions. Besides, the codes of methods used to calculate the density are provided.

Keywords Density · Thermodynamics · Correlations · Methods · Saline water

2.1 Introduction

Density is the reciprocal of the specific volume; it is a measure of how much mass there is in a given volume. The density of saline water is an important property for flow characteristics and energy calculations. That is why a lot of density methods and correlations have been reported in the literature. The saline water density is found to be a function of temperature, pressure, and salinity (Millero et al. 2008). When salt is dissolved in water, the density of the solution increases because the mass increases. When the same amount of water is heated, its density decreases due to the increase in volume, and the opposite happens with the increase in pressure (Bryden 1973; Isdale and Morris 1986).

The methods used to estimate the density of saline water are based on thermodynamic properties, empirical correlation, and detailed methods (especially when dealing with multiple components). A salinity range of 0–250 g/kg, a pressure range of 0.1–100 MPa, and a temperature range of 0–180 °C were considered by different studies to calculate the saltwater density. Multiple ion correlations and methods over a wide range of salinity were also investigated.

N. A. Qasem et al., *Thermodynamic and Thermophysical Properties of Saline Water*, Springer Water, https://doi.org/10.1007/978-3-031-35193-8_2

Millerio and Poisson (Huang 2009; Poisson et al. 1982) proposed a density correlation obtained from experimental data with a considerable accuracy of ± 0.00333 kg/m^3. Later, this work was used to generate the international equation of state for saline water. The development of the equation of state was a big milestone in determining the density of saline water (Millero et al. 2008). It depends on temperature and pressure over a wider range of salinity. However, it is not the only equation to cover the effect of temperature and pressure. There are other correlations reported in the literature as well. Sun et al. (2008a) reported a really comprehensive correlation that encompasses the effect of temperature (0–180 °C), pressure (1–100 MPa), and salinity (0–42 g/kg). Gebhart & Mollendorf (1977) also developed a correlation to evaluate the density of saline water as a function of temperature, pressure, and salinity. Chil (Lam et al. 2008) considered salinity as an independent parameter to determine salinity. A slight variation in the composition of saline water and its effect on density was modeled by Schmidt et al. (2018). Chebyshev (Isdale and Morris 1986) also suggested a density correlation as a function of temperature and salinity. Some types of saline water, such as ocean water, may also have a considerable amount of dissolved air; therefore, Bigg (1967) investigated the effects of dissolved air on density calculations.

2.2 Thermodynamic Approach

The thermodynamic method relies upon specific volume to estimate the density as given in Eq. (2.1).

$$\rho = \frac{1}{v} \tag{2.1}$$

where ρ (in kg/m^3) is the density and v in (m^3/kg) is the specific volume. It is basically given by the reciprocal of the pressure derivative of the Gibbs function at constant absolute salinity and in situ temperature as (IOC et al. 2010):

$$\rho = \rho(S_A, T, P) = \left(\left. \frac{\partial g}{\partial P} \right|_{S_A,T} \right)^{-1} \tag{2.2}$$

Here, S_A is absolute salinity, T is the in situ temperature that can be converted to a conservative temperature (IOC et al. 2010), P is the pressure, and g is the Gibbs function. Note that the absolute salinity (in g/kg) is higher than that of practical salinity by about 0.165–0.195 g/kg in the open ocean. Thus, if the practical salinity is not used in Eq. (2.2), a density error of 0.12–0.15 kg/m^3 would result.

Generally, the thermodynamic approach is cumbersome and needs a lot of details, so researchers proposed some methods and correlations to simplify the saltwater density estimation for multicomponent saline water and NaCl seawater.

2.3 Multiple Components Models

Multiple component models were proposed to be general for all types of saltwater under different salinity values. Major constituents of the aqueous ionic solutions are cations, anions, and water. The density of an ionic solution can be defined as the ratio of the average molar mass and volume of the aqueous solution. In this scenario, Lam et al. (2008) proposed the following model.

$$\rho = \frac{\sum_{i=1}^{n} x_i M_i + \left(1 - \sum_{i=1}^{n} x_i\right) M_w}{\sum_{i=1}^{n} v_i x_i + v_w\left(1 - \sum_{i=1}^{n} x_i\right) + \sum_{i=1}^{n} \alpha_i x_i\left(1 - \sum_{i=1}^{n} x_i\right)} \tag{2.3}$$

where x_i is the mole fraction of anionic species, M_w is the molar mass of water (in kg/mol), M_i is the molar mass of ionic species (in kg/mol), and n is the number of moles. The model is valid and can be used with confidence at a given temperature based on the availability of empirical parameters v_i and α_i. The values of these empirical parameters at 298.15 K are listed in Table 2.1. The model is used for all salinity values, and it was tested for a pressure of 0.1 MPa with a standard deviation of ± 0.08%.

For the same conditions above with variable temperature, Laliberté and Cooper (2004) developed a density model; that is expressed as:

$$\rho = \frac{1}{W_w V_w + \sum_{i=1}^{n} W_i V_i} \tag{2.4}$$

where W_w, V_w, W_i, and V_i are the mass fraction of water, the specific volume of water, the mass fraction of the ith component of solute, and the apparent specific volume of the ith species, respectively. The apparent specific volume (in m^3/kg) can be modeled by the following equation:

$$V_i = \frac{(W_i + c_2 + c_3 T)}{(c_o W_i + c_1)\exp\left(0.000001(T + c_4)^2\right)} \tag{2.5}$$

Here, T is the temperature in °C, and the constants c_o, c_1, c_2, c_3, and c_4 for a variety of electrolytes may be obtained from Table 2.2. This method can be used for a temperature less than the boiling point, P = 0.1 MPa, with a standard deviation of less than ± 1.39 kg/m^3.

Table 2.1 Ionic volumes and values of α_i at 298.15 K for Lam et al. (2008) density model

Ion	v_i^o, cm^3/mol	α_i, cm^3/mol	Ion	v_i^o, cm^3/mol	α_i, cm^3/mol
Cation			Anion		
Ba^{2+}	18.3023	− 5.1552	Br^-	26.0512	− 9.4688
Ca^{2+}	25.1865	− 22.3772	BrO_3^-	13.4392	13.1391
Cd^{2+}	9.8123	1.4568	CO_3^{2-}	44.8606	− 61.8929
Co^{2+}	7.7433	− 10.5132	Cl^-	23.5130	− 14.6221
Cs^+	34.4534	− 2.6230	ClO_3^-	12.6994	12.6882
Cu^{2+}	30.4567	− 34.6929	ClO_4^-	46.1001	− 13.2870
Fe^{3+}	11.8134	− 7.3003	HCO_3^-	7.6564	7.2686
Hg^{2+}	15.3070	15.1201	$H_2PO_4^-$	23.0551	4.5946
K^+	27.0279	− 8.0763	HSO_4^-	76.6467	− 52.8580
La^{3+}	31.1914	− 32.9643	I^-	36.5122	− 9.8899
Li^+	8.4351	0.0655	IO_3^-	66.4296	− 47.5567
Mg^{2+}	19.4707	− 18.9749	MoO_4^{2-}	81.2460	− 65.2988
Mn^{2+}	0.0000	6.6409	N_3^-	23.6464	− 5.1772
NH_4^+	24.6421	2.8238	NO_3^-	36.3809	− 16.0250
Na^+	14.6883	− 5.9081	OH^-	17.2940	− 29.3781
Ni^+	17.2824	− 25.0728	SO_3^{2-}	59.1677	− 63.8203
Pb^{2+}	94.0885	− 86.8831	SO_4^{2-}	69.4141	− 65.8818
Rb^+	30.2399	− 5.7210			
Sr^{2+}	38.5117	− 35.5831			
Tl^+	10.9815	10.3291			
Zn^{2+}	40.1139	− 34.5036			

Table 2.2 Values of constants c_0 to c_4 for Laliberté correlation (Cisternas and Lam 1989)

Compound	c_0, kg/m^3	c_1, kg/m^3	c_2, dimensionless	c_3, 1/°C	c_4, °C	Vapp, at W = 0.01 and T = 25 °C/m^3/kg
$AlCl_3$	221.68	160.90	0.15125	0.00250	1500.0	0.0001340
$Al_2(SO_4)_3$	− 0.0017202	0.0018967	− 0.030904	0.00408	3804.2	0.0000185
$BaCl_2$	− 0.0030518	0.00526670	4.1785	0.06827	3971.9	0.0001299
$CaCl_2$	− 0.63254	0.93995	4.2785	0.04831	3180.9	0.0002025
$CdCl_2$	− 0.090879	0.29116	7.3827	− 0.03185	− 3477	0.0001513
$CdSO_4$	-1.0440×10^{-7}	6.1070×10^{-8}	− 0.003761	0.00410	5007.7	0.0000182
$CoCl_2$	-8.0924×10^{-8}	8.0261×10^{-8}	410.24	9.1808	5619.8	0.0001169
$CoSO_4$	− 118.36	1368.1	0.01304	− 0.00014	− 294.0	0.0000132
$CrCl_3$	3.1469	232.160	0.20191	0.00250	1500.0	0.0001155
$Cr_2(SO_4)_3$	1.0045	1.7679	− 0.085017	0.00250	1500.0	− 0.0006873
$CuCl_2$	1868.5	1137.20	0.07185	0.00256	575.7	0.0000880
$CuSO_4$	-1.9827×10^{-7}	1.0883×10^{-7}	− 0.12506	0.00383	4936.8	− 0.0000037
$FeCl_2$	98.654	199.51	0.33639	0.00384	1650.1	0.00001334
$FeSO_4$	-3.6737×10^{-6}	1.2465×10^{-4}	− 0.062861	0.00156	3943.0	− 0.0000159
$FeCl_3$	− 1333.8	4369.2	1.5298	0.00709	829.21	0.0001901
$Fe_2(SO_4)_3$	0.47444	− 0.64624	− 713.10	− 25.569	4023.2	0.0001610
HCl	− 80.061	255.42	118.42	1.0164	2619.5	0.0005186
HCN	255.82	283.11	0.66888	0.00620	891.0	0.0012616
HNO_3	12.993	− 23.579	− 3.6070	0.00794	− 2427	0.0004521

(continued)

Table 2.2 (continued)

Compound	c_0, kg/m^3	c_1, kg/m^3	c_2, dimensionless	c_3, 1/°C	c_4, °C	Vapp, at W = 0.01 and T = 25 °C/m^3/kg
H_3PO_4	1358.3	– 4327.7	– 4.5950	0.00438	– 912	0.0004720
H_2SO_4	89.891	224.48	0.82285	0.00684	1571.5	0.0003482
KCl	– 0.46782	4.30800	2.3780	0.02204	2714.0	0.0003769
K_2CO_3	– 1.4313	2.49170	1.1028	0.01311	2836.0	0.0001621
KNO_3	7.5436	26.38800	1.2396	0.01165	2214.0	0.0003872
KOH	194.85	407.31	0.14542	0.00204	1180.9	0.0001178
K_2SO_4	-2.6681×10^{-5}	3.0412×10^{-5}	0.97118	0.01981	4366.1	0.0002071
LiCl	17.807	32.011	1.3951	– 0.00623	– 2131	0.0004588
Li_2SO_4	0.0014730	0.0026934	0.17699	0.00413	3640.7	0.0001565
$MgCl_2$	– 0.00051500	0.0013444	0.58358	0.00858	3843.6	0.0001910
$MgSO_4$	3.9412×10^{-7}	1.4425×10^{-6}	– 0.05372	0.00206	4563.3	0.0000039
$MnCl_2$	0.000001869	0.00004545	1.5758	– 0.01077	– 4369	0.0001832
$MnSO_4$	0.0032447	0.057246	0.05136	0.00214	3287.8	0.0000344
NaBr	109.770	513.04	1.54540	0.01101	1618.1	0.0002394
NaCl	– 0.00433	0.06471	1.01660	0.01462	3315.6	0.0003065
$NaClO_3$	0.014763	0.024913	1.2924	– 0.00761	– 3454	0.0003465
Na_2CO_3	0.012755	0.014217	– 0.091456	0.00213	3342.4	– 0.0000233
NaF	2.8191×10^{-6}	2.1777×10^{-7}	– 0.041483	0.00021	4586.9	– 0.0000613
$NaHCO_3$	-9.4794×10^{-8}	1.5657×10^{-7}	0.9912	0.02264	4900.2	0.0002939

(continued)

Table 2.2 (continued)

Compound	c_0, kg/m^3	c_1, kg/m^3	c_2, dimensionless	c_3, 1/°C	c_4, °C	Vapp, at W = 0.01 and T = 25 °C/m^3/kg
NaH_2PO_4	208.77	641.05	0.78893	0.00455	1198.4	0.0003177
Na_2HPO_4	1096.7	937.57	0.01424	− 0.00055	− 860.2	0.0000054
$NaHSO_3$	6.1384×10^{-6}	1.3029×10^{-6}	0.13635	− 0.00146	− 4472	0.0002066
NaI	626.15	1858.2	1.7387	0.01050	1203.3	0.0002386
Na_2MoO_4	− 2.0813	4.8446	4.4342	− 0.02081	− 2942	0.0001638
$NaNO_2$	78.365	298.00	0.96246	0.0021999	1500.0	0.0003360
$NaNO_3$	49.209	94.737	0.77927	0.0075451	1819.2	0.0003423
NaOH	385.55	753.47	− 0.10938	0.0006953	542.88	− 0.000078
Na_3PO_4	1015.6	1533.7	− 0.15180	0.0001366	173.71	− 0.000086
Na_2SO_3	1.5197×10^{-5}	4.3766×10^{-7}	0.10296	− 0.0015271	− 4500.9	0.0002528
$Na_2S_2O_3$	0.84462	− 1.5142	− 42.949	0.19335	− 3425.9	0.0002400
Na_2SO_4	-1.2095×10^{-7}	4.3474×10^{-7}	0.15364	0.0072514	4731.5	0.0001189
NH_3	0.12693	0.10470	1.0302	− 0.0050803	− 2973.7	0.0014430
NH_4Cl	6.56150	89.772	4.9024	− 0.016574	− 2089.3	0.0007061
NH_4NO_3	1397.3	1124.4	0.65598	0.0014106	176.41	0.0005916
$(NH_4)_2SO_4$	− 123.22	452.59	3.2898	0.016292	1692.4	0.0004301
$NiCl_2$	-1.3900×10^{-6}	4.1879×10^{-6}	0.77734	− 0.0066936	− 4638.1	0.0000851
$NiSO_4$	− 0.03894	0.22109	− 0.14443	0.0009867	3073.8	− 0.000033
$SrCl_2$	1.3534×10^{-6}	-7.4877×10^{-7}	− 1.9356	0.010704	− 4882.1	0.0001281
$ZnCl_2$	1943.6	304.34	− 0.01375	0.0011543	573.79	0.0000542
$ZnSO_4$	18.378	35.927	− 0.08919	0.0010773	2066.3	− 0.000018

2.4 Correlations and Data

Correlations are the easiest way to calculate the saltwater density as they are fitted from the experimental data. Most of the available density correlations in the literature are for seawater. Rowe and Chou (1970) proposed a density correlation for petroleum reservoir brines and seawater as follows:

$$\rho = \frac{1000}{\left(A - (BP) - \left(CP^2\right) + (DS) + \left(ES^2\right) + (FSP) - \left(GS^2P\right) - \left(HSP^2\right)\right)} \tag{2.6}$$

where,

$$A = 5.916365 - 0.01035794T + 0.9270048 \times 10^{-5}T^2 - 1127.522T^{-1} + 100674.1T^{-2}$$

$$B = 0.5204914 \times 10^{-2} - 0.10482101 \times 10^{-4}T + 0.8328532 \times 10^{-8}T^2 - 1.1702939T^{-1} + 102.2783T^{-2}$$

$$C = 0.11854 \times 10^{-7} - 0.6599143 \times 10^{-10}T$$

$$D = -2.5166 + 0.0111766T - 0.170552 \times 10^{-4}T^2$$

$$E = 2.84851 - 0.0154305T + 0.22398 \times 10^{-4}T^2$$

$$F = -0.0014814 + 0.82969 \times 10^{-5}T - 0.12469 \times 10^{-7}T^2$$

$$G = 0.0027141 - 0.15391 \times 10^{-4}T + 0.22655 \times 10^{-7}T^2$$

$$H = 0.62158 \times 10^{-6} - 0.40075 \times 10^{-8}T + 0.65972 \times 10^{-11}T^2$$

The density here is in kg/m^3, T is the temperature in K, P is the pressure in kg/cm^2, and S is the salinity in kg/kg. This correlation (Eq. 2.6) is valid for the salinity of 0–250 g/kg, the pressure of 0.1–34.5 MPa, and the temperature of 0–176.5 °C at an accuracy of $\pm$ 0.2%.

El-Dessouky and Ettouney (2002) suggested a seawater density correlation being valid for the salinity of 0–160 g/kg, pressure of 0.1 MPa, and temperature of 10–180 °C as:

$$\rho = 1000(A_1F_1 + A_2F_2 + A_3F_3 + A_4F_4) \tag{2.7}$$

where,

$$
\begin{aligned}
A_1 &= 4.032219G_1 + 0.115313G_2 + 3.26 \times 10^{-4}G_3 \\
A_2 &= -0.108199G_1 + 1.571 \times 10^{-3}G_2 - 4.23 \times 10^{-4}G_3 \\
A_3 &= -0.012247G_1 + 1.74 \times 10^{-3}G_2 - 9 \times 10^{-6}G_3 \\
A_4 &= 6.92 \times 10^{-4}G_1 - 8.7 \times 10^{-5}G_2 - 5.3 \times 10^{-5}G_3 \\
F_1 &= 0.5 \\
F_2 &= A \\
F_3 &= 2A^2 - 1 \\
F_4 &= 4A^3 - 3A \\
G_1 &= 0.5 \\
G_2 &= B \\
G_3 &= 2B^2 - 1 \\
A &= \frac{2T - 200}{160} \\
B &= \frac{\left(\frac{2S}{1000} - 150\right)}{150}
\end{aligned}
$$

The density is in kg/m^3, T is taken in °C, and S is in ppm. The accuracy of this correlation was not provided by the source. However, it is within ± 0.35% of the experimental data of Leyendekkers (1980), as checked by Generous et al. (2020).

For salinity in kg/kg, temperature in °C, and pressure in MPa; density of seawater can be determined by Nayar et al. (2016) correlation as:

$$\rho = K \times \rho_{S1} \tag{2.8}$$

And,

$$
\begin{aligned}
\rho_{s1} = {} & a_1 + (a_2T) + \left(a_3T^2\right) + \left(a_4T^3\right) + \left(a_5T^4\right) + (b_1S) + (b_2ST) \\
& + \left(b_3ST^2\right) + \left(b_4ST^3\right) + \left(b_5S^2T^2\right)
\end{aligned} \tag{2.9}
$$

$$
\begin{aligned}
K = \exp\Big[& (P - P_o) \times \left(C_1 + (C_2T) + \left(C_3T^2\right) + \left(C_4T^3\right) + \left(C_5T^4\right) + \left(C_6T^5\right)\right) \\
& + \left(S \times \left(d_1 + (d_2T) + \left(d_3T^2\right)\right)\right) + \left(\frac{(P - P_o)}{2}\right) \times \left(C_7 + (C_8T) + \left(C_9T^3\right) + (d_4S)\right)\Big]
\end{aligned} \tag{2.10}
$$

where the constants are listed in Table 2.3. The operating conditions are salinity of 0–56 g/kg, pressure of 0–12 MPa, and temperature of 10–180 °C to have an accuracy of ± 0.14%.

Table 2.3 Constant values of Eqs. (2.9) and (2.10)

Constant	Value	Constant	Value
a_1	9.999×10^2	C_3	5.693×10^{-8}
a_2	2.034×10^{-2}	C_4	-3.7263×10^{-10}
a_3	-6.162×10^{-3}	C_5	1.4465×10^{-12}
a_4	2.261×10^{-5}	C_6	-1.7058×10^{-15}
a_5	-4.657×10^{-8}	C_7	-1.3389×10^{-6}
b_1	8.020×10^2	C_8	4.8603×10^{-9}
b_2	-2.001	C_9	-6.8039×10^{-13}
b_3	1.677×10^{-2}	d_1	-1.1077×10^{-6}
b_4	-3.060×10^{-5}	d_2	5.5584×10^{-9}
b_5	-1.613×10^{-5}	d_3	-4.2539×10^{-11}
C_1	5.0792×10^{-4}	d_4	8.3702×10^{-9}
C_2	-3.4168×10^{-6}		

Poisson et al. (1982) proposed a density correlation as follows:

$$(\rho - \rho_W) \times 10^3 = S\left(\frac{h}{35} + (S - 35)\, g\right) \tag{2.11}$$

where the subscript w denotes pure water and,

$$\begin{aligned}(\rho_W) \times 10^3 = {} & 999.842594 + 6.793952 \times 10^{-2}T \\ & - 9.095290 \times 10^{-3}T^2 + 1.001685 \times 10^{-4}T^3 \\ & - 1.120083 \times 10^{-6}T^4 + 6.536332 \times 10^{-9}T^5\end{aligned} \tag{2.12}$$

$$h = 28.265 - (0.11813T) + \left(1.937 \times 10^{-3}T^2\right) - \left(1.339 \times 10^{-5}T^3\right) \tag{2.13}$$

$$\begin{aligned}g = {} & \left(-0.3107 \times 10^{-3}\right) + \left(0.2120 \times 10^{-4}S\right) + \left(0.1427 \times 10^{-4}T\right) \\ & - \left(0.5423 \times 10^{-6}S^2\right) - \left(0.1953 \times 10^{-6}ST\right) \\ & - \left(0.3613 \times 10^{-6}T^2\right) + \left(0.524 \times 10^{-8}S^3\right) \\ & + \left(0.310 \times 10^{-8}S^2T\right) + \left(0.130 \times 10^{-9}ST^2\right) \\ & + \left(0.50 \times 10^{-8}T^3\right)\end{aligned} \tag{2.14}$$

Here, T is the temperature in °C, and S is salinity in g/kg. The operating limits are salinity of 5–70 g/kg, pressure of 0.1 MPa, and temperature of 0–90 °C at a standard deviation of ± 0.0032 kg/m^3. Another seawater correlation for salinity of less than 150 g/kg, temperature of 0–180 °C, and pressure of 0.1 MPa was suggested by Chebyshev (Isdale and Morris 1986) as:

Table 2.4 Constant values of Eq. (2.15)

j	i = 0	i = 1	i = 2
0	3.97260	– 0.084913	– 0.005318
1	0.115079	0.002695	0.000682
2	0.000109	– 0.000330	– 0.000011

$$\rho = \sum_{j=1}^{02} b_{ij} t_i \left(\frac{2S - 150}{150}\right) t_j \left(\frac{2T - 250}{110}\right) \tag{2.15}$$

where b_{ij} = Chebyshev Coefficients, t_i and t_j = Chebyshev Polynomials, $t_0 = 1$, $t_1 = x$, $t_2 = 2x^2 - 1$, $t_{r+1} = 2xt_r - t_{r-1}$, and the constants are given in Table 2.4. Notably, the values of Chebyshev polynomials in terms of "x" are determined by using known density values of pure water and equating it with the density of saline water. It is accurate up to $\pm$ 0.02%.

Sharqawy et al. (2011) suggested a density correlation in kg/m^3 as a function of temperature in °C, and salinity in kg/kg, as given by:

$$\begin{aligned}\rho = {} & a_1 + (a_2T) + \left(a_3T^2\right) + \left(a_4T^3\right) + \left(a_5T^4\right) \\ & + \left(b_1S + (b_2ST) + \left(b_3ST^2\right) + \left(b_4ST^3\right) + \left(b_5S^2T^2\right)\right)\end{aligned} \tag{2.16}$$

where the constants are listed in Table 2.5. The correlation is valid for salinity < 160 g/kg, pressure = 0.1 MPa, and temperature 0–90 °C to achieve an accuracy of $\pm$ 0.0003 kg/m^3.

For a smaller range of salinity (5–70 g/kg), pressure of 0.1 MPa, and temperature of 0–90 °C, Millero and Huang proposed the following correlation (Huang 2009):

$$\rho - \rho_W = AS_A + BS_A^{1.5} + CS_A^2 \tag{2.17}$$

Table 2.5 Constant values of Eq. (2.16)

Constants	Values
a_1	9.999×10^2
a_2	2.034×10^{-2}
a_3	-6.162×10^{-3}
a_4	2.261×10^{-5}
a_5	-4.657×10^{-8}
b_1	8.020×10^2
b_2	-2.001
b_3	1.677×10^{-2}
b_4	-3.060×10^{-5}
b_5	-1.613×10^{-5}

where,

$$S_A = \frac{35.16504}{35} S \tag{2.18}$$

$$\begin{aligned}\rho_W = {} & 999.842594 + 6.793952 \times 10^{-2}T - 9.095290 \times 10^{-3}T^2 \\ & + 1.001685 \times 10^{-4}T^3 - 1.120083 \times 10^{-6}T^4 \\ & + 6.536332 \times 10^{-9}T^5\end{aligned} \tag{2.19}$$

$$\begin{aligned}A = {} & 0.824493 - 4.0899 \times 10^{-3}T + 7.6438 \times 10^{-5}T^2 \\ & - 8.2467 \times 10^{-7}T^3 + 5.3875 \times 10^{-9}T^4\end{aligned} \tag{2.20}$$

$$B = -5.72466 \times 10^{-3} + 1.0227 \times 10^{-4}T - 1.6546 \times 10^{-6}T^2 \tag{2.21}$$

and $C = 4.8314 \times 10^{-4}$.

On top of the mentioned correlations, the equation of state of saline water is an empirical thermodynamic relation (nonlinear) that relates seawater density to a number of state variables, like temperature, salinity, and pressure (Millero 2010). The dependence of the density on temperature, salinity, and pressure can be expressed as:

$$d\rho = \frac{\partial \rho}{\partial T} dT - \frac{\partial \rho}{\partial S} dS + \frac{\partial \rho}{\partial P} dP \tag{2.22}$$

This equation could be written in terms of thermal expansion coefficient, Haline contraction coefficient, and adiabatic compressibility as:

$$d\rho = -\alpha \rho dT + \varepsilon \rho dS + K \rho dP \tag{2.23}$$

where the thermal expansion coefficient can be defined as:

$$\alpha = -\frac{1}{\rho} \frac{\partial \rho}{\partial T} \tag{2.24}$$

Haline contraction coefficient is expressed as:

$$\varepsilon = \frac{1}{\rho} \frac{\partial \rho}{\partial S} \tag{2.25}$$

And, the adiabatic compressibility can be defined as:

$$K = \frac{1}{\rho} \frac{\partial \rho}{\partial P} \tag{2.26}$$

Also, the international equation of state was proposed but with a lower range of salinity (0–40 g/kg) at pressure of 0.01–100 MPa and temperature of 0–40 °C (UNESCO 1981a). For temperature in °C, pressure in bar, and salinity in g/kg, the international equation of state for saline water density is expressed as:

$$\rho = \frac{K}{P}\left[a_o + (a_1T) + (a_2T^2) + (a_3T^3) + (a_4T^4) + (a_5T^5) + (B \times S) + (C \times S^{1.5}) + (D \times S^2)\right] \tag{2.27}$$

where $B = b_o + (b_1T) + (b_2T^2) + (b_3T^3) + (b_4T^4)$, $C = C_o + (C_1T) + (C_2T^2)$, and $D = 4.8314 \times 10^{-4}$, $K = K_{Atm} + (A_k \times P) + (B_k \times P^2)$, $K_{Atm} = A_w + ((i_o) + (i_1T) + (i_2T^2)) \times S + (j_oS^{1.5})$, $A_w = h_o + (h_1T) + (h_2T^2) + (h_3T^3)$, $B_k = B_w + ((m_oT) + (m_1T) + (m_2T^2)) \times S$, and $B_w = K_o + (K_1T) + (K_2T^2)$. The constants are given in Table 2.6. This method has an accuracy of ± 1.0%.

At a salinity of < 40 g/kg, pressure of 0.1–100 MPa, and temperature of 0–374 °C, Sun et al. suggested the following correlation for seawater density considering the temperature in °C, pressure in MPa, and salinity in g/kg:

$$\rho = \sum_{i=1}^{18} m_in_i - \sum_{i=1}^{06} X_iY_i \tag{2.28}$$

The constants of Eq. (2.28) are listed in Table 2.7. It shows a mean square error of ± 2.5%.

Table 2.6 Constant values of Eq. (2.27)

Constants	Values	Constants	Values
a_o	999.842594	h_o	3.23990
a_1	6.793953×10^{-2}	h_1	1.43713×10^{-3}
a_2	-9.095290×10^{-3}	h_2	1.16092×10^{-4}
a_3	1.001685×10^{-4}	h_3	-5.77905×10^{-7}
a_4	$-1.1200083 \times 10^{-6}$	i_o	2.28380×10^{-3}
a_5	6.536332×10^{-9}	i_1	-1.09810×10^{-5}
b_o	8.2449×10^{-1}	i_2	-1.60780×10^{-6}
b_1	-4.0899×10^{-3}	j_o	1.91075×10^{-4}
b_2	7.6483×10^{-5}	k_o	8.50935×10^{-5}
b_3	-8.2467×10^{-7}	k_1	-6.12293×10^{-6}
b_4	5.3875×10^{-9}	k_2	5.27870×10^{-8}
C_o	-5.7246×10^{-3}	m_o	-9.9348×10^{-7}
C_1	1.0227×10^{-4}	m_1	2.0816×10^{-8}
C_2	-1.6546×10^{-6}	m_2	9.1697×10^{-10}

Table 2.7 Constant values of Eq. (2.28)

m_i	n_i	X_i	Y_i
9.9920571×10^{2}	1	7.99992230×10^{1}	S
9.5390097×10^{-2}	T	2.4093650×10^{-3}	ST
$-7.6186636 \times 10^{-3}$	T^2	-2.580527×10^{-5}	ST^2
3.1305828×10^{-5}	T^3	6.8560840×10^{-8}	ST^3
$-6.1737704 \times 10^{-8}$	T^4	6.2976110×10^{-4}	P S
4.3368858×10^{-1}	P	-9.362637×10^{-7}	P^2S
2.5495667×10^{-5}	PT^2		
$-2.8988021 \times 10^{-7}$	PT^3		
$9.5784313 \times 10^{-10}$	PT^4		
1.7627497×10^{-3}	P^2		
$-1.2312703 \times 10^{-4}$	P^2T		
1.3659381×10^{-6}	P^2T^2		
4.0454583×10^{-9}	P^2T^3		
$-1.4673241 \times 10^{-5}$	P^3		
8.8391585×10^{-7}	P^3T		
1.1021321×10^{-9}	P^3T^2		
$4.2472611 \times 10^{-11}$	P^3T^3		
$-3.9591772 \times 10^{-14}$	P^3T^4		

UNESCO (1981b) suggested a seawater correlation in terms of salinity (0–42 g/kg) and temperature (0–40 °C) at 0.1 MPa with an accuracy of ± 1%, as follows.

$$\rho = a_o + (a_1T) + (a_2T^2) + (a_3T^3) + (a_4T^4) + (a_5T^5) + (B \times S) + (C \times S^{1.5}) + (D \times S^2) \quad (2.29)$$

where $B = b_o + (b_1T) + (b_2T^2) + (b_3T^3) + (b_4T^4)$, $C = C_o + (C_1T) + (C_2T^2)$, and $D = 4.8314 \times 10^{-4}$. And the constants are illustrated in Table 2.8.

Table 2.8 Constant values of Eq. (2.29)

Constants	Values	Constants	Values
a_o	999.842594	b_1	-4.0899×10^{-3}
a_1	6.793953×10^{-2}	b_2	7.6483×10^{-5}
a_2	-9.095290×10^{-3}	b_3	-8.2467×10^{-7}
a_3	1.001685×10^{-4}	b_4	5.3875×10^{-9}
a_4	$-1.1200083 \times 10^{-6}$	C_o	-5.7246×10^{-3}
a_5	6.536332×10^{-9}	C_1	1.0227×10^{-4}
b_o	8.2449×10^{-1}	C_2	-1.6546×10^{-6}

Sun et al. (2008b) suggested a modified correlation for seawater and brackish water for a salinity of < 40 g/kg, pressure of 0.1–100 MPa, and temperature of 0–374 °C as:

$$\rho = \sum_{i=1}^{18} m_i n_i - \sum_{i=1}^{09} X_i Y_i \tag{2.30}$$

where constant values are listed in Table 2.9. This method could achieve an accuracy of ± 1%.

Also, Poisson et al. (1982) proposed a modified seawater density correlation at only 35 g/kg for P = 0.1 MPa and T = 0–30 °C, as:

$$(\rho - \rho_W) = 28.255 - 0.1181T + 1.937 \times 10^{-3}T^2 - 1.399 \times 10^{-5}T^3 \tag{2.31}$$

where pure water density is considered as:

$$\begin{aligned}\rho_w = {} & 999.842594 + 6.793952 \times 10^{-2}T - 9.095290 \times 10^{-3}T^2 \\ & + 1.001685 \times 10^{-4}T^3 - 1.120083 \times 10^{-6}T^4 \\ & + 6.536332 \times 10^{-9}T^5\end{aligned} \tag{2.32}$$

Table 2.9 Constant values of Eq. (2.30)

m_i	n_i	X_i	Y_i
9.9920571×10^2	1	-8.23082×10^1	S
9.5390097×10^{-2}	T	1.779319×10^{-3}	S^2
$-7.6186636 \times 10^{-3}$	T^2	-5.881556×10^{-5}	S^3
3.1305828×10^{-5}	T^3	6.1531992×10^{-7}	S^4
$-6.1737704 \times 10^{-8}$	T^4	2.808321×10^{-3}	S T
4.3368858×10^{-1}	P	-2.671617×10^{-5}	ST^2
2.5495667×10^{-5}	PT^2	8.120625×10^{-4}	P S
$-2.8988021 \times 10^{-7}$	PT^3	-1.0347×10^{-6}	P^2S
$9.5784313 \times 10^{-10}$	PT^4	-8.83528×10^{-6}	PST
1.7627497×10^{-3}	P^2		
$-1.2312703 \times 10^{-4}$	p^2T		
1.3659381×10^{-6}	p^2T^2		
4.0454583×10^{-9}	P^2T^3		
$-1.4673241 \times 10^{-5}$	p^3		
8.8391585×10^{-7}	p^3T		
$-1.1021321 \times 10^{-9}$	P^3T^2		
$4.2472611 \times 10^{-11}$	P^3T^3		
$-3.9591772 \times 10^{-14}$	P^3T^4		

T is taken in °C. The standard deviation of this correlation is about ± 0.002 kg/m³. Tanaka et al. (2001) proposed a density correlation for pure water at 0.1 MPa and 0–180 °C. It is expressed by:

$$\rho_W = a_5\left(1 - \frac{(T + a_1)^2(T + a_2)}{a_3(T + a_4)}\right) \tag{2.33}$$

where $a_1 = -\,3.983035 \pm 0.00067$, $a_2 = 301.797$, $a_3 = 522{,}528.9$, $a_4 = 69.34881$, and $a_5 = 999.974950 \pm 0.00084T$. The accuracy of this correlation is reported to be temperature-dependent (Tanaka et al. 2001) as $(\Delta\rho T^2) \times 10^3 = 0.8394 - 0.00128T + 0.000110T^2 - 0.00000609T^3 + 0.000000116T^4$.

2.5 Case Study

The density of saline water is a function of salinity, temperature, and pressure. The density usually increases with an increase in salinity, as shown in Fig. 2.1. The density values were obtained by using the correlations and methods of Millero and Huang (2009), the international equation of state (UNESCO 1981a), El-Dessouky and Ettouney (2002), UNESCO (1981b), Nayar et al. (2016), Sun et al. (2008a), Poisson et al. (1982), Sharqawy et al. (2011), Sun et al. (2008b) (modified), Lam et al. (2008), Rowe and Chou (1970), and Laliberté and Cooper (2004) are compared to the available NaCl experimental data of Leyendekkers (1980), as shown in Fig. 2.1, against the salinity at 20 °C and 0.1 MPa. It can be seen from the figure that the density values generated by these correlations mostly overlap under their proposed range of operation. However, the Laliberté and Cooper (2004) model expects slightly lower density values than the other correlations and models within a difference of ± 3% from the experimental data (for $S < 150$ ppt). The other correlations and models have a difference of ± 1% from the experimental data.

Concerning the influenced density by temperature, the density of liquids always decreases with an increase in temperature under the same pressure. The dependence of temperature (0–100 °C) on the saltwater density values is investigated using the correlations of Millero and Huang (2009), El-Dessouky and Ettouney (2002), UNESCO (1981b), Sun et al. (2008a), Poisson et al. (1982) (modified), Sharqawy et al. (2011), Sun et al. (2008b) (modified), Poisson et al. (1982), Rowe and Chou (1970), international equation of state (UNESCO 1981a), and Nayar et al. (2016) against the experimental data of Isdale and Morris (1986) at 35 g/kg and 0.1 MPa, as shown in Fig. 2.2. The density values generated by all these correlations are close to each other, i.e., within ± 2% difference, except for the Nayar et al. (2016) correlation that deviates more than ± 2% from the experimental data after T = 60 °C.

The effect of pressure on the density and other properties is important for some applications, such as reverse osmosis. The density of saline water slightly decreases with an increase in pressure. The effect of pressure (< 100 MPa) on the calculated

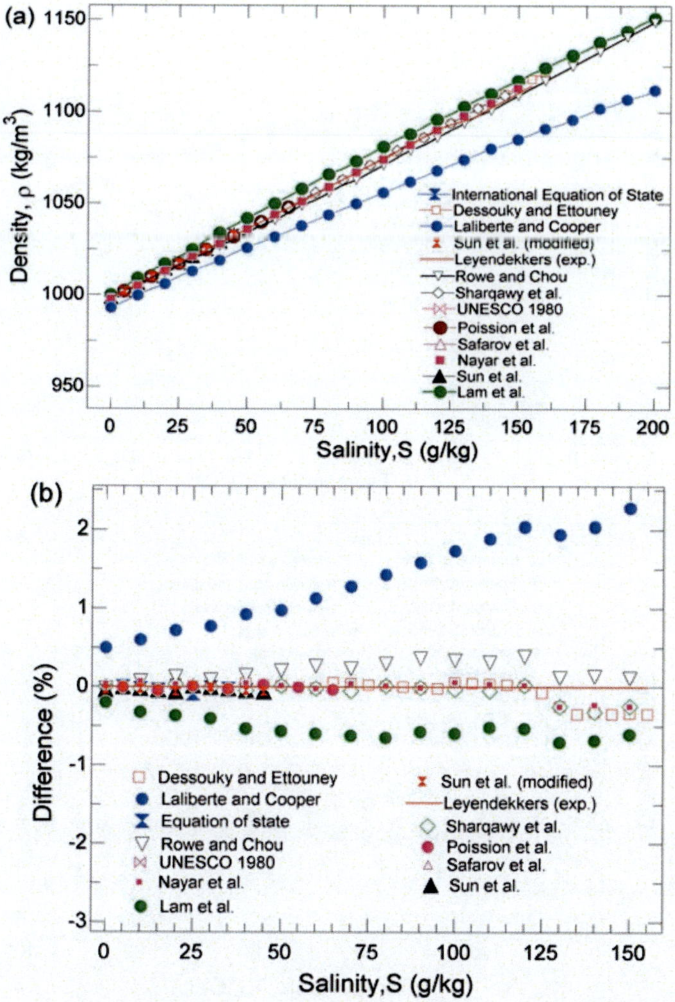

Fig. 2.1 Saline water density as a function of salinity: **a** salinity versus density and **b** the difference from the experimental data of Leyendekkers (1980) for Safarov et al. (2012), International equation of state (UNESCO 1981a), El-Dessouky and Ettouney (2002), Nayar et al. (2016), Sun et al. (2008a), Poisson et al. (1982), Sharqawy et al. (2011), Sun et al. (2008b) (modified), Lam et al. (2008), Rowe and Chou (1970), UNESCO (1981b), and Laliberté and Cooper (2004) methods at 20 °C and 0.1 MPa

density values is also plotted against the data of the international equation of state (UNESCO 1981a), as shown in Fig. 2.3. The international equation of state is one of the most accurate equations (UNESCO 1981b) that can be used to calculate density up to 1000 MPa. It successfully introduces the idea of the compressibility factor to consider the effect of pressure on saline water density. The other investigated correlations are those of Sun et al. (2008b) (modified), Rowe and Chou (1970), and

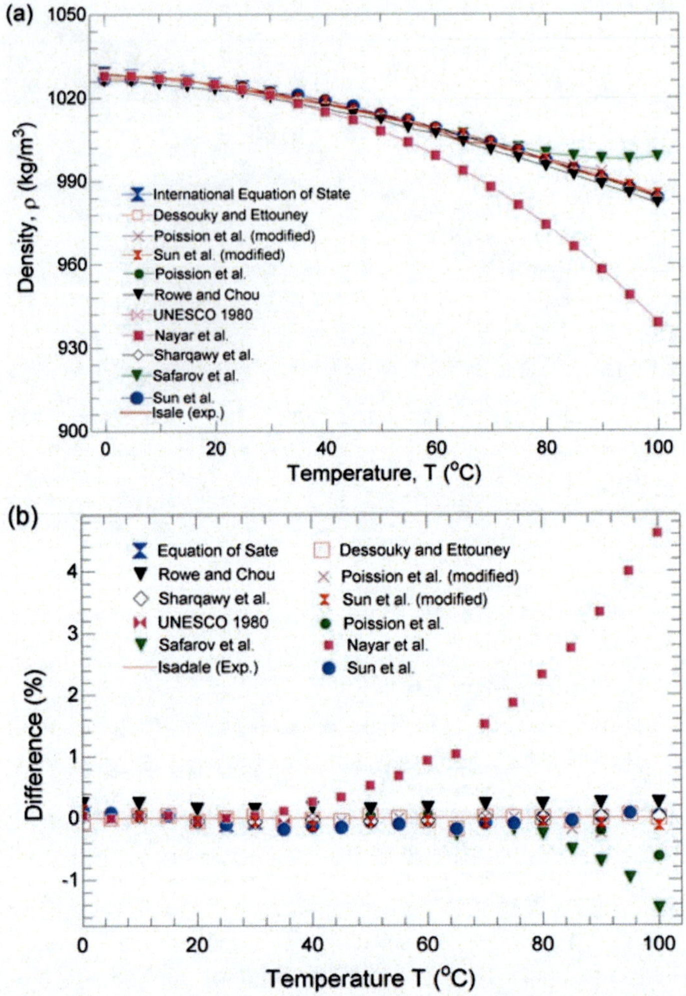

Fig. 2.2 Saline water density as a function of temperature: **a** temperature versus density and **b** the difference from the experimental data of Isdale and Morris (1986) for the international equation of state, Safarov et al. (2012), El-Dessouky and Ettouney (2002), Sun et al. (2008a), Poisson et al. (1982) (modified), Sharqawy et al. (2011), Sun et al. (2008b) (modified), Poisson et al. (1982), UNESCO (1981b), Rowe and Chou (1970), and Nayar et al. (2016) methods at 0.1 MPa and 35 g/kg

Nayar et al. (2016) at 15 °C and 35 g/kg. Figure 2.3 shows similar trends for the density values calculated by these correlations in the entire investigated pressure range to those of the international equation of state within a 0.3% difference. It is important to note that the effect of pressure on density values is not highly significant. For example, the difference between the density at 0.1 and 100 MPa is less than 4.4%. Thus, pressure dependence can be neglected.

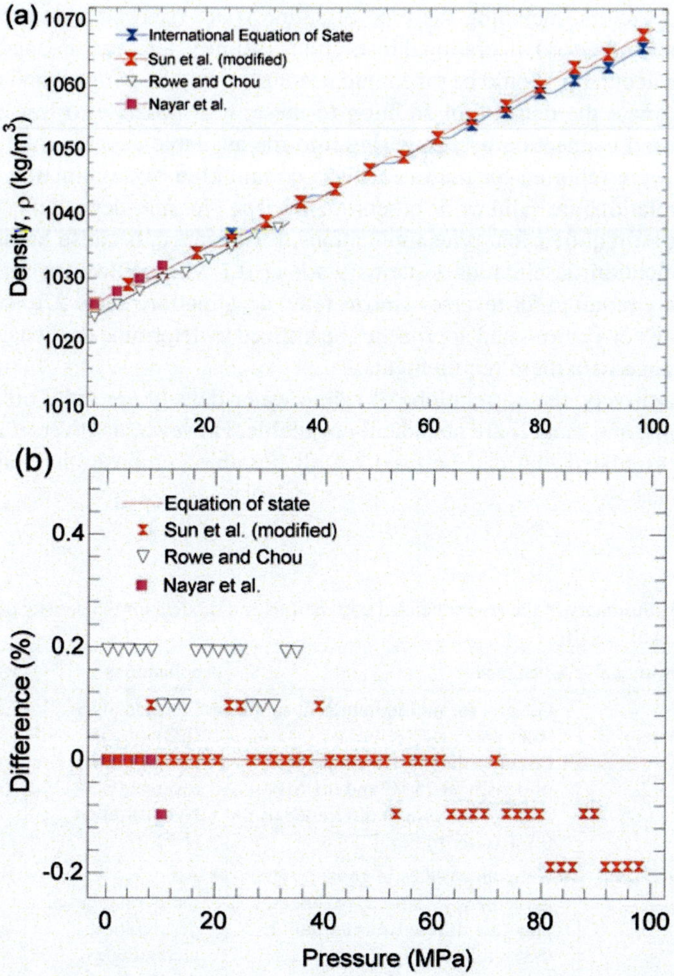

Fig. 2.3 Saline water density as a function of temperature: **a** pressure versus density and **b** the difference from the data of international equation of state (UNESCO 1981a) for Sun et al. (2008b) (modified), Rowe and Chou (1970), and Nayar et al. (2016) at 15 °C and 35 g/kg

Notably, seawater can be considered an aqueous solution of NaCl due to its dominant contribution. And because the experimental data are abundant for NaCl, the properties comparison is restricted for NaCl data since they are abundant. The multicomponent models are represented by handling Na and Cl as components.

Several factors, including the type of saline water (seawater, multicomponent, high saline, and so on), operational restrictions (salinity, temperature, and pressure), and fitting accuracy, should be taken into account when choosing a good correlation to approximate the density. In addition to these, it is possible to test and choose the suggested connection by comparing it to the available experimental data, most of which were reported for seawater. For a certain kind of desalination procedure, some correlations are valid while others are not. The pressure-dependent correlations might not be required for some applications, for instance, because humidification-dehumidification desalination systems work at 0.1 MPa. However, high-pressure qualities are required for reverse osmosis (RO) desalination. Table 2.10 suggests the appropriate correlations and approaches be utilized to determine the density of saline in accordance with these requirements.

Comparatively, the correlations of saline water density are quite old. The data, correlations, and models are abundantly available. However, a universal model may still be formulated that can be used for all possible compositions and operating conditions.

Table 2.10 Summary of the recommended correlations and models for the density of saline water (Generous et al. 2020)

Correlation-model	Advantages	Applications	References
Lam et al. (multi-species) (Eq. 2.3)	• General for multicomponent saline solution • Provides the most accurate values of density at 25 °C and 0.1 MPa within a maximum difference of ± 0.7% (refer to Fig. 2.1)	• Electrodialysis (ED) • Electrodialysis reversal (EDR) • Nanofiltration (NF)	Lam et al. (2008)
Nayar et al. (Eq. 2.8)	• Generates values in good agreement with the experimental data (the difference is around ± 0.3%, refer to Fig. 2.1) • It is a function of temperature, pressure, and salinity • Highly accurate in temperature and pressure ranges of 0–60 °C and 0–100 MPa, respectively	• All membrane-based desalination technologies • All thermal-based desalination technologies	Nayar et al. (2016)
Rowe and Chou (Eq. 2.6)	• Accurate within ± 0.7% difference (refer to Fig. 2.1) within its proposed range of operation • Valid in the temperature range of 0–100 °C with a difference of less than ± 0.5% (refer to Fig. 2.2) • Valid in the pressure range of 0–34.5 MPa with less than ± 0.3% difference (refer to Fig. 2.3)	• All membrane-based desalination technologies • All thermal-based desalination technologies • High-grade water recycling • Wave-powered desalination	Rowe and Chou (1970), Standing (1977)

(continued)

Table 2.10 (continued)

Correlation-model	Advantages	Applications	References
Equation of state (Eq. 2.27)	• Comprehensive in terms of temperature, pressure, and salinity • Valid in the salinity range of 0–42 g/kg with less than a difference of ± 0.5% (refer to Fig. 2.1) • Has a difference of less than ± 0.5% (refer to Fig. 2.2) within a temperature range of 0–40 °C • The best correlation in terms of accuracy in the pressure range of 0–100 MPa (refer to Fig. 2.3)	• All membrane-based desalination technologies • All thermal-based desalination technologies • High-grade water recycling	UNESCO (1981a)

2.6 EES Codes

The EES codes for the correlations discussed above:

Lam et al. Model

```
"Concentrations in mol/kg"
"Please note that 1: H2O, 2: Cl−, 3: Na+, 4: Mg++, 5: SO4−−, 6: Ca++, 7: K+"
n[1] = 53.6
{n[2] = 0.546
n[3] = 0.469
n[4] = 0.0528
n[5] = 0.0282
n[6] = 0.0103
n[7] = 0.0102}
n_t = sum(n[i], i = 1, 7) "total no. of moles"

"Mole Fractions"

Duplicate i = 1, 7
X[i] = n[i]/n_t
End

"Molar Masses in Kg/mol"
m[1] = 18/1000
m[2] = 35.453/1000
m[3] = 23/1000
```

```
m[4] = 24.305/1000
m[5] = 96/1000
m[6] = 40/1000
m[7] = 39.1/1000

"Summation of mole fraction of all ions"

X_ions = sum(X[i], i = 2, 7)

"Mole Fraction of Water"

X_water = 1 – X_ions

"Volume of Species in m3/mol"
v[1] = 18/10^6
v[2] = 23.5130/10^6
v[3] = 14.6883/10^6
v[4] = 19.4707/10^6
v[5] = 69.4141/10^6
v[6] = 25.1865/10^6
v[7] = 27.0279/10^6

"Compensation Factor for Ions-water molecules interaction in m3/mol"
alpha[2] = – 14.6221/10^6
alpha[3] = – 5.9081/10^6
alpha[4] = – 18.8749/10^6
alpha[5] = – 65.8818/10^6
alpha[6] = – 22.3772/10^6
alpha[7] = – 8.0763/10^6

"Average Molar Mass of Solution"

M_av = (sum(x[i] * m[i], i = 2, 7)) + (X_water * m[1])

"Molar Volume of Solution"
```

```
V = (sum(v[i] * x[i], i = 2, 7)) + (v[1] * X_water) + (sum(alpha[i] * x[i] * X_water, i = 2, 7))

"Density"
rho = M_av/V
```

Laliberté and Cooper Density Model

```
T = 20
rho_H2O = (((((−2.8054253/10^10 * T + 1.0556302/10^7) * T − 4.6170461/10^5 * T) * T − 0.0079870401) * T + 16.945176) * T + 999.83952)/(1 + 0.01687985 * T)

V_H2O = 1/rho_H2O

"let for example, we have NaCl. 1 represent Na and 2 represent Cl"
n_Na = 0.469 "moles/kg of water"
n_Cl = 0.546 "moles/kg of water"

"masses"

{m[1] = 0.469 * 23/1000 "kg"
m[2] = 0.546 * 35.453/1000 "kga"
m_s = sum(m[i], i = 1, 2)}
{m_s = 35.154/1000}

m_t = m_s + 1

W_H2O = 1/m_t
W[1] = m_s/m_t

V_NaCl = (W[1] + C2 + C3 * T)/((Co + C1) * exp(0.000001 * (T + C4)^2))

Co = − 0.00433
C1 = 0.06471
C2 = 1.01660
C3 = 0.014624
C4 = 3315.6

rho_m = 1/(W_H2O * V_H2O + W[1] * V_NaCl)
```

Rowe and Chou Correlation

“Validity of temperature upto 180C, pressure of 345 bars, and salinity of 250 g/m^3”
“accuracy within 0.2%”

“Inputs”

“Temperature in K”

T = 293.15 “K”

“absolute pressure in kg/cm^2”

P = 1.029913 “kg/cm^2”

“Salinity kg/kg of saline water”

Cs = 0.035

“Specific Volume in g/cm^3”

V = A – (B * P) – (C * P^2) + (D * Cs) + (E * Cs^2) – (F * Cs * P) – (G * Cs^2 * P) – (H * Cs * P^2)

“Constants”

A = 5.916365 – (0.01035794 * T) + (0.9270048/10^5 * T^2) – (1127.522/T) + (100674.1/T^2)
B = (0.5204914/10^2) – (0.10482101/10^4 * T) + (0.8328532/10^8 * T^2) – (1.1702939/T) + (102.2783/T^2)

C = (0.11854/10^7) – (0.6599143/10^10 * T)
D = – 2.5166 + (0.0111766 * T) – (0.170552/10^4 * T^2)
E = 2.84851 – (0.0154305 * T) + (0.223982/10^4 * T^2)
F = – 0.0014814 + (0.82969/10^5 * T) – (0.12469/10^7 * T^2)

```
G = 0.0027141 - (0.15391/10^4 * T) + (0.22655/10^7 * T^2)
H = (0.62158/10^6) - (0.40075/10^8 * T) + (0.65972/10^11 * T^2)

"specific volume in m^3/kg"

V_SI = (V * 1000)/10^6

"Density in SI units (kg/m^3)"

rho = 1/V_SI
```

El-Dessouky and Ettouney

```
"Correlation for density of saline water (kg/m3) in the salinity range of 0 < S <
160000 PPM and temperature range of 10 < T < 180 Co is given below"

S = 35000 "PPM"
"Salinity in PPM for multi-species can be calculated by using same procedure as
reported in the my work related to concentration earlier"

T = 20 "Degree C"

rho = 10^3 * (A_1 * F_1 + A_2 * F_2 + A_3 * F_3 + A_4 * F_4)

"where"

A_1 = (4.032219 * G_1) + (0.115313 * G_2) + (3.26/10^4 * G_3)
A_2 = (-0.108199 * G_1) + (1.571/10^3 * G_2) - (4.23/10^4 * G_3)
A_3 = (-0.012247 * G_1) + (1.74/10^3 * G_2) + (9/10^6 * G_3)
A_4 = (6.92/10^4 * G_1) - (8.7/10^5 * G_2) - (5.3/10^5 * G_3)
F_1 = 0.5
F_2 = A
F_3 = 2 * A^2 - 1
F_4 = (4 * A^3) - (3 * A)
G_1 = 0.5
G_2 = B
G_3 = 2 * B^2 - 1
A = (2 * T - 200)/160
B = ((2 * S/1000) - 150)/150

"Where "S" is salinity in PPM and "T" is temperature of saline water in Degree
C"
```

Nayar et al.

```
"Density as function of temperature, pressure and salinity"
"Salinity range 0–150 g/kg, pressure 1–10 bar"

T = 20 "C"
S = 0.035 "kg/kg"
P_o = 0.1 "MPa"
P = 0 "MPa"

rho_T_S = a1 + (a2 * T) + (a3 * T^2) + (a4 * T^3) + (a5 * T^4) + (b1 * S) +
(b2 * S * T) + (b3 * S * T^2) + (b4 * S * T^3) + (b5 * S^2 * T^2)

a1 = 9.999 * 10^2
a2 = 2.034/10^2
a3 = – 6.162/10^3
a4 = – 2.261/10^5
a5 = – 4.657/10^8

b1 = 8.020 * 10^2
b2 = – 2.001
b3 = 1.677/10^2
b4 = – 3.060/10^5
b5 = – 1.613/10^5

"Compressibility Factor, K"

K_o = (P – P_o) * ((C1 + C2 * T + C3 * T^2 + C4 * T^3 + C5 * T^4 + C6 *
T^5) + (S * (d1 + d2 * T + d3 * T^2))) + (((P^2 – P_o^2)/2) * (C7 + C8 * T
+ C9 * T^3 + d4 * S))

K = exp(K_o)

"Constants"

C1 = 5.0792/10^4
C2 = – 3.4168/10^6
C3 = 5.693/10^8
C4 = – 3.7263/10^10
```

```
C5 = 1.4465/10^12
C6 = – 1.7058/10^15
C7 = – 1.3389/10^6
C8 = 4.8603/10^9
C9 = – 6.8039/10^13

d1 = – 1.1077/10^6
d2 = 5.5584/10^9
d3 = – 4.2539/10^11
d4 = 8.3702/10^9

rho_T_S_P = rho_T_S * K
```

Poisson et al. Correlation

```
"Temperature"

T = 20 "C"

"Salinity in g/kg"

S = 35 "g/kg"

"Correlation Reported Poisson and Brunet"

(rho – rho_o) * 10^3 = S * (((1/35) * h_t) + ((S – 35) * g_t_s))

"h as a function of T"

h_t = 28.265 – (0.11923 * t) + (1.99189/10^3 * t^2) – (1.5065/10^5 * t^3)

"g as a function of S and t"

g_t_s = (–0.3057/10^3) + (0.2054/10^4 * S) + (0.1440/10^4 * t) – (0.5073/
10^6 * S^2) – (0.1966/10^6 * S * t) – (0.3710/10^6 * t^2) + (0.4076/10^8 *
S^3) + (0.311/10^8 * S^2 * t) + (0.1253/10^9 * S * t^2) + (0.5209/10^8 * t^3)

"density of water"
```

rho_o * 10^3 = (999.842594) + (6.793952/10^2 * t) − (9.095290/10^3 * t^2) + (1.001685/10^4 * t^3) − (1.120083/10^6 * t^4) + (6.536332/10^9 * t^5)
rho_f = rho * 1000

Sharqawy et al. Correlation

"Correlation is valid from salinity range of 0 to 0.16 kg/kg of saline water and temperature range of 0 to 180 C"

T = 20 "C"
S = 0.035 "kg/kg of saline Water"

rho = a_1 + (a_2 * T) + (a_3 * T^2) + (a_4 * T^3) + (a_5 * T^4) + ((b_1 * S) + (b_2 * S * T) + (b_3 * S * T^2) + (b_4 * S * T^3) + (b_5 * S^2 * T^2))

a_1 = 9.999 * 10^2
a_2 = 2.034/10^2
a_3 = − 6.162/10^3
a_4 = 2.261/10^5
a_5 = − 4.657/10^8

b_1 = 8.020 * 10^2
b_2 = − 2.020
b_3 = 1.677/10^2
b_4 = − 3.060/10^5
b_5 = − 1.613/10^5

Millero and Huang Correlation

"Temperature"

{T = 20}

"Salinity in g/kg"

S = 35 "g/kg"

"Correlation Reported by F. J. Millero"

rho − rho_o = A * S_a + B * S_a^1.5 + C * S_a^2

"Absolute Salinity Correlation (S_a)"

S_a = (35.16504/35) * S

"Correlations for Constants A, B, C"

A = (0.824493) – (4.0899/10^3 * t) + (7.6438/10^5 * t^2) – (8.2467/10^7 * t^3) + (5.3875/10^9 * t^4)

B = (−5.72466/10^3) + (1.0227/10^4 * t) – (1.6546/10^6 * t^2)

C = 4.8314/10^4

"Correlation for density of pure water as function of temperature"

rho_o = (999.842594) + (6.793952/10^2 * t) – (9.095290/10^3 * t^2) + (1.001685/10^4 * t^3) – (1.120083/10^6 * t^4) + (6.536332/10^9 * t^5)

"this correlation is valid for salinity range of 5–70 g/kg and temperature range of 273.15 K to 363.15 K"

International Equation of State

"Most Versatile Model for Density of Saline Water as Function of Temperature, Pressure, and Salinity"

"Density of Sea Water as function Salinity and Temperature"

"Correlation is valid from salinity range of 0 to 42 g/kg of saline water and temperature range of 0 to 40 C"

T = 20
S = 35 "g/kg of saline Water"
P = 10.1325

rho_S_T = a_o + (a_1 * T) + (a_2 * T^2) + (a_3 * T^3) + (a_4 * T^4) + (a_5 * T^5) + (B * S) + (C * S^1.5) + (D * S^2)

```
B = b_o + (b_1 * T) + (b_2 * T^2) + (b_3 * T^3) + (b_4 * T^4)
C = c_o + (c_1 * T) + (c_2 * T^2)
D = 4.8314/10^4

b_o = 8.2449/10^1
b_1 = – 4.0899/10^3
b_2 = 7.6483/10^5
b_3 = – 8.2467/10^7
b_4 = 5.3875/10^9

c_o = – 5.7246/10^3
c_1 = 1.0227/10^4
c_2 = – 1.6546/10^6
a_o = 999.842594
a_1 = 6.793953/10^2
a_2 = – 9.095290/10^3
a_3 = 1.001685/10^4
a_4 = – 1.120083/10^6
a_5 = 6.536332/10^9

"Let us now find Compressibility Factor at atmospheric pressure"

K_atm = K_w + (F * S) + (G * S)
K_w = e_o + (e_1 * T) + (e_2 * T^2) + (e_3 * T^3) + (e_4 * T^4)
F = f_o + (f_1 * T) + (f_2 * T^2) + (f_3 * T^3)
G = g_o + (g_1 * T) + (g_2 * T^2)

e_o = 19652.210
e_1 = 148.420600
e_2 = – 2.327105
e_3 = 1.360477/10^2
e_4 = – 5.155288/10^5

f_o = 54.674600
f_1 = – 0.603459
f_2 = 1.099870/10^2
f_3 = – 6.167000/10^5

g_o = 7.9440/10^2
g_1 = 1.6483/10^2
g_2 = – 5.3009/10^4
```

```
"Let us now find Compressibility Factor at any Given Pressure"

K = K_atm + (A_k * P) + (B_k * P^2)
A_k = A_w + (i_o + (i_1 * T) + (i_2 * T^2)) * S + (j_o * S^1.5)
A_w = h_o + (h_1 * T) + (h_2 * T^2) + (h_3 * T^3)

h_o = 3.23990
h_1 = 1.43713/10^3
h_2 = 1.16092/10^4
h_3 = – 5.77905/10^7

i_o = 2.28380/10^3

i_1 = – 1.09810/10^5

i_2 = – 1.60780/10^6

j_o = 1.91075/10^4

B_k = B_w + (m_o + (m_1 * T) + (m_2 * T^2)) * S
B_w = k_o + (k_1 * T) + (k_2 * T^2)

k_o = 8.50935/10^5
k_1 = – 6.12293/10^6
k_2 = 5.27870/10^8

m_o = – 9.9348/10^7
m_1 = 2.0816/10^8
m_2 = 9.1697/10^10

"Density as a function of Temperature, Pressure and Salinity"

Rho_T_P_S = Rho_S_T/(1 – (P/K))
```

Sun et al. Correlation

```
"Correlation for Density as a Function of Temperature, Pressure, and Salinity"

"Please note that here in the model m, n, x, and y are notations to represent terms"

"Variable to be defined by the user"
```

```
P = 0.1 "MPA"
T = 20 "C"
S = 35 "g/kg"

m[1] = 9.9920571 * 10^2
m[2] = 9.5390097/10^2
m[3] = - 7.6186636/10^3
m[4] = 3.1305828/10^5
m[5] = - 6.1737704/10^8
m[6] = 4.3368858/10^1
m[7] = 2.5495667/10^5
m[8] = - 2.8988021/10^7
m[9] = 9.5784313/10^10
m[10] = 1.7627497/10^3
m[11] = - 1.2312703/10^4
m[12] = 1.3659381/10^6
m[13] = 4.0454583/10^9
m[14] = - 1.4673241/10^5
m[15] = 8.8391585/10^7
m[16] = - 1.1021321/10^9
m[17] = 4.2472611/10^11
m[18] = - 3.9591772/10^14

n[1] = 1
n[2] = t
n[3] = t^2
n[4] = t^3
n[5] = t^4
n[6] = p
n[7] = p * t^2
n[8] = p * t^3
n[9] = p * t^4
n[10] = p^2
n[11] = p^2 * t
n[12] = p^2 * t^2
n[13] = p^2 * t^3
n[14] = p^3
n[15] = p^3 * t
n[16] = p^3 * t^2
n[17] = p^3 * t^3
n[18] = p^3 * t^4

x[1] = S
x[2] = S * t
```

```
x[3] = S * t^2
x[4] = S * t^3
x[5] = p * S
x[6] = p^2 * S

y[1] = – 7.99992230/10^1
y[2] = 2.40936500/10^3
y[3] = – 2.58052775/10^5
y[4] = 6.85608405/10^8
y[5] = 6.29761106/10^4
y[6] = – 9.36263713/10^7

"Relation for Saline Water Density as a Function of Temperature, Pressure, and Salinity"

rho = sum(m[i] * n[i], i = 1, 18) – sum(x[i] * y[i], i = 1, 6)
```

UNESCO-1981

```
"Density of Sea Water as function Salinity and Temperature"

"Correlation is valid from salinity range of 0 to 42 g/kg of saline water and temperature range of 0 to 40 C"

T = 20 "C"
S = 35 "g/kg of saline Water"

rho = a_o + (a_1 * T) + (a_2 * T^2) + (a_3 * T^3) + (a_4 * T^4) + (a_5 * T^5)
+ (B * S) + (C * S^1.5) + (D * S^2)

B = b_o + (b_1 * T) + (b_2 * T^2) + (b_3 * T^3) + (b_4 * T^4)

C = c_o + (c_1 * T) + (c_2 * T^2)

D = 4.8314/10^4

b_o = 8.2449/10^1
b_1 = – 4.0899/10^3
b_2 = 7.6483/10^5
b_3 = – 8.2467/10^7
b_4 = 5.3875/10^9
```

```
c_o = – 5.7246/10^3
c_1 = 1.0227/10^4
c_2 = – 1.6546/10^6
a_o = 999.842594
a_1 = 6.793953/10^2
a_2 = – 9.095290/10^3
a_3 = 1.001685/10^4
a_4 = – 1.120083/10^6
a_5 = 6.536332/10^9
```

Sun et al. Correlation (Modified)

```
"Modified-Sun Correlation for Density as a Function of Temperature, Pressure, and Salinity"

"Please note that here in the model m, n, x, and y are notations to represent terms"

"Variable to be defined by the user"
P = 0.1 "MPA"
{T = 20} "C"
S = 35 "g/kg"

m[1] = 9.9920571 * 10^2
m[2] = 9.5390097/10^2
m[3] = – 7.6186636/10^3
m[4] = 3.1305828/10^5
m[5] = – 6.1737704/10^8
m[6] = 4.3368858/10^1
m[7] = 2.5495667/10^5
m[8] = – 2.8988021/10^7
m[9] = 9.5784313/10^10
m[10] = 1.7627497/10^3
m[11] = – 1.2312703/10^4
m[12] = 1.3659381/10^6
m[13] = 4.0454583/10^9
m[14] = – 1.4673241/10^5
m[15] = 8.8391585/10^7
m[16] = – 1.1021321/10^9
m[17] = 4.2472611/10^11
m[18] = – 3.9591772/10^14

n[1] = 1
n[2] = t
n[3] = t^2
```

```
n[4] = t^3
n[5] = t^4
n[6] = p
n[7] = p * t^2
n[8] = p * t^3
n[9] = p * t^4
n[10] = p^2
n[11] = p^2 * t
n[12] = p^2 * t^2
n[13] = p^2 * t^3
n[14] = p^3
n[15] = p^3 * t
n[16] = p^3 * t^2
n[17] = p^3 * t^3
n[18] = p^3 * t^4

x[1] = S
x[2] = S^2
x[3] = S^3
x[4] = S^4
x[5] = S * t
x[6] = S * t^2
x[7] = p * S
x[8] = p^2 * S
x[9] = p * S * t

y[1] = – 8.23082313/10^1
y[2] = 1.77931996/10^3
y[3] = – 5.88155619/10^5
y[4] = 6.15319929/10^7
y[5] = 2.80832192/10^3
y[6] = – 2.67161731/10^5
y[7] = 8.12062506/10^4
y[8] = – 1.03473343/10^6
y[9] = – 8.83528781/10^6

"Relation for Saline Water Density as a Function of Temperature, Pressure, and Salinity"

rho = sum(m[i] * n[i], i = 1, 18) – sum(x[i] * y[i], i = 1, 9)
```

Poisson et al. (Modified)

"Correlation for density of saline water reported by A. Poisson, C. Brunet and J. C. Brun-Cottan is given below"
"Temperature at which density is to be determined"

T = 15 "C"

"Density Correlaions"
"Please note that temperature is degree C"

(rho − rho_o) * 10^3 = 28.255 − (0.1181 * t) + (1.937/10^3 * t^2) − (1.399/10^5 * t^3)

"Density of pure water as a function of temperature is given as"

rho_o * 10^3 = (999.842594) + (6.793952/10^2 * t) − (9.095290/10^3 * t^2) + (1.001685/10^4 * t^3) − (1.120083/10^6 * t^4) + (6.536332/10^9 * t^5)

"This correlation is valid for temperature range of 0 to 30 C and salinity range of 0 to 40 parts per thousands"
"Pressure is atmospheric or close to atmospheric"
rho_f = rho * 1000

References

Bigg PH (1967) Density of water in SI units over the range 0–40 °C. Br J Appl Phys 18(4):521–525. https://doi.org/10.1088/0508-3443/18/4/315

Bryden HL (1973) New polynomials for thermal expansion, adiabatic temperature gradient and potential temperature of sea water. Deep-Sea Res Oceanogr Abs 20(4):401–408. https://doi.org/10.1016/0011-7471(73)90063-6

Cisternas LA, Lam EJ (1989) A analytic correlation of vapour pressure of aqueous and non-aqueous solutions of single and mixed electrolytes. Fluid Phase Equilib 53(C):243–249. https://doi.org/10.1016/0378-3812(89)80092-5

El-Dessouky HT, Ettouney HM (2002) Fundamentals of salt water desalination. https://doi.org/10.1016/b978-0-444-50810-2.x5000-3

Gebhart B, Mollendorf JC (1977) A new density relation for pure and saline water. Deep-Sea Res 24(9):831–848. https://doi.org/10.1016/0146-6291(77)90475-1

Generous MM, Qasem NAA, Qureshi BA, Zubair SM (2020) A comprehensive review of saline water correlations and data-part I: thermodynamic properties. Arab J Sci Eng 45(11):8817–8876. https://doi.org/10.1007/s13369-020-05019-y

IOC, SCOR, IAPSO (2010) The international thermodynamic equation of seawater—2010: calculation and use of thermodynamic properties. Manuals and guides no. 56, June. Intergovernmental Oceanographic Commission, p 196

Isdale JD, Morris R (1986) Physical properties of sea water: density. Numer Data Funct Relat Sci Technol 3a(3):237–259. https://doi.org/10.1007/10201933_43

Laliberté M, Cooper WE (2004) Model for calculating the density of aqueous electrolyte solutinos. J Chem Eng Data 49(5):1141–1151. https://doi.org/10.1021/je0498659

Lam EJ, Alvarez MN, Galvez ME, Alvarez EB (2008) A model for calculating the density of aqueous multicomponent electrolyte solutions. J Chil Chem Soc 53(1):1404–1409. https://doi.org/10.4067/S0717-97072008000100015

Leyendekkers JV (1980) Prediction of the heat capacities of seawater and other multicomponent solutions from the Tammann–Tait–Gibson model. Mar Chem 9(1):25–35. https://doi.org/10.1016/0304-4203(80)90004-3

Millero FJ (2010) History of the equation of state of seawater. Oceanography 23(3):18–33. https://doi.org/10.5670/oceanog.2010.21

Millero FJ, Huang F (2009) The density of seawater as a function of salinity (5 to 70 g kg^{-1}) and temperature (273.15 to 363.15 K). Ocean Sci 27(1):91–110. https://doi.org/10.1017/S0373463300025200

Millero FJ, Waters J, Woosley R, Huang F, Chanson M (2008) The effect of composition on the density of Indian Ocean waters. Deep Sea Res Part 1 Oceanogr Res Pap 55(4):460–470. https://doi.org/10.1016/j.dsr.2008.01.006

Nayar KG, Sharqawy MH, Banchik LD, Lienhard JH (2016) Thermophysical properties of seawater: a review and new correlations that include pressure dependence. Desalination 390:1–24. https://doi.org/10.1016/j.desal.2016.02.024

Numbere D, Brigham WE, Standing MB (1977) Correlation for physical properties of petroleum reservoir brines. United States. https://doi.org/10.2172/6733264

Poisson A, Brunet C, Brun-Cottan JC (1982) Density of standard seawater solutions at atmospheric pressure. Mar Geodesy 5(4):363–365. https://doi.org/10.1080/15210608209379434

Rowe AM, Chou JCS (1970) Pressure-volume-temperature-concentration relation of aqueous NaCI solutions. J Chem Eng Data 15(1):61–66. https://doi.org/10.1021/je60044a016

Safarov J, Berndt S, Millero F, Feistel R, Heintz A, Hassel E (2012) (p, ρ, T) properties of seawater: extensions to high salinities. Deep Sea Res Part 1 Oceanogr Res Pap 65:146–156. https://doi.org/10.1016/j.dsr.2012.03.010

Schmidt H, Seitz S, Hassel E, Wolf H (2018) The density-salinity relation of standard seawater. Ocean Sci 14(1):15–40. https://doi.org/10.5194/os-14-15-2018

Sharqawy MH, Lienhard VJH, Zubair SM (2011) Thermophysical properties of seawater: a review of existing correlations and data. Desalin Water Treat 29(1–3):355. https://doi.org/10.5004/dwt.2011.2947

Standing MB (1977) Correlations for physical properties of petroleum reservoir brines. https://doi.org/10.2172/6733264

Sun H, Feistel R, Koch M, Markoe A (2008a) New equations for density, entropy, heat capacity, and potential temperature of a saline thermal fluid. Deep Sea Res Part 1 Oceanogr Res Pap 55(10):1304–1310. https://doi.org/10.1016/j.dsr.2008.05.011

Sun H, Koch M, Markoe A, Feistel R (2008b) New equation of density, potential temperature, and entropy of saline thermal fluid

Tanaka M, Girard G, Davis R, Peuto A, Bignell N (2001) Recommended table for the density of water between 0 °C and 40 °C based on recent experimental reports. Metrologia 38(4):301–309. https://doi.org/10.1088/0026-1394/38/4/3

Gebhart B, Mollendorf JC (1977) A new density relation for pure and saline water. Deep-Sea Res 24(9):831–848. https://doi.org/10.1016/0146-6291(77)90475-1

UNESCO (1981a) Background papers and supporting data on the international equation of state of seawater 1980. UNESCO technical papers in marine science, vol 37

UNESCO (1981b) Background papers and supporting data on the practical salinity scale 1978. UNESCO technical papers in marine science, vol 37, pp 1–144

Chapter 3
Specific Heat Capacity

Abstract The specific heat capacity of saline water is an important thermodynamic property for the estimation of the energy equation, thermal energy capacity, and industrial processes, including desalination. This chapter discusses multicomponent models, correlation, and data to estimate specific heat capacity at various salinity, temperature, and pressure values. The estimated specific heat capacity values from the correlations and methods are compared and assessed using experimental data from seawater. Based on their accuracy and operating conditions, some correlations and methods are recommended. In addition, the codes for the methods used to calculate specific heat capacity are provided.

Keywords Specific heat capacity · Correlations · Methods · Energy · Saline water

3.1 Introduction

The specific heat capacity of saline water is vital to help in evaluating energy consumption for treatment and desalination processes. Thus, the objective of this chapter is to summarize the procedures and correlations reported in the literature to calculate (isobaric) specific heat capacity of saline water. As saline water is a mixture of some species, the heat capacity of a mixture is equal to the sum of individual constituents provided that its heat capacity in the mixture behaves similarly as it behaves in the pure form (Randall and Rossini 1929). The heat capacity of a mixture can also be additive if the deviation of both substances is exactly the reciprocal of each other. Unfortunately, this is a rare phenomenon in the real world. Therefore, the properties are not directly additive, and the calculation of heat capacity is required to be formulated.

The (isobaric) specific heat capacity of saline water depends on the temperature, pressure, and salinity. It has been found that increasing the salinity decreases the amount of energy required to raise the unit temperature of saline water. Therefore, isobaric heat capacity decreases with increasing the salinity level and vice versa (Fisher and Dial 2005). Darros-Barbosa et al. (2003) investigated the effect of temperature on the saline water heat capacity. The specific heat capacity of saline

N. A. A. Qasem et al., *Thermodynamic and Thermophysical Properties of Saline Water*, Springer Water, https://doi.org/10.1007/978-3-031-35193-8_3

water is shown to be increased with increasing temperature. The pressure is also an important factor that could influence the heat capacity. The increase in pressure results in decreasing the specific heat capacity. However, the heat capacity was found to be a weak function of pressure (e.g., increasing pressure from 0 to 100 MPa was found to depreciate the heat capacity value only by 0.1 J/kg K (Zemaitis et al. 1986)).

The models and correlations reported in the literature are compiled in correlations and data. The range of operating conditions, accuracy, and saline water type is illustrated. This covers a salinity range from 0 to 180 g/kg, a temperature ranges from 0 to 200 °C, and a pressure range from 0 to 100 MPa. Saline water with ionic impurities can also be considered a multi-species ionic solution. The heat capacity of saline water can be calculated by adding the heat capacity of water in the solution and the heat capacity of salt ions (Leduc and Desnoyers 2006). Leyendekkers (1980) developed an approach based on this concept and used Tammann–Tait–Gibson (TTG) model to calculate heat capacity at any given pressure. As summarized in this chapter, a similar model, reported by Bromley et al. (1970), has used the concept of ionic strength (I) and partial and apparent heat capacity concepts. This model is efficient in calculating the isobaric heat capacity of solutions having a variety of ionic species.

3.2 Multiple Component Models

Multiple components models were proposed to be general for all types of saltwater under different salinity values. Major constituents of the aqueous ionic solutions are cations, anions, and water. Bromley's (1970) method, based on Debye Huckel theory, was extended and generalized for multi-ionic solutions. Specific heat values at saturation were experimentally determined for 80–200 °C (see Appendix 1), and these were used to calculate the required specific heat values. Data from 2 to 80 °C (Bromley et al. 1967) (see Appendix 2) was also combined to extend the temperature range of the model.

$$C_p = C_{pw}(1 - x) + x(\Phi C_{ps}) \tag{3.1}$$

where C_p is the heat capacity of the solution in cal/g °C, C_{pw} is the partial heat capacity of water in the solution in cal/(g °C), and ΦC_{ps} is the apparent heat capacity of salts in the solution in cal/g °C.

$$C_{pw} = C_p^* - \frac{A}{2}\left(\frac{\sigma I^{0.5}}{3}\right)\left(\frac{2I^{1.5}}{1000}\right) + \left(\frac{B}{1000}\right)\frac{\left(2\sum_{i=1}^{n} m_i M_i\right)}{\sum_{i=1}^{n} m_i}\left(\frac{\sum_{i=1}^{n} m_i}{2}\right)^2 \tag{3.2}$$

Here, C_p^* is the heat capacity of pure water in kJ/kg K, M_i is the molar mass of ions in g/mol, and A and B are constants (refer to Table 3.1). Ionic strength (I) can be calculated as:

$$I = \frac{1}{2}\sum_{i=1}^{n} m_i Z_i^2 \tag{3.3}$$

$$\Phi C_{ps} = C_p^o - \frac{A}{2}\left(\frac{\sum_{i=1}^{n} m_i Z_i^2}{\sum_{i=1}^{n} m_i M_i}\right) I^{0.5}\left(\frac{1}{1+I^{0.5}} - \frac{\sigma I^{0.5}}{3}\right) - B\frac{\sum_{i=1}^{n} m_i}{2} \tag{3.4}$$

where C_p^o is the heat capacity of salt at infinite dilution, and

$$\frac{\sigma I^{0.5}}{3} = \frac{1}{I^{0.5}}\left(1 + I^{0.5} - \frac{1}{1+I^{0.5}} - 2\ln\left(1+I^{0.5}\right)\right) \tag{3.5}$$

The method can be used in a salinity range of 0–120 g/kg, temperature range of 0–200 °C, and pressure of 0.1 MPa. The accuracy is reported to be within 0.0042 kJ/kg K, but it is ± 3% when compared with the experimental data of Chou (1968) (see Appendix 3).

Table 3.1 Constants used in modified Debye–Hückel calculation of heat capacities (Bromley 1973)

T (°C)	A	C_p^α	C_p^-	B
0	7.498	1.00804	– 0.76760	0.22811
5	8.136	1.00434	– 0.61757	0.16929
10	8.737	1.00194	– 0.51709	0.13140
15	9.309	1.00040	– 0.44876	0.10679
20	9.862	0.99947	– 0.40269	0.09112
25	10.405	0.99893	– 0.37021	0.08057
30	10.945	0.99866	– 0.34659	0.07305
35	11.492	0.99860	– 0.32950	0.06751
40	12.055	0.99869	– 0.31662	0.06287
45	12.643	0.99889	– 0.30641	0.05844
50	13.264	0.99919	– 0.29853	0.05403
55	13.927	0.99959	– 0.29301	0.04961
60	14.640	1.00008	– 0.28951	0.04501
65	15.414	1.00065	– 0.28771	0.04001
70	16.255	1.00132	– 0.28842	0.03496
75	17.175	1.00208	– 0.29133	0.02966
80	18.180	1.00295	– 0.29726	0.02445

3.3 Correlations and Data

Correlations are the easiest way to calculate the saltwater heat capacity as they are fitted from the experimental data. Most of the available specific heat capacity correlations in the literature are for seawater. Nayar et al. (2016) correlation can be used to calculate the specific heat capacity of seawater. For temperature (T), in °C, pressure (P), in MPa, and salinity (S), in g/kg,

$$C_p(S, T, P) = C_p(T, S) + (P - P_o)\left(a_1 + a_2T + a_3T^2 + a_4T^3\right) + \left(S\left(a_5 + a_6T + a_7T^2 + a_8T^3\right)\right) \quad (3.6)$$

C_p in terms of temperature and salinity can be expressed as:

$$C_p(T, S) = C_p(T) + AS + BS^{1.5} \quad (3.7)$$

C_p in terms of temperature is expressed as:

$$C_p(T) = C_o + C_1T + C_2T^2 + C_3T^3 + C_4T^4 \quad (3.8)$$

Constants A and B are as given below:

$$A = d_o + d_1T + d_2T$$

$$B = b_o + b_1T + b_2T$$

Values of the constants used for this correlation are provided in Table 3.2.

The correlation can be used in a salinity range of 0–180 g/kg, temperature range of 0–180 °C, and pressure of 0.1–1 MPa with an accuracy of ± 1%. However, it is

Table 3.2 Constant for Nayar et al. (2016) correlation

Constants	Values	Constants	Values
a_1	-3.1118	b_2	5.148×10^{-5}
a_2	0.0157	C_o	4217.4
a_3	5.1014×10^{-5}	C_1	-3.72030283
a_4	-1.0302×10^{-6}	C_2	0.1412855
a_5	0.0107	C_3	-2.654387×10^{-3}
a_6	-3.9716×10^{-5}	C_4	2.093236×10^{-5}
a_7	3.2088×10^{-8}	d_o	-7.643575
a_8	1.0119×10^{-9}	d_1	0.1072763
b_o	0.1770383	d_2	-1.38385×10^{-3}
b_1	-4.07718×10^{-3}	b_2	5.148×10^{-5}

valid too for the salinity range of 0–42 g/kg, temperature range of 0–180 °C, and pressure of 0.1–12 MPa with an accuracy of ± 1%.

Ghallab and Elnahas (2016) also proposed a correlation for seawater and brackish water in terms of temperature (T), in °C, pressure (P), in MPa, and salinity (S), in g/kg,

$$C_p = \sum_{i=1}^{28} X_i Y_i Z_i \tag{3.9}$$

X, Y, and Z are expressed in terms of salinity and temperature in Table 3.3. The correlation is valid in the salinity range of 0–180 g/kg, temperature range of 0–180 °C, and pressure of 0.1 MPa with an accuracy of ± 1%.

El-Dessouky and Ettouney (2002) proposed a correlation for specific heat capacity calculation of seawater in terms of temperature (T), in °C, and salinity (S), in g/kg,

$$C_p = \left(A + BT\, CT^2 + DT^3\right) \times 10^{-3} \tag{3.10}$$

The constants are provided below,

$$A = 4206.8 - 6.6197S + 1.2288 \times 10^{-2}S^2$$

$$B = -1.1262 + 5.4178 \times 10^{-2}S - 2.2719 \times 10^{-4}S^2$$

$$C = 1.2026 \times 10^{-2} - 5.3566 \times 10^{-4}S + 1.8960 \times 10^{-6}S^2$$

$$D = 6.8777 \times 10^{-7} + 1.517 \times 10^{-6}S - 4.4268 \times 10^{-9}S^2$$

The correlation can be used in a salinity range of 20–160 g/kg, temperature range of 20–80 °C, and pressure of 0.1. The accuracy is not mentioned by the source. However, it is within ± 2.2% of the experimental data of Chou (1968).

Chou (1971) proposed two correlations for the specific heat capacity calculation of seawater. His first correlation in terms of temperature (T), in °C, and salinity (S), in g/kg, is provided in Eq. (3.11), and the constants are provided in Table 3.4.

$$C_p = A + BT + CT^2 + \frac{D}{T} + \left(E + FT + GT^2 + \frac{H}{T}\right)S + \left(1 + JT + KT^2 + \frac{L}{T}\right) \tag{3.11}$$

The first correlation can be used in a salinity range of 0–120 g/kg, temperature range of 0–200 °C, and pressure of 0.1 MPa with an accuracy of ± 2.3% of the experimental data of Chou (1968).

Table 3.3 Data for Ghallab and Elnahas (2016) correlation

X_i	Y_i	Z_i
4.21	S^0	T^0
− 6.66	S^1	T^0
14.38	S^2	T^0
− 40.4	S^3	T^0
379.7	S^4	T^0
− 1692	S^5	T^0
2920	S^6	T^0
1.18×10^{-3}	S^0	T^1
5.62×10^{-2}	S^1	T^1
-2.7×10^{-1}	S^2	T^1
2.5×10^{-1}	S^3	T^1
− 1.38	S^4	T^1
2.434	S^5	T^1
1.50×10^{-5}	S^0	T^2
-5.77×10^{-4}	S^1	T^2
2.45×10^{-3}	S^2	T^2
-3×10^{-4}	S^3	T^2
1.40×10^{-3}	S^4	T^2
-5.73×10^{-8}	S^0	T^3
1.79×10^{-6}	S^1	T^3
-7.29×10^{-6}	S^2	T^3
-4.61×10^{-7}	S^3	T^3
5.28×10^{-10}	S^0	T^4
-4.25×10^{-10}	S^1	T^4
2.47×10^{-10}	S^2	T^4
-2.27×10^{-12}	S^0	T^5
6.72×10^{-13}	S^1	T^5
3.75×10^{-15}	S^0	T^6

Table 3.4 Constants for the first Chou (1971) correlation

Constants	Values	Constants	Values
A	1.7875111	G	$-0.16410434 \times 10^{-7}$
B	$-0.33318151 \times 10^{-2}$	H	− 1.4014369
C	0.43603294×10^{5}	I	$0.59886571 \times 10^{-4}$
D	− 53.950447	J	$-0.23457641 \times 10^{-6}$
E	$0.45477025 \times 10^{-2}$	K	$0.27592503 \times 10^{-9}$
F	0.7704077×10^{-6}	L	$-0.35977131 \times 10^{-2}$

The second Chou (1971) correlation in terms of temperature (T) in °C and salinity (S) in g/kg is given as,

$$\begin{aligned} C_p = {} & 1.5651785 - 0.3608443 \times 10^{-2}T + 0.57252694 \times 10^{-5}T^2 \\ & - \left((0.02458979 - 0.14470982 \times 10^{-3}T + 0.22311467 \times 10^{-6}T^2)S\right) \\ & + \left((0.15246268 \times 10^{-3} - 0.94150958 \times 10^{-6}T + 0.1460181 \times 10^{-8}T^2)S^2\right) \end{aligned} \tag{3.12}$$

The specific heat capacity is in cal/g °C. The second correlation can be used in a salinity range of 0–120 g/kg, temperature range of 0–80 °C, and pressure of 0.1 MPa. Accuracy is within ± 2.3% of the experimental data of Chou (1968).

Bromley et al. (1967) proposed three correlations for the specific heat capacity of seawater. His first correlation in terms of temperature (T) in °C and salinity (S) in g/kg is given below:

$$\begin{aligned} C_p = {} & 1.0049 - 0.016210S + 3.5261 \times 10^3S^2 \\ & - \left((3.2506 - 1.4795S + 0.027454S^2) \times 10^{-4}T\right) \\ & + \left((3.8013 - 1.2084S + 0.06121S^2) \times 10^{-6}T^2\right) \end{aligned} \tag{3.13}$$

The specific heat capacity is in cal/g °C. The correlation can be used in a salinity range of 0–120 g/kg, temperature range of 2–80 °C, and pressure of 0.1 MPa. The maximum deviation from experimental values at the 90% confidence limit is 0.001 cal/g °C (Bromley et al. 1967). Bromley's second correlation, in terms of temperature (T) in °C and salinity (S) in g/kg, is given below:

$$\begin{aligned} C_p = {} & 1.007 - 0.014880S + \left(2.8618 \times 10^{-4}S^2\right) \\ & - \left((0.14054 - 0.48497S + 0.027454S^2) \times 10^{-4}T\right) \end{aligned} \tag{3.14}$$

The specific heat capacity is in cal/g °C. The second correlation can be used in a salinity range of 0–120 g/kg, temperature range of 2–80 °C, and pressure of 0.1 MPa. The maximum deviation from experimental values at the 90% confidence limit is 0.003 cal/g °C (Bromley et al. 1967). Bromley's third correlation, in terms of temperature (T) in °C and salinity (S) in wt%, is given below:

$$\begin{aligned} C_p = {} & 0.9948 - 0.011412S + (0.42997 + 0.15192S) \\ & - \left((3.2506 - 1.4795S + 0.027454S^2) \times 10^{-4}T\right) \end{aligned} \tag{3.15}$$

The specific heat capacity is in cal/g °C. The correlation can be used in a salinity range of 2–120 g/kg, temperature range of 2–80 °C, and pressure of 0.1 MPa. The maximum deviation from experimental values at the 90% confidence limit is 0.004 cal/g °C (Bromley et al. 1967).

Table 3.5 Constants for Jamieson et al. (1969) correlation for the specific heat capacity of seawater

Constants	Values	Constants	Values
a_1	5.328	c_1	9.6×10^{-6}
a_2	-9.76×10^{-2}	c_2	-1.927×10^{-6}
a_3	4.04×10^{-4}	c_3	8.23×10^{-9}
b_1	-6.913×10^{-3}	d_1	2.5×10^{-9}
b_2	7.351×10^{-4}	d_2	1.666×10^{-9}
b_3	-3.15×10^{-6}	d_3	-7.125×10^{-12}

Jamieson et al. (1969) proposed a correlation for the calculation of specific heat capacity of seawater in terms of temperature (T), in K, and salinity (S), in kg/kg, is given in Eq. (3.16), and constants are summarized in Table 3.5.

$$\begin{aligned} C_p(T, S) = {} & \left(a_1 + a_2 S + a_3 S^2\right) + \left(b_1 + b_2 S^2 + b_3 S^2\right) T \\ & + \left(c_1 + c_2 S + c_3 S^2\right) T^2 + \left(d_1 S + d_2 S + d_3 S^2\right) T^3 \end{aligned} \quad (3.16)$$

The data was extracted from the graphs provided and is given in Appendix 4. The correlation can be used in a salinity range of 0–180 g/kg, temperature range of 0–180 °C, and pressure of 0.1 MPa with reproducibility within ± 1%.

The extended Millero method for the specific heat of seawater is derived from UNESCO (1981), for temperature (T), in °C, pressure (P), in bar, and salinity (S), in g/kg,

$$C_p(S,\ T,\ P) = C_{p1} + \Delta C_{p1} + \Delta C_{p2} \quad (3.17)$$

C_{p1}, ΔC_{p1} and ΔC_{p2} can be calculated from the following expressions,

$$\begin{aligned} \Delta C_{p1} = {} & \left(d_o + d_1 T + d_2 T^2 + d_3 T^3 + d_4 T^4\right) P \\ & + \left(e_o + e_1 T + e_2 T^2 + e_3 T^3 + e_4 T^4\right) P^2 \\ & + \left(f_o + f_1 T + f_2 T^2 + f_3 T^3\right) P \end{aligned} \quad (3.18)$$

$$\begin{aligned} \Delta C_{p2} = {} & \left(\left(g_o + g_1 T + g_2 T^2 + g_3 T^3 + g_4 T^4\right) S + \left(h_o + h_1 T + h_2 T^2\right) S^{1.5}\right) P \\ & + \left(\left(I_o + I_1 T + I_2 T^2 + I_3 T^3\right) S + J S^{1.5}\right) P^2 \\ & + \left(\left(k_o + k_1 T + k_2 T^2\right) S + L T S^{1.5}\right) P^3 \end{aligned} \quad (3.19)$$

$$C_{p1} = C_{p_T} + AS + BS^{1.5} \quad (3.20)$$

C_{p_T} as a function of temperature is expressed in Eq. (3.21). The constant for this correlation is provided in Table 3.6.

$$C_{p_T} = C_o + C_1 T + C_2 T^2 + C_3 T^3 + C_4 T^4 \quad (3.21)$$

A and B in Eq. (3.21) are calculated as,

$$A = a_o + a_1 T + a_2 T$$

$$B = b_o + b_1 T + b_2 T$$

The correlation can be used in a salinity range of 0–40 g/kg, temperature range of 0–40 °C, and pressure of 0.1–100 MPa with an accuracy of ±2.26% of the experimental data of Chou (1968).

Millero correlation obtained from UNESCO (1981) is valid for seawater only. It is in terms of temperature (T) in °C, and salinity (S), in g/kg. The main correlation is summarized in Eq. (3.22), and relevant constant values are provided in Table 3.7.

$$C_p(T, S) = C_p(T) + AS + BS^{1.5} \tag{3.22}$$

Table 3.6 Constant for the extended Millero method for the specific heat of seawater is derived from UNESCO (1981) report

Constants	Values	Constants	Values
a_o	-7.643575	f_o	-5.422×10^{-8}
a_1	0.1072763	f_1	2.6380×10^{-9}
a_2	-1.38385×10^{-3}	f_2	-6.5637×10^{-11}
b_o	0.1770383	f_3	6.136×10^{-13}
b_1	-4.07718×10^{-3}	g_o	4.9247×10^{-3}
b_2	5.148×10^{-5}	g_1	-1.28315×10^{-4}
c_o	4217.4	g_2	9.802×10^{-7}
c_1	-3.72030283	g_3	2.5941×10^{-8}
c_2	0.1412855	g_4	-2.9179×10^{-10}
c_3	-2.654387×10^{-3}	h_o	-1.2331×10^{-4}
c_4	2.093236×10^{-5}	h_1	-1.517×10^{-6}
d_o	-4.9592×10^{-1}	h_2	3.122×10^{-8}
d_1	1.45747×10^{-2}	I_o	-2.9558×10^{-6}
d_2	-3.13885×10^{-4}	I_1	1.17054×10^{-7}
d_3	2.0357×10^{-6}	I_2	-2.3905×10^{-9}
d_4	1.7168×10^{-8}	I_3	1.8448×10^{-11}
e_o	2.4931×10^{-4}	J	9.971×10^{-8}
e_1	-1.08645×10^{-5}	k_o	5.540×10^{-10}
e_2	2.87533×10^{-7}	k_1	-1.7682×10^{-11}
e_3	-4.0027×10^{-9}	k_2	3.513×10^{-13}
e_4	2.2956×10^{-11}	L	-1.4300×10^{-12}

Table 3.7 The constant for Millero correlation was obtained from UNESCO (1981)

Constants	Values	Constants	Values
a_o	-7.643575	c_o	4217.4
a_1	0.1072763	c_1	-3.72030283
a_2	-1.38385×10^{-3}	c_2	0.1412855
b_o	0.1770383	c_3	-2.654387×10^{-3}
b_1	-4.07718×10^{-3}	c_4	2.093236×10^{-5}
b_2	5.148×10^{-5}		

Specific heat as a function of temperature can be expressed as,

$$C_p(T) = c_o + c_1 T + c_2 T^2 + cT^3 + c_4 T^4 \tag{3.23}$$

A and B in Eq. (3.22) can be expressed as,

$$A = a_o + a_1 T + a_2 T$$

$$B = b_o + b_1 T + b_2 T$$

The correlation can be used in a salinity range of 0–40 g/kg, temperature range of 0–35 °C, and pressure of 0.1 MPa with an accuracy of ± 0.5 J/kg °C.

Sun et al. (2008) proposed a correlation for seawater and brackish water in terms of temperature (T) in °C, pressure (P) in MPa, and salinity (S) in g/kg as in Eq. (3.24). The relevant constants are summarized in Table 3.8.

$$C_p = \sum_{i=1}^{10} m_i n_i - \sum_{i=1}^{10} X_i Y_i \tag{3.24}$$

The correlation can be used in a salinity range of 0–40 g/kg, temperature range of 0–374 °C, and pressure of 0.1–100 MPa with an accuracy of ±4.62%.

Cox and Smith (1959) proposed a correlation for seawater in terms of temperature (T), in °C and salinity (S), in g/kg,

$$C_p = A - 0.005075 - 0.000014S^2 \tag{3.25}$$

where A is isobaric heat capacity for pure water (refer to No. # 16). The correlation can be used in a salinity range of 0–40 g/kg, temperature range of − 2 to 30 °C, and pressure of 0.1 MPa with an accuracy of ± 0.003 J/g.

Table 3.8 Constants for Sun et al. (2008) correlation for seawater and brackish water

m_i	n_i	X_i	Y_i
4.19284306	1	$5.020186422 \times 10^{3}$	S
$-2.2732541 \times 10^{-4}$	T	$-9.9612292 \times 10^{-6}$	S^2
$2.36862694 \times 10^{-6}$	T^2	$6.814886633 \times 10^{-8}$	S^3
$1.6700924 \times 10^{-10}$	T^4	$-2.6046203 \times 10^{-5}$	ST
$-3.9782283 \times 10^{-3}$	P	$4.585292916 \times 10^{-8}$	ST^2
$3.22914232 \times 10^{-5}$	PT	$7.64172529 \times 10^{-10}$	ST^3
$-1.0725210 \times 10^{-9}$	PT^3	$-3.6488019 \times 10^{-8}$	S^2T
$1.91296765 \times 10^{-5}$	P^2	$2.49629547 \times 10^{-10}$	S^3T
$-4.1758292 \times 10^{-7}$	P^2T	$1.186419227 \times 10^{-6}$	SP
$2.30627396 \times 10^{-9}$	P^2T^2	$4.34585797 \times 10^{-9}$	STP

Millero et al. (1973) proposed a correlation (see Appendix 5 for data) for seawater in terms of chlorinity (Cl) in ‰ and temperature (T) in °C,

$$C_p = C_{p_o} + A\,Cl + B\,Cl^{1.5} \tag{3.26}$$

Salinity is related to chlorinity by Eq. (3.27).

$$S = 1.81578\,Cl \tag{3.27}$$

C_{p_w} is the specific heat of pure water and can be determined from Eq. (3.28).

$$\begin{aligned} C_{p_w} = {} & 4.2174 - 3.720283 \times 10^{-3}T + 1.412855 \times 10^{-4}T^2 \\ & - 2.654387 \times 10^{-6}T^3 + 2.093236 \times 10^{-8}T^4 \end{aligned} \tag{3.28}$$

Constants A and B in Eq. (3.26) can be calculated as:

$$A = 13.81 - 0.1938T + 0.0025T^2$$

$$B = 0.43 - 0.0099T + 0.00013T^2$$

The correlation can be used in a salinity range of 0–40 g/kg, temperature range of 5–35 °C, and pressure of 0.1 MPa with an accuracy of ± 0.0013 J/g °C from the data of Cox and Smith (1959).

A correlation for pure water (Wagner and Pruß 2002) in terms of temperature (T) in °C and pressure (P) in MPa is provided in Eq. (3.29). All the relevant constants for the specific heat of pure water are given in Table 3.9.

$$C_p = \sum_{i=1}^{10} m_i n_i \tag{3.29}$$

Table 3.9 Constants for the calculation of specific heat of pure water

m_i	n_i
4.19284306	1
$-2.27325412 \times 10^{-4}$	T
$2.36862694 \times 10^{-6}$	T^2
$1.67009248 \times 10^{-10}$	T^4
$-3.97822834 \times 10^{-3}$	P
$3.22914232 \times 10^{-5}$	PT
$-1.07252107 \times 10^{-9}$	PT^3
$1.91296765 \times 10^{-5}$	P^2
$-4.1758292 \times 10^{-7}$	P^2T
$2.30627396 \times 10^{-9}$	P^2T^2

The correlation is valid for the temperature range of 0–1374 °C and pressure of 0.1–100 MPa with an accuracy of $\pm 0.01\%$ (best fit to the data of Wagner and Pruß 2002).

3.4 Case Study

The specific heat capacity of saline water is a function of salinity, temperature, and pressure. It decreases with an increase in salinity. The methods of Millero (extended) (UNESCO 1981), Sun et al. (2008), Bromley et al. (1970) model, Bromley et al. (1967), Ghallab and Elnahas (2016), El-Dessouky and Ettouney (2002), Millero et al. (1973), Nayar et al. (2016) and Cox and Smith (1959) are plotted along with the experimental data of Chou (1968) at 25 °C and 0.1 MPa, as shown in Fig. 3.1. It can be seen from Fig. 3.1a that the specific heat capacity values calculated by the correlations closely follow the experimental data. However, the Bromley model (Bromley 1973) has slightly higher values along with the salinity values. On the other hand, the Bromley correlation (Bromley et al. 1967) generates close values to the experimental data.

Figure 3.1b shows the difference in the specific heat capacity estimated by the correlations against the experimental data of Chou (1968). It can be shown that Millero (extended) (UNESCO 1981), Bromley (1973) model, Bromley et al. (1967), Ghallab and Elnahas (2016), and Nayar et al. (2016) correlations could expect the correct values of C_p within $\pm$ 3% of their proposed range of operation while the difference of the other correlations is within $\pm$ 10%.

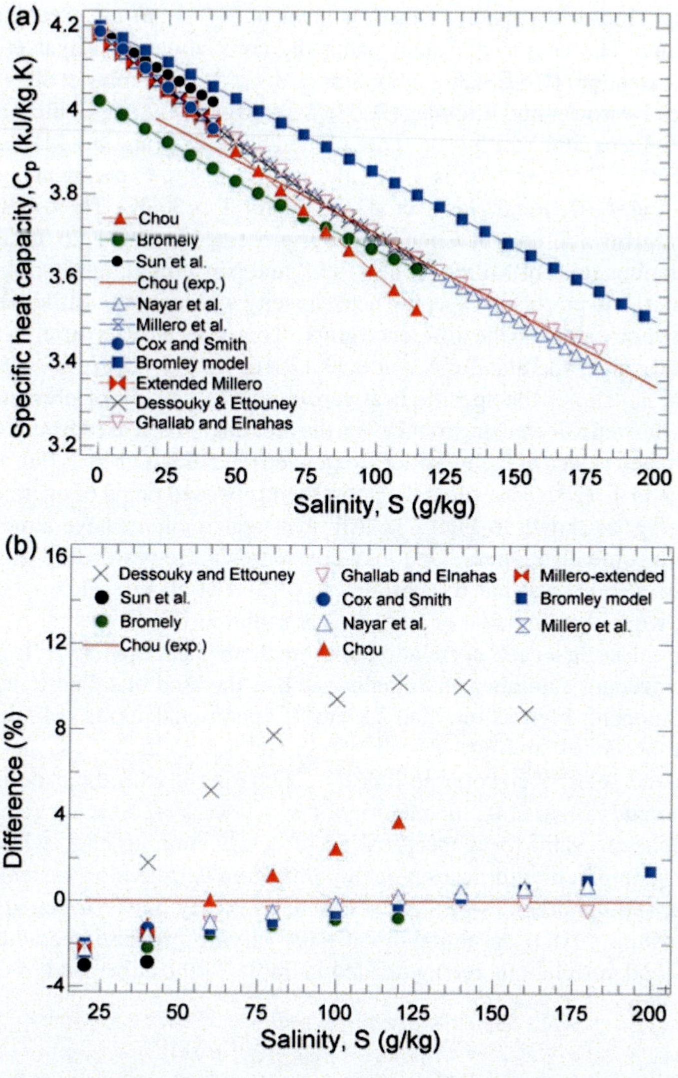

Fig. 3.1 **a** Salinity versus isobaric heat capacity and **b** the difference against the experimental data of Chou (1968) for the methods of Millero (extended) (UNESCO 1981), Sun et al. (2008), Chou (1971), Bromley et al. (1970) model, Bromley et al. (1967), Ghallab and Elnahas (2016), El-Dessouky and Ettouney (2002), Millero et al. (1973), Nayar et al. (2016), and Cox and Smith (1959) at 0.1 MPa and 25 °C

The specific heat capacity values (shown in Fig. 3.2a) are weak functions of temperature. The estimated values using the correlations of Nayar et al. (2016), Millero (extended) (UNESCO 1981), Sun et al. (2008), Bromley et al. (1967), Chou (1971), El-Dessouky and Ettouney (2002), Millero et al. (1973), Ghallab and Elnahas (2016), and Cox and Smith (1959) are plotted at 30 g/kg and 0.1 MPa. Figure 3.2a shows that all the correlations give values close to the experimental data except Millero et al. (1973) and Nayar et al. (2016) for T > 80 °C. The difference based on the experimental data of Chou (1968) is presented in Fig. 3.2b. It confirms that all correlations (except Millero et al. (1973) and Nayar et al. (2016) for T > 80 °C) could accurately predict the specific heat capacity within ± 3%. Millero et al. (1973) results are not visible on the difference graph. This is because its difference is beyond the plotted range, and also for Nayar et al. (2016) for T > 80 °C.

Figure 3.3 shows the specific heat capacity as a function of pressure. The heat capacity showed a decreasing trend when the pressure is increased from 0 to 100 MPa. The estimated values of C_p by Millero (extended) (UNESCO 1981), Sun et al. (2008), and Nayar et al. (2016) are plotted against their proposed range of operation at 25 °C and 35 g/kg, as shown in Fig. 3.3a. All investigated curves have almost the same trend with some differences. One may consider Millero (extended) as a reference; thus, Fig. 3.3b shows that the correlation of Sun et al. (2008) is accurate within ± 1.5%, while that of Nayar et al. (2016) is within ± 1%.

When selecting a good correlation to estimate the heat capacity, it is important to take into account a number of variables, such as the kind of saline water (seawater, multicomponent, high saline, and so forth), operational limits (salinity, temperature, and pressure), and fitting accuracy. In addition to these, it is also feasible to evaluate and select the proposed link by contrasting it with the experimental data that are already accessible, the majority of which were reported for seawater. Some connections are valid for a specific kind of desalination process, while others are not. For example, humidification-dehumidification desalination systems operate at 0.1 MPa, and the pressure-dependent correlations may not be necessary. However, reverse osmosis (RO) desalination calls for varying pressure characteristics. The methods and correlations recommended in Table 3.10 can be used to calculate the heat capacity.

3.5 EES Codes

The EES codes for the correlations discussed above:

Bromley Method Based on Debye Huckel Theory That Was Extended and Generalized for Multi-ionic Solutions

"Model equations"

"C_p in cal/(g.C)"

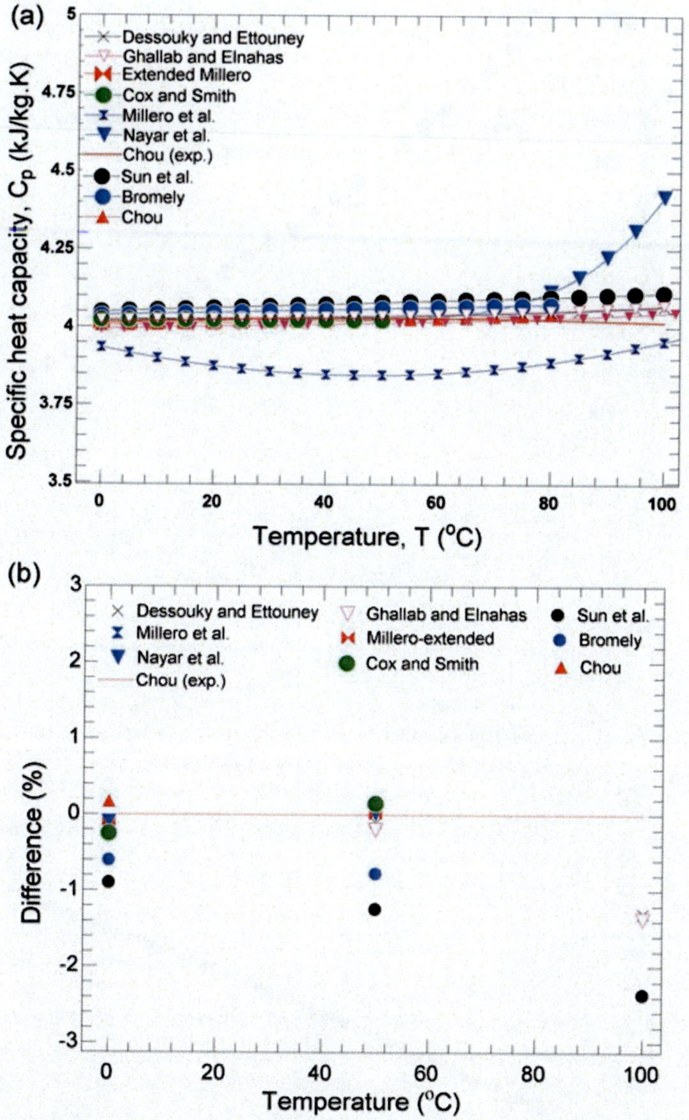

Fig. 3.2 **a** Temperature versus specific heat capacity and **b** the difference against the experimental data for the methods of Chou (1968) for Nayar et al. (2016), Millero (extended) (UNESCO 1981), Sun et al. (2008), Bromley et al. (1967), Chou (1971), El-Dessouky and Ettouney (2002), Millero et al. (1973), Ghallab and Elnahas (2016), and Cox and Smith (1959) at 0.1 MPa and 30 g/kg

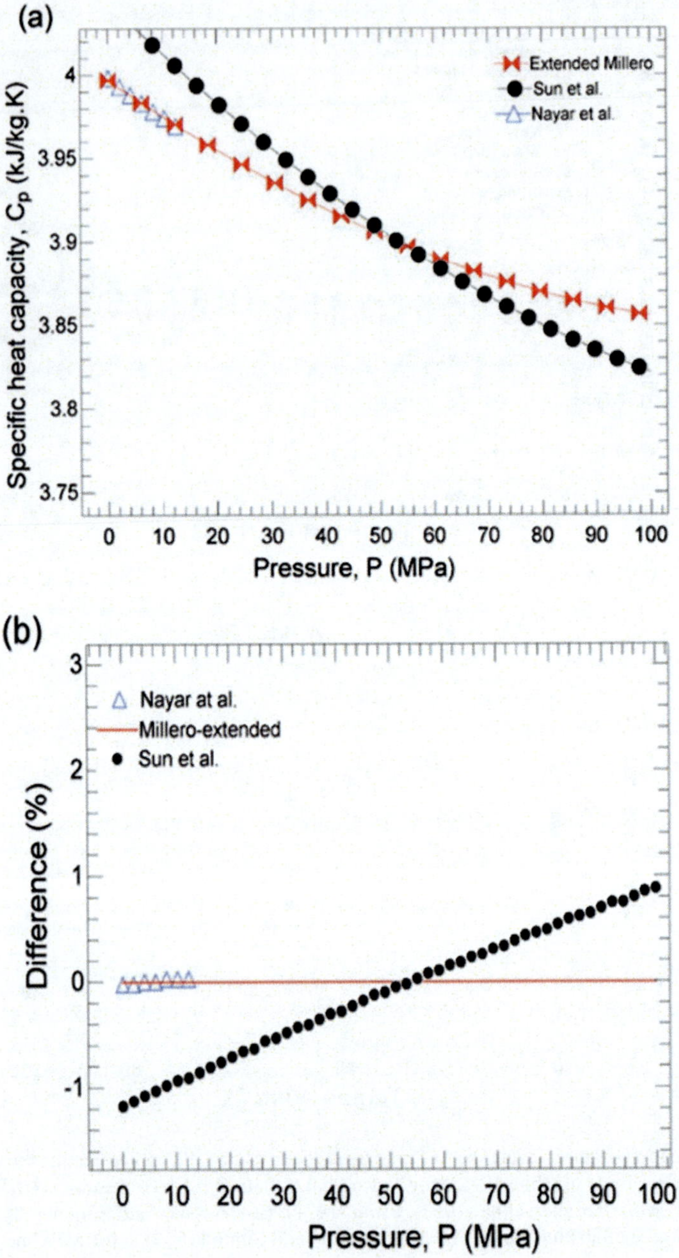

Fig. 3.3 **a** Pressure versus specific heat capacity and **b** the difference against the data of Millero (extended) (UNESCO 1981) for Sun et al. (2008) and Nayar et al. (2016) correlations at 25 °C and 35 g/kg

Table 3.10 Summary of the recommended correlations and models for the heat capacity of saline water (Generous et al. 2020)

Correlation-model	Advantages	Applications	References
Bromley multicomponent model (Eq. 3.1)	• General for multicomponent saline solution • Valid in a salinity range of 0–200 g/kg with a difference of ± 2% (refer to Fig. 3.1)	• Electrodialysis (ED) • Electrodialysis reversal (EDR) • Nanofiltration (NF) • Membrane distillation (MD) • Seawater greenhouse • High-grade water recycling • Wave-powered desalination	Bromley et al. (1970)
Ghallab and Elnahas (Eq. 3.9)	• Valid in a salinity range of 0–180 g/kg within a difference of ± 2.5% (refer to Fig. 3.1) • Valid in a temperature range of 0–100 °C within a difference of ± 1.3% (refer to Fig. 3.2)	• All thermal-based desalination technologies • High-grade water recycling • Wave-powered desalination	Ghallab and Elnahas (2016)
Millero-extended (Eq. 3.17)	• It can be used up to a salinity of 40 g/kg within difference of ± 2 % (refer to Fig. 3.1) • Valid in the pressure range of 0–100 MPa within a negligible difference (refer to Fig. 3.3)	• All membrane-based desalination technologies • All thermal-based desalination technologies • Seawater greenhouse • High-grade water recycling • Wave-powered desalination	Millero et al. (1973)
Bromley (Eq. 3.15)	• It can be used for a salinity of 120 g/kg within a difference of ± 3% (refer to Fig. 3.1) • Covers a broader temperature range of 2–80 °C within a difference of ± 1.6% (refer to Fig. 3.2)	• All thermal-based desalination technologies	Bromley et al. (1967)
Sun et al. (Eq. 3.24)	• It can be used up to a salinity of 40 g/kg within a difference of ± 3.5% (refer to Fig. 3.1) • Cover a wide range of temperatures (0–100 °C) with a difference of ± 2.5% (refer to Fig. 3.2) • It can be applied to high-pressure conditions (0–100 MPa) with a difference of ± 1.2% (refer to Fig. 3.3)	• All membrane-based desalination technologies, especially forward and reverse osmosis • All thermal-based desalination technologies • Seawater greenhouse • High-grade water recycling • Wave-powered desalination	Sun et al. (2008)

(continued)

Table 3.10 (continued)

Correlation-model	Advantages	Applications	References
Nayar et al. (Eq. 3.6)	• One of the most accurate correlations that is valid for a salinity range of 0–180 g/kg within a difference of ± 2.5% (refer to Fig. 3.1) • It can be conveniently applied to the temperature range of 0–80 °C within a difference of ± 0.5% (refer to Fig. 3.2) • It can be applied to high-pressure conditions (0–12 MPa) within a difference of ± 0.5% (refer to Fig. 3.3)	• All membrane-based desalination technologies • All thermal-based desalination technologies • Seawater greenhouse • High-grade water recycling • Wave-powered desalination	Nayar et al. (2016)

```
C_p = C_p_H2O * (1 − X) + X * (PhiC_p)

"PhiC_p"

PhiC_p = C_p_o + (A/2) * (p/q) * I^0.5 * (((1/(1 + I^0.5)) − ((sigma * I^0.5)/
3))) − (B * r/2)

C_p_o = − 0.37021
A = 10.405
"Concentrations in mol/kg"
"Please note that 1: H2O, 2: Cl−, 3: Na+, 4: Mg++, 5: SO4−−, 6: Ca++, 7: K+"
m[1] = 53.6
m[2] = 0.546 * 7.4
m[3] = 0.469 * 7.4
m[4] = 0.0528 * 7.4
m[5] = 0.0282 * 7.4
m[6] = 0.0103 * 7.4
m[7] = 0.0102 * 7.4

"Charge of each species"
z[2] = − 1
z[3] = + 1
z[4] = + 2
```

```
z[5] = − 2
z[6] = + 2
z[7] = + 1

p = sum(m[i] * z[i]^2, i = 2, 7)

"Molecular weight of ions in g/mol"
w[1] = 18
w[2] = 35.453
w[3] = 23
w[4] = 24.305
w[5] = 96
w[6] = 40
w[7] = 39.1

m_t = sum(m[i] * w[i], i = 1, 7)
m_s = sum(m[i] * w[i], i = 2, 7)
X = m_s/m_t

q = sum(m[i] * w[i], i = 2, 7)

I = p/2

((sigma * I^0.5)/3) = (1/I^1.5) * (1 + I^0.5 − (1/(1 + I^0.5)) − (2 * ln(1 +
I^0.5)))

r = sum(m[i], i = 2, 7)

"lets find C_p_H2O"

C_p_H2O = C_p_s − ((A/2) * ((sigma * I^0.5)/3) * ((2 * I^1.5)/1000)) + (B/
1000) * ((2 * q)/r) * (r/2)^2

C_p_s = 0.99893
B = 0.08057

"Conversion to SI units from cal/gC to kJ/kg.C"

C_p_SI = C_p * 4.18
```

Nayar et al. Correlation

```
"temperature 0–180 C"
"pressure 0.1 to 1 MPa"
"Salinity 0–180 g/kg"

"Temperature"

T = 25 "C"

"Salinity"

S = 35 "g/kg"

"Pressure"
P_o = 0.1
P = 0

C_p_tsp = (C_p_st + (((P – P_o) * (a1 + a2 * T + a3 * T^2 + a4 * T^3 + (S *
(a5 + a6 * T + a7 * T^2 + a8 * T^3))))))/1000

C_p_st = C_p_t + (A * S) + (B * S^1.5)

C_p_t = C_o + (C1 * T) + (C2 * T^2) + (C3 * T^3) + (C4 * T^4)

"Constants"

A = d_o + (d1 * T) + (d2 * T^2)

B = b_o + (b1 * T) + (b2 * T^2)

a1 = – 3.1118
a2 = 0.0157
a3 = 5.1014/10^5
a4 = – 1.0302/10^6
a5 = 0.0107
a6 = – 3.9716/10^5
a7 = 3.2088/10^8
a8 = 1.0119/10^9
```

```
C_o = 4217.4
C1 = − 3.72030283
C2 = 0.1412855
C3 = − 2.654387/10^3
C4 = 2.093236/10^5

d_o = − 7.643575
d1 = 0.1072763
d2 = − 1.38385/10^3

b_o = 0.1770383
b1 = − 4.07718/10^3
b2 = 5.148/10^5
```

Ghallab and Elnahas Correlation

```
"Temperature"

T = 25 "C"

"Salinity"

S = 0.035 "kg/kg"

X[1] = 4.21 * S^0 * T^0
X[2] = − 6.66 * S^1 * T^0
X[3] = 14.38 * S^2 * T^0
X[4] = − 40.4 * S^3 * T^0
X[5] = 379.7 * S^4 * T^0
X[6] = − 1692 * S^5 * T^0
X[7] = 2920 * S^6 * T^0
X[8] = − 1.18/10^3 * S^0 * T^1
X[9] = 5.62/10^2 * S^1 * T^1
X[10] = − 2.7/10^1 * S^2 * T^1
X[11] = 2.5/10^1 * S^3 * T^1
X[12] = − 1.38 * S^4 * T^1
X[13] = 2.434 * S^5 * T^1
X[14] = 1.50/10^5 * S^0 * T^2
X[15] = − 5.77/10^4 * S^1 * T^2
X[16] = 2.45/10^3 * S^2 * T^2
X[17] = − 3/10^4 * S^3 * T^2
X[18] = 1.40/10^3 * S^4 * T^2
X[19] = − 5.73/10^8 * S^0 * T^3
X[20] = 1.79/10^6 * S^1 * T^3
```

```
X[21] = − 7.29/10^6 * S^2 * T^3
X[22] = − 4.61/10^7 * S^3 * T^3
X[23] = 5.28/10^10 * S^0 * T^4
x[24] = − 4.25/10^10 * S^1 * T^4
X[25] = 2.47/10^10 * S^2 * T^4
X[26] = − 2.27/10^12 * S^0 * T^5
X[27] = 6.72/10^13 * S^1 * T^5
X[28] = 3.75/10^15 * S^0 * T^6

C_p = sum(X[i], i = 1, 28) "kJ/kg.C"
```

El-Dessouky and Ettouney Correlation

```
"temperature 20–180C"
"Salinity 20–160 g/kg"

T = 25 "C"
S = 35 "g/kg"

"Specific heat capacity at constant pressure can be given by the following correlation"

C_p = (A + (B * T) + (C * T^2) + (D * T^3))/10^3

"Constants"

A = 4206.8 – (6.6197 * S) + (1.2288/10^2 * S^2)
B = − 1.1262 + (5.4178/10^2 * S) – (2.2719/10^4 * S^2)
C = (1.2026/10^2) – (5.3566/10^4 * S) + (1.8960/10^6 * S^2)
D = (6.8777/10^7) + (1.517/10^6 * S) – (4.4268/10^9 * S^2)
```

(a) ***Chou Correlation***

```
"Chou Correlation for 0 to 200C"

"Temperature Range: 0–80C"
"Pressure Range: 1 bar"
"Salinity Range: 0–120 g/kg"
"Accuracy: within 0.00097"
```

```
T = 298.15 "C"
S = 35.154 "g/kg"

"C_P in cal/g.C"

C_p = A + B * T + C * T^2 + D/T + (E + F * T + G * T^2 + H/T) * S + (I
+ J * T + K * T^2 + L/T)

A = 1.7875111
B = - 0.33318151/10^2
C = 0.43603294/10^5
D = - 53.950447
E = 0.45477025/10^2
F = 0.7704077/10^6
G = - 0.16410434/10^7
H = - 1.4014369
I = 0.59886571/10^4
J = - 0.23457641/10^6
K = 0.27592503/10^9
L = - 0.35977131/10^2

"C_p in kJ/kg.C"

C_p_SI = C_p * 4.184
```

(b) ***Another Chou Correlation***

```
"Chou Correlation for 0 to 80C"

"Temperature Range: 0–80C"
"Pressure Range: 1 bar"
"Salinity Range: 0–120 g/kg"
"Accuracy: within 0.00065"

T = 298.15 "C"
```

S = 35.154 “g/kg”

“C_P in cal/g.C”

C_p = 1.5651785 – (0.3608443/10^2 * T) + ((0.57252694/10^5 * T^2) – ((0.02458979) – (0.14470982/10^3 * T) + (0.22311467/10^6 * T^2)) * S) + ((0.15246268/10^3) – (0.94150958/10^6 * T) + (0.1460181/10^8 * T^2)) * S^2

“C_p in kJ/kg.C”

C_p_SI = C_p * 4.184

Bromley Correlations

(a): For temperature (T) in °C and salinity (S) in g/kg,

“Temperature Range: 0–120C”
“Pressure Range: 1 bar”
“Salinity Range: 0–120 g/kg”
“Accuracy: within 0.001”

T = 25 “C”
S = 35.154 “g/kg”

C_p = 1.0049 – (0.016210 * S) + (3.5261/10^4 * S^2) – (((3.2506) – (1.4795 * S) + (0.027454 * S^2)) * T/10^4) + (((3.8013) – (1.2084 * S) + (0.06121 * S^2)) * T^2/10^6)

C_p_SI = C_p * 4.184
(b): For temperature (T) in °C and salinity in mass %,

“Temperature Range: 0–120C”
“Pressure Range: 1 bar”
“Salinity Range: 0–120 g/kg”
“Accuracy: within 0.003”

T = 25 “C”
S = 3.5 “% salinity”

C_p = 1.007 – (0.014880 * S) + (2.8618/10^4 * S^2) – (((0.14054) – (0.48497 * S) + (0.027454 * S^2))/10^4 * T)
C_p_SI = C_p * 4.184

(c): For temperature in °C and salinity in % mass,

```
"Temperature Range: 0–120C"
"Pressure Range: 1 bar"
"Salinity Range: 0–120 g/kg"
"Accuracy: within 0.004"

T = 25 "C"
S = 3.5 "% salinity"

C_p = 0.9948 – (0.011412 * S) + ((0.42997 + (0.15192 * S))/10^4 * T)
C_p_SI = C_p * 4.184
```

Jamieson et al. Correlation

```
"Temperature Range: 0–160C"
"Pressure Range: 1.01 bar"
"Salinity Range: 0–120 g/kg"
"Accuracy: within 1%"

T = 298 "K"
S = 3.5 "kg/kg"

"C_p as function of temperature, and salinity"

C_p = (a1 + a2 * S + a3 * S^2) + (b1 + b2 * S + b3 * S^2) * T + (c1 + c2 *
S + c3^S^2) * T^2 + (d1 + d2 * S + d3 * S^2) * T^3

a1 = 5.328
a2 = – 9.76/10^2
a3 = 4.04/10^4

b1 = – 6.913/10^3
b2 = 7.351/10^4
b3 = – 3.15/10^6

c1 = 9.6/10^6
c2 = – 1.927/10^6
c3 = 8.23/10^9
```

```
d1 = 2.5/10^9
d2 = 1.666/10^9
d3 = − 7.125/10^12
```

Extended Millero Method from UNESCO Report 1981

```
"Temperature Range: 0–40C"
"Pressure Range: 1000 bar"
"Salinity Range: 0–40 g/kg"
"Standar Deviation: 0.062 J/kgC"

T = 25 "C"
S = 35.154 "g/kg"
P = 1.01 "bar"

"C_p as function of temperature, salinity and pressure"
C_p_pst = C_p_st + delC_p_tp + delC_p_stp
"Determination of C_p_st"

C_p_st = C_p_t + (A * S) + (B * S^1.5)
C_p_t = C_o + (C1 * T) + (C2 * T^2) + (C3 * T^3) + (C4 * T^4)

"Constants"

A = a_o + (a1 * T) + (a2 * T^2)
B = b_o + (b1 * T) + (b2 * T^2)

C_o = 4217.4
C1 = − 3.72030283
C2 = 0.1412855
C3 = − 2.654387/10^3
C4 = 2.093236/10^5

a_o = − 7.643575
a1 = 0.1072763
a2 = − 1.38385/10^3

b_o = 0.1770383
b1 = − 4.07718/10^3
```

b2 = 5.148/10^5

"Correlation for delC_p_tp"

delC_p_tp = (d_o + (d1 * T) + (d2 * T^2) + (d3 * T^3) + (d4 * T^4)) * P + (e_o + (e1 * T) + (e2 * T^2) + (e3 * T^3) + (e4 * T^4)) * P^2 + (f_o + (f1 * T) + (f2 * T^2) + (f3 * T^3)) * P

"Constants"

d_o = − 4.9592/10^1
d1 = 1.45747/10^2
d2 = − 3.13885/10^4
d3 = 2.0357/10^6
d4 = 1.7168/10^8

e_o = 2.4931/10^4
e1 = − 1.08645/10^5
e2 = 2.87533/10^7
e3 = − 4.0027/10^9
e4 = 2.2956/10^11

f_o = − 5.422/10^8
f1 = 2.6380/10^9
f2 = − 6.5637/10^11
f3 = 6.136/10^13

"Correlation for delC_p_stp"

delC_p_stp = ((g_o + (g1 * T) + (g2 * T^2) + (g3 * T^3) + (g4 * T^4)) * S + (h_o + (h1 * T) + (h2 * T^2)) * S^1.5) * P + ((I_o + (I1 * T) + (I2 * T^2) + (I3 * T^3)) * S + J * S^1.5) * P^2 + (((k_o + (k1 * T) + (k2 * T^2)) * S + L * T * S^1.5)) * P^3

"Constants"

g_o = 4.9247/10^3
g1 = − 1.28315/10^4

```
g2 = 9.802/10^7
g3 = 2.5941/10^8
g4 = – 2.9179/10^10

h_o = – 1.2331/10^4
h1 = – 1.517/10^6
h2 = 3.122/10^8

I_o = – 2.9558/10^6
I1 = 1.17054/10^7
I2 = – 2.3905/10^9
I3 = 1.8448/10^11

J = 9.971/10^8

k_o = 5.540/10^10
k1 = – 1.7682/10^11
k2 = 3.513/10^13

L = – 1.4300/10^12

"Unit Converstion to kJ/kg.C"

C_p_SI = C_p_pst/1000
```

Millero Correlation Obtained from UNESCO 1981

```
"Temperature Range: 0–35C"
"Pressure Range: 1 bar"
"Salinity Range: 0–40 g/kg"
"Standard Deviation: 0.5 J/kgC"

T = 25 "C"
S = 35.154 "g/kg"

"C_p as a function of temperature and salinity"

C_p_st = C_p_t + (A * S) + (B * S^1.5)
C_p_t = C_o + (C1 * T) + (C2 * T^2) + (C3 * T^3) + (C4 * T^4)

"Constants"

A = a_o + (a1 * T) + (a2 * T^2)
```

```
B = b_o + (b1 * T) + (b2 * T^2)

C_o = 4217.4
C1 = - 3.72030283
C2 = 0.1412855
C3 = - 2.654387/10^3
C4 _ 2.093236/10^5

a_o = - 7.643575
a1 = 0.1072763
a2 = - 1.38385/10^3

b_o = 0.1770383
b1 = - 4.07718/10^3
b2 = 5.148/10^5

"C_p in kJ/kg C"

C_p_SI = C_p_st/1000
```

Sun et al. Correlation Using Temperature (T) in °C, Pressure (P) in MPa, and Salinity (S) in g/kg

```
"Salinity 0–40 g/kg"
"Temperature 0–374 C"
"Pressure 0–100 MPa"
"Error R^2 by regression analysis gives 0.9974"

T = 25 "C"
P = 0.1 "MPa"
S = 35.154 "g/kg"

m[1] = 1
m[2] = t
m[3] = t^2
m[4] = t^4
m[5] = p
m[6] = p * t
m[7] = p * t^3
m[8] = p^2
m[9] = p^2 * t
m[10] = p^2 * t^2

n[1] = 4.19284306
```

```
n[2] = – 2.27325412/10^4
n[3] = 2.36862694/10^6
n[4] = 1.67009248/10^10
n[5] = – 3.97822834/10^3
n[6] = 3.22914232/10^5
n[7] = – 1.07252107/10^9
n[8] = 1.91296765/10^5
n[9] = – 4.1758292/10^7
n[10] = 2.30627396/10^9

x[1] = s
x[2] = s^2
x[3] = s^3
x[4] = s * t
x[5] = s * t^2
x[6] = s * t^3
x[7] = s^2 * t
x[8] = s^3 * t
x[9] = s * p
x[10] = s * t * p

y[1] = 5.020186422/10^3
y[2] = – 9.96122929/10^6
y[3] = 6.814886633/10^8
y[4] = – 2.604620356/10^5
y[5] = 4.585292916/10^8
y[6] = 7.641725299/10^10
y[7] = – 3.648801938/10^8
y[8] = 2.496295470/10^10
y[9] = 1.186419227/10^6
y[10] = 4.345857976/10^9

C_p = sum(m[i] * n[i], i = 1, 10) – (sum(x[i] * y[i], i = 1, 10))
```

Cox and Smith Correlation

```
"Temperature Range: – 2 - 30C"
"Pressure Range: 1 bar"
"Salinity Range: 0–40 g/kg"
"Accuracy: within 0.003 J/g"

T = 25 "C"
```

S = 35 "g/kg"

"C_P in kJ/kgC"

C_p = A – (0.005075 * S) – (0.000014 * S^2)

"here A is specific heat of pure water in kJ/kg.C"

A = 4.188 "kJ/kg.C"

Millero Correlation in Terms of Chlorinity

"Temperature Range: 0–40C"
"Pressure Range: 1 bar"
"Salinity Range: 0–40 g/kg"
"Accuracy: 0.0013"

T = 25 "C"
S = 35.154 "g/kg"

"S is linked with cholorinity"

S = 1.81578 * Cl

"C_P in kJ/kgC"

C_p = (C_p_o + (A * Cl) + (B * Cl^1.5))/10^3

"here C_p_o is specific heat of pure water in kJ/kg.C"

C_p_o = 4.2174 – (3.720283/10^3 * T) + (1.412855/10^4 * T^2) – (2.654387/10^6 * T^3) + (2.093236/10^8 * T^4)

"Constants"

A = 13.81 – (0.1938 * T) + (0.0025 * T^2)
B = 043 – (0.0099 * T) + (0.00013 * T^2)

Pure Water Correlation

"Pure Water"
"Temperature 0–374 C"

```
"Pressure 0–100 MPa"

T = 25 "C"
P = 0.1 "MPa"

m[1] = 1
m[2] = t
m[3] = t^2
m[4] = t^4
m[5] = p
m[6] = p * t
m[7] = p * t^3
m[8] = p^2
m[9] = p^2 * t
m[10] = p^2 * t^2

n[1] = 4.19284306
n[2] = – 2.27325412/10^4
n[3] = 2.36862694/10^6
n[4] = 1.67009248/10^10
n[5] = – 3.97822834/10^3
n[6] = 3.22914232/10^5
n[7] = – 1.07252107/10^9
n[8] = 1.91296765/10^5
n[9] = – 4.1758292/10^7
n[10] = 2.30627396/10^9

"Please note that this correlation is for isobaric heat capacity. it is not valid for isochoric heat capacity"

C_p = sum(m[i] * n[i], i = 1, 10)
```

Appendix 1

See Table 3.11.

Appendix 2

See Table 3.12.

Table 3.11 Specific heat capacities (cal/g °C) of sea salt solutions (80–200 °C) (Bromley et al. 1970)

T (°C)	S (%)							
	1.02	2	3.34	5	6.75	8.5	10.24	12
80	0.9902	0.9781	0.9637	0.9444	0.9264	0.9095	0.8928	0.8760
	0.9906	0.9785	0.9642	0.9451	0.9266	0.9093	0.8925	0.8761
90	0.9923	0.9803	0.9658	0.9463	0.9280	0.9110	0.8936	0.8776
	0.9922	0.9811	0.9651	0.9466	0.9279	0.9106	0.8937	0.8774
100	0.9947	0.9832	0.9680	0.9477	0.9302	0.9124	0.8950	0.8792
	0.9948	0.9829	0.9678	0.9481	0.9303	0.9120	0.8953	0.8785
110	0.9979	0.9856	0.9703	0.9508	0.9321	0.9142	0.8969	0.8804
	0.9981	0.9858	0.9706	0.9501	0.9322	0.9134	0.8964	0.8804
120	1.0013	0.9886	0.9735	0.9528	0.9346	0.9162	0.8991	0.8823
	1.0011	0.9892	0.9736	0.9529	0.9342	0.9160	0.8982	0.8813
130	1.0044	0.9921	0.9767	0.9559	0.9367	0.9189	0.9004	0.8837
	1.0049	0.9926	0.9772	0.9562	0.9370	0.9182	0.9002	0.8834
140	1.0090	0.9969	0.9802	0.9597	0.9395	0.9212	0.9026	0.8854
	1.0092	0.9967	0.9812	0.9594	0.9399	0.9206	0.9023	0.8857
150	1.0137	1.0004	0.9830	0.9629	0.9430	0.9237	0.9048	0.8870
	1.0137	1.0008	0.9828	0.9632	0.9429	0.9242	0.9055	0.8866
160	1.0204	1.0052	0.9878	0.9673	0.9464	0.9272	0.9075	0.8893
	1.0200	1.0059	0.9874	0.9674	0.9468	0.9284	0.9076	0.8899
170	1.0263	1.0123	0.9932	0.9724	0.9509	0.9313	0.9108	0.8921
	1.0268	1.0115	0.9930	0.9721	0.9515	0.9308	0.9114	0.8931
180	1.0334	1.0194	0.9994	0.9781	0.9563	0.9347	0.9157	0.8959
	1.0338	1.0196	0.9998	0.9778	0.9563	0.9357	0.9153	0.8966
190	1.0421	1.0268	1.0063	0.9846	0.9617	0.9404	0.9202	0.9005
	1.0418	1.0268	1.0073	0.9850	0.9622	0.9404	0.9204	0.9006
200	1.0520	1.0359	1.0160	1.9919	0.9686	0.9454	0.9252	0.9055
	1.0513	1.0359	1.0154	0.9917	0.9689	0.9468	0.9251	0.9057

Appendix 3

See Table 3.13.

Appendix 4

See Table 3.14.

Table 3.12 Specific heat capacities (cal/g °C) of sea water solutions (2–80 °C) (Bromley et al. 1967)

T °C	C_p (cal/g °C)	T °C	C_p (cal/g °C)	T °C	C_p (cal/g °C)	T °C	C_p (cal/g °C)	T °C	C_p (cal/g °C)
S = 1.12%		S = 1.7% cont.		S = 3.36% cont.		S = 5.01% cont.		S = 10.1% cont.	
9	0.9866	11.2	0.9771	65	0.9613	64.9	0.9437	80	0.8929
12.3	0.9854	14.5	0.9763	65	0.962	79.7	0.9453	2.1	0.8777
15.3	0.9854	17.6	0.9762	80	0.9643	S = 6.67%		5.3	0.8793
18.4	0.9853	20.7	0.9766	16	0.9568			8.1	0.8808
35.1	0.9848	32.5	0.9772	19	0.9574	8.7	0.9156	11.3	0.8814
49.9	0.9857	44.4	0.9772	55	0.9611	12.2	0.9158	14.3	0.8827
60	0.9864	59.9	0.9785	58	0.9613	18.9	0.9176	17.4	0.8835
75.5	0.9896	70	0.9805	65	0.9621	29.7	0.9203	20.4	0.8851
78.6	0.9893	S = 2.23%		75	0.9634	40.1	0.9227	22.4	0.8851
9.8	0.9861			3.5	0.9553	49.9	0.924	25.5	0.8853
12.8	0.9855	8.9	0.9701	6.5	0.9558	60.3	0.9264	28.1	0.8867
16	0.985	11.8	0.9701	11.1	0.9556	69.4	0.9266	31	0.8872
19	0.9848	14.8	0.9701	14.2	0.9559	80.4	0.9279	34.1	0.889
25.1	0.9852	22	0.9703	17.4	0.9562	15	0.9167	40.1	0.8893
28.1	0.985	25.1	0.9706	20.4	0.957	18.5	0.9192	52	0.8907
30.7	0.9846	34.8	0.9713	30.2	0.9579	34.5	0.9227	60	0.8913
33.7	0.9846	49.9	0.9727	40.1	0.959	36	0.9226	70	0.8921
50	0.9854	64.9	0.9736	50	0.9597	55	0.9254	10	0.8814
S = 1.65%		79.8	0.9766	64.9	0.9614	68	0.9267	24.9	0.8867
		S = 3%		80.1	0.9635	71	0.9272	50.7	0.8891
10.7	0.9782			S = 3.36%[a]		S = 8.32%		80	0.8924
19.8	0.9776	2.7	0.9587					12.8	0.8833
35	0.9778	8.1	0.9586	7.9	0.956	8	0.8971	79	0.8955
49.9	0.9788	11.1	0.9591	10.8	0.9559	11	0.8984	S = 11.4%	
64.5	0.9796	40	0.9621	11	0.9556	14	0.8997		
80	0.9833	64.9	0.9656	17	0.9567	20.1	0.9015		
S = 1.69%		S = 3.36%		20	0.9566	34.9	0.9053	2.1	0.8649
				21.1	0.9575	48.9	0.908	5.3	0.8666
3.3	0.9782	22	0.9579	26.2	0.9571	64.7	0.9099	8.5	0.8684
6.3	0.978	25	0.9587	35.7	0.9588	79.8	0.9123	11.5	0.8694
9.3	0.9778	35	0.9586	45.5	0.9599	S = 9.96%		14.7	0.8701
12.5	0.9772	55	0.9616	65.7	0.9627			17.8	0.8716
15.5	0.9772	65	0.9621	S = 3.40%[b]		10	0.882	20.8	0.872

(continued)

Table 3.12 (continued)

T °C	C_p (cal/g °C)	T °C	C_p (cal/g °C)	T °C	C_p (cal/g °C)	T °C	C_p (cal/g °C)	T °C	C_p (cal/g °C)
18.5	0.9771	65	0.9619			13.1	0.8833	23.9	0.873
21.5	0.9773	80	0.9646	7.6	0.9536	16.1	0.8844	29.5	0.8749
24.4	0.977	15.5	0.956	10.7	0.9546	19.1	0.885	45.5	0.878
27.5	0.9773	16.9	0.9562	19.4	0.9551	22.1	0.8858	54.7	0.8781
29.9	0.9775	19.9	0.9574	34.3	0.9573	25.3	0.8868	64.7	0.8797
33	0.9778	59	0.9613	50.1	0.9586	39.1	0.8896	79.8	0.8813
40.1	0.9776	55	0.9601	61.8	0.961	60	0.8923	S = 11.74%	
60.1	0.9797	75	0.9634	70.2	0.9616	80	0.8957		
13.1	0.9176	80	0.9643	79.1	0.9641	S = 10.1%			
16.3	0.9768	80	0.9636	S = 5.01%				11.8	0.8658
19.3	0.977	22.5	0.9568					14.8	0.8677
22.2	0.9767	23.5	0.9578			10.1	0.8801	17.8	0.8688
40.1	0.9779	50	0.9603	8.9	0.9345	16.7	0.8831	20.9	0.8695
55.1	0.979	70	0.9623	11.9	0.935	19.8	0.8836	23.8	0.8703
80	0.9828	25	0.9567	15.2	0.9355	29.2	0.8864	35.8	0.8735
S = 1.7%		35	0.9582	18.2	0.9368	44.8	0.8881	49.8	0.8755
		45	0.9578	33.5	0.9398	60	0.8905	65	0.8762
8.1	0.9772	47	0.9575	49.6	0.9418	69.8	0.8911	80.1	0.8788

[a] Atlantic (Wrightsville Beach, NC) sea water

[b] Equatorial Pacific sea water

Table 3.13 Experimental data on specific heat capacity from Chou (1968)

T °C	Salt %	C_p cal/g °C
25	24.5	0.79
50	1.16	0.984
50	2.84	0.965
50	6.81	0.924
50	12.75	0.871
50	24.5	0.787
75	1.16	0.988
75	2.84	0.969
75	6.81	0.936
75	12.75	0.869
75	24.5	0.78
6	24.49	0.805

(continued)

Table 3.13 (continued)

T	Salt	C_p
°C	%	cal/g °C
6	17.78	0.826
6	13.95	0.853
6	7.5	0.91
6	5.13	0.934
6	3.88	0.948
6	3.14	0.958
6	2.11	0.968
6	1.59	0.978
20	24.49	0.807
20	17.78	0.828
20	13.95	0.858
20	7.5	0.913
20	5.13	0.937
20	3.88	0.951
20	3.14	0.96
20	2.11	0.97
20	1.59	0.978
20	1.07	0.983
20	0.8	0.989
33	24.49	0.81
33	17.78	0.829
33	13.95	0.861
33	7.5	0.916
33	5.13	0.94
33	3.88	0.953
33	3.14	0.961
33	2.11	0.972
33	1.59	0.978
57	24.49	0.815
57	17.78	0.841
57	13.95	0.87
57	7.5	0.917
20	0.2	0.997
20	0.4	0.994
20	0.6	0.991

(continued)

Table 3.13 (continued)

T	Salt	C_p
°C	%	cal/g °C
20	0.8	0.989
20	1	0.986
20	1.2	0.984
20	1.4	0.981
20	2	0.974
20	3	0.962
20	4	0.951
20	6	0.929
20	8	0.909
20	10	0.891
20	12	0.873
20	14	0.857
20	16	0.842
20	18	0.828
20	20	0.816
20	22	0.804
20	24	0.793
20	26	0.783
25	0.2	0.996
25	0.4	0.994
25	0.6	0.991
25	0.8	0.989
25	1	0.986
25	1.2	0.984
25	1.4	0.981
25	2	0.974
25	3	0.963
25	4	0.952
25	6	0.93
25	8	0.91
25	10	0.892
25	12	0.875
25	14	0.859
25	16	0.843
25	18	0.829

(continued)

Table 3.13 (continued)

T	Salt	C_p
°C	%	cal/g °C
25	20	0.817
25	22	0.805
25	24	0.794
25	26	0.783
15	0.06	0.999
25	0.06	0.997
35	0.06	0.997
45	0.06	0.997
15	0.29	0.996
25	0.29	0.994
35	0.29	0.994
45	0.29	0.995
15	0.41	0.994
25	0.41	0.993
35	0.41	0.992
45	0.41	0.994
15	0.58	0.992
25	0.58	0.99
35	0.58	0.99
45	0.58	0.992
15	1.72	0.977
25	1.72	0.976
35	1.72	0.976
45	1.72	0.978
15	4	0.949
25	4	0.95
35	4	0.951
45	4	0.952
15	5.69	0.93
25	5.69	0.932
35	5.69	0.933
45	5.69	0.935

Table 3.14 Average of experimental values of specific for various temperatures used by Jamieson et al. (1969)

S (g/kg)	T (°C)	C_p (kJ/kg K)	S (g/kg)	T (°C)	C_p (kJ/kg K)
33.57	169.792	3.426	110.88	159.775	1.817
	159.900	3.317		149.691	1.824
	150.186	3.404		139.607	1.831
	139.545	3.317		130.084	1.809
	120.324	2.990		119.631	1.685
	109.684	2.881		109.548	1.663
	99.976	2.830		99.841	1.591
	90.081	2.771		89.759	1.532
	79.806	2.872		79.680	1.409
	70.283	2.865		79.856	1.677
	60.573	2.843		69.959	1.662
	50.303	2.835		60.062	1.662
	40.406	2.806		49.978	1.661
	29.949	2.798		40.269	1.610
	20.425	2.798		29.814	1.566
	10.343	2.768		20.104	1.537
65.55	170.194	2.760		10.209	1.500
	160.108	2.796	153.46	159.610	1.303
	149.840	2.737		149.899	1.303
	139.948	2.614		139.628	1.317
	119.788	2.410		129.359	1.273
	110.081	2.316		119.464	1.207
	99.999	2.264		109.757	1.142
	89.732	2.192		99.488	1.098
	80.200	2.387		89.779	1.061
	70.116	2.372		79.884	0.996
	60.032	2.372			
	50.322	2.364			
	40.240	2.313			
	29.971	2.284			
	20.074	2.262			
	10.180	2.189			

Table 3.15 Experimental values of specific heat capacity used for correlation by Millero et al. (1973)

Cl (‰)	Temperature (°C)			
	5°	15°	25°	35°
0.000	4.2019	4.1855	4.1793	4.1779
0.552	4.1948	4.1793	4.1737	4.1725
1.3805	4.1847	4.1703	4.1652	4.1644
2.762	4.1679	4.1556	4.1513	4.1511
5.524	4.1360	4.1262	4.1243	4.1254
8.379	4.1015	4.0972	4.0968	4.0994
11.037	4.0723	4.0708	4.0718	4.0758
13.797	4.0428	4.0439	4.0464	4.0508
16.519	4.0134	4.0184	4.0215	4.0269
17.951	3.9986	4.0049	4.0088	4.0154
19.375	3.9840	3.9913	3.9964	4.0030
20.733	3.9699	3.9786	3.9848	3.9915
22.143	3.9568	3.9659	3.9728	3.9795

Appendix 5

See Table 3.15.

References

Bromley LA (1973) Thermodynamic properties of strong electrolytes in aqueous solutions. AIChE J 19(2):313–320. https://doi.org/10.1002/aic.690190216

Bromley LA, Desaussure VA, Cupp JC, Wright JS (1967) Heat capacities of sea water solutions at salinities of 1 to 12% and temperatures of 2° to 80°. J Chem Eng Data 12(2):202–206. https://doi.org/10.1021/je60033a013

Bromley LA, Diamond AE, Salami E, Wilkins DG (1970) Heat capacities and enthalpies of sea salt solutions to 200 °C. J Chem Eng Data 15(2):246–253. https://doi.org/10.1021/je60045a038

Chou JCS (1968) Thermodynamic properties of aqueous sodium chloride solutions from 32 to 350 °F. Georgia Institute of Technology

Chou JCS (1971) Thermal properties of sea water. Technical report 49 (issue 49). Hawaii University, Water Resources Research Center

Cox RA, Smith ND (1959) The specific heat of sea water. Proc R Soc Lond Ser A Math Phys Sci 252(1268):51–62. https://doi.org/10.1098/rspa.1959.0136

Darros-Barbosa R, Balaban MO, Teixeira AA (2003) Temperature and concentration dependence of heat capacity of model aqueous solutions. Int J Food Prop 6(2):239–258. https://doi.org/10.1081/JFP-120017845

El-Dessouky HT, Ettouney HM (2002) Fundamentals of salt water desalination. https://doi.org/10.1016/b978-0-444-50810-2.x5000-3

Fisher FH, Dial OE (2005) Equation of state of water and seawater. J Acoust Soc Am 45(1):325–325. https://doi.org/10.1121/1.1971921

Generous MM, Qasem NAA, Qureshi BA, Zubair SM (2020) A comprehensive review of saline water correlations and data-part I: thermodynamic properties. Arab J Sci Eng 45(11):8817–8876. https://doi.org/10.1007/s13369-020-05019-y

Ghallab AO, Elnahas MH (2016) Determination of physical properties of saline water and different pressure and concentrations. Sci Proc Int Sci Conf 1(2):109–112

Jamieson DT, Tudhope JS, Morris R, Cartwright G (1969) Physical properties of sea water solutions: heat capacity. Desalination 7(1):23–30. https://doi.org/10.1016/S0011-9164(00)80271-4

Leduc P-A, Desnoyers JE (2006) Apparent molal heat capacities and volumes of tetrabutylammonium carboxylates and related solutes in water at 25 °C. Can J Chem 51(17):2993–2998. https://doi.org/10.1139/v73-445

Leyendekkers JV (1980) Prediction of the heat capacities of seawater and other multicomponent solutions from the Tammann–Tait–Gibson model. Mar Chem 9(1):25–35. https://doi.org/10.1016/0304-4203(80)90004-3

Millero FJ, Perron G, Desnoyers JE (1973) Heat capacity of seawater solutions from 5° to 35 °C and 0.5 to 22‰ chlorinity. J Geophys Res 78(21):4499–4507. https://doi.org/10.1029/JC078i021p04499

Nayar KG, Sharqawy MH, Banchik LD, Lienhard JH (2016) Thermophysical properties of seawater: a review and new correlations that include pressure dependence. Desalination 390:1–24. https://doi.org/10.1016/j.desal.2016.02.024

Randall M, Rossini FD (1929) Heat capacities in aqueous salt solutions. J Am Chem Soc 51(2):323–345. https://doi.org/10.1021/ja01377a001

Sun H, Feistel R, Koch M, Markoe A (2008) New equations for density, entropy, heat capacity, and potential temperature of a saline thermal fluid. Deep Sea Res Part 1 Oceanogr Res Pap 55(10):1304–1310. https://doi.org/10.1016/j.dsr.2008.05.011

UNESCO (1981) Background papers and supporting data on the international equation of state of seawater 1980. UNESCO technical papers in marine science, vol 37

Wagner W, Pruß A (2002) Revised release on the (IAPWS) formulation 1995 for the thermodynamic properties of ordinary water substance for general and scientific use. J Phys Chem Ref Data 31(2):387–535. https://doi.org/10.1158/0008-5472.CAN-09-0634

Zemaitis JF, Clark DM, Rafal M, Scrivner NC (1986) Handbook of aqueous electrolyte thermodynamics. Wiley. https://doi.org/10.1002/9780470938416

Chapter 4
Specific Enthalpy

Abstract The specific enthalpy of saline water is one of the critical thermodynamic properties used in estimating the thermal energy of desalination and salt separation processes. This chapter explains the thermodynamic approaches, multicomponent models, correlation, and data used to estimate specific enthalpy at various salinity and temperature values. A new correlation for NaCl-specific enthalpy with high accuracy is suggested for a salinity of 0–260 g/kg and ambient conditions. Moreover, the evaluated specific enthalpy values from the correlations and methods are compared and assessed using experimental data of the specific enthalpy of seawater. Based on their accuracy and operating conditions, some correlations and methods are recommended. Furthermore, the codes used for estimating the investigated specific enthalpy are provided.

Keywords Specific enthalpy · Thermodynamics · Correlations · Methods · Energy · Saline water

4.1 Introduction

Specific enthalpy of saline water is a critical thermodynamic property due to its direct use in evaluating energy consumption and performance, using the 1st and 2nd laws of thermodynamics for saline water treatment and desalination processes. Therefore, the enthalpy of saline water is of considerable importance for the design and energy analysis of desalination technologies.

Most of the laboratory experiments on saline water properties are based on the NaCl solution. Chou and Rowe (1969) tested the enthalpy of NaCl solution in the temperature range of 32–350 °F and pressure range of 0–5000 psig. Tremendous efforts were made by Millero and his fellow scientists (1983) to investigate the enthalpy of saline water and other thermodynamic properties in the 1970s. The enthalpy at atmospheric pressure and 25 °C was proposed to be derived from the heat of dilution experiments. Similarly, free enthalpy was proposed to be derived from freezing point depression measurements, and the fundamental equations for the heat capacity and enthalpy were available in 1980 (Feistel 1993).

N. A. A. Qasem et al., *Thermodynamic and Thermophysical Properties of Saline Water*, Springer Water, https://doi.org/10.1007/978-3-031-35193-8_4

Salinity, temperature, and pressure affect the enthalpy of saline water. Therefore, many researchers studied the effect of these variables on enthalpy. The specific enthalpy up to a pressure of 100 MPa, salinity in a range of 0–40 g/kg, and temperature in a range of 0–20 °C were addressed by Kukulka et al. (1987). A specific enthalpy correlation, as a function of salinity up to 120 g/kg, was proposed by Feistel (2008) from the experimental data of freezing points, vapor pressure, and mixing heats.

4.2 Thermodynamic Approach

For multi-component saline water, the total enthalpy is related to the apparent and partial enthalpies as follows (Singh and Bromley 1973):

$$h = x^{\phi}h = x\bar{h}_s + (1 - x)\bar{h}_w \tag{4.1}$$

where "h" is the enthalpy of the solution, ${}^{\phi}h$, is the apparent enthalpy of the salts in the solution; $\bar{h}_s$ is the partial enthalpy of the salts in solution; $\bar{h}_w$ is the partial enthalpy of the water in solution; x is approximately the mass fraction of salts in solution—accurately, it is 0.001 times the salinity in parts per thousand (0.01 times the salinity in percent). The extended Pitzer model (Pitzer et al. 1984; Pitzer 1987; Thiel and Lienhard 2014) is considered the most reliable model to determine the specific enthalpy for multi-component solutions.

4.3 Multiple Components Models

Multiple components models were proposed to be general for all types of saltwater under different salinity values. Major constituents of the aqueous ionic solutions are cations, anions, and water. The extended Pitzer model (Pitzer 1987; Pitzer et al. 1984; Thiel and Lienhard 2014) is a multi-component model that can be used for all possible concentration levels.

$$\frac{G^{ex}}{n_w RT} = \sum_i m_i(\ln \gamma_i + (1 - \phi)) \tag{4.2}$$

$$L = G^{ex} - T\left(\frac{\partial G^{ex}}{\partial T}\right)_{P,m} \tag{4.3}$$

$${}^{\phi}L = L/n_2 \tag{4.4}$$

and

$$\mathrm{h} = {}^{\phi}L$$

The multi-component activity coefficient and osmotic coefficient are as described in Part II (Qasem et al. 2019) based on Thiel and Lienhard (2014). n_2 is the moles of solute. The model was initially tested for aqueous NaCl, and values of specific enthalpy are given in Appendix 1, taken from Pitzer et al. (1984), Pitzer (1987) and Thiel and Lienhard (2014), for a significant variation of temperature (0–300 °C), pressure (1–100 MPa), and molality (0.1–6.0 mol/kg). It is important to note that specific entropy is a weak function of pressure and temperature limitation depends upon measured parameters. Nonetheless, the complete data set was provided for the convenience of checking all correlations provided in the next section against the work of Pitzer.

Bromley (1973) model can be used for specific enthalpy calculations for multiple components and various electrolytes. It should be kept in mind that specific enthalpy has been reported to be a weak function of pressure.

$$
\begin{aligned}
{}^{\phi}L = {} & \frac{\nu}{2} A_H |z_+ z_-| \sqrt{I} \left[\frac{1}{1+\sqrt{I}} - \frac{\sigma\left(\sqrt{I}\right)}{3} \right] \\
& - 2.303 \nu R T^2 I \left\{ \left[\frac{1}{(1+I)^2} - \frac{\psi(aI)}{2} \right] (0.6|z_+ z_-|) + \frac{1}{2} \right\} \frac{dB}{dT}
\end{aligned} \tag{4.5}
$$

and

$$
h = {}^{\phi}L
$$

where

$$
\sigma\left(\sqrt{I}\right) = \frac{3}{\left(\sqrt{I}\right)^3} \left[1 + \sqrt{I} - \frac{1}{1+\sqrt{I}} - 2\ln\left(1+\sqrt{I}\right) \right]
$$

$$
\psi(aI) = \frac{2}{aI} \left[\frac{1+2aI}{(1+aI)^2} - \frac{\ln(1+aI)}{aI} \right]; \quad a = \frac{1.5}{|z_+ z_-|}
$$

Here, $A_H = 688$ cal/mol at T = 25 °C. B values are for solutions and ions, and the effect of temperature was listed by Kukulka et al. (1987).

4.4 Correlations and Data

Correlations are the easiest way to calculate the saltwater heat capacity as they are fitted from the experimental data. Most of the available specific enthalpy correlations in the literature are for seawater. Nayar et al. (2016) proposed a correlation which

Table 4.1 Constant for Nayar et al. (2016) correlation for specific enthalpy of seawater which is also reported in IAPWS-2008 (Cooper et al. 2008)

Constants	Values	Constants	Values
a_1	996.7767	b_4	-1.44606×10^7
a_2	-3.2406	b_5	7.82607×10^3
a_3	0.0127	b_6	-4.41733×10^1
a_4	-4.7723×10^{-5}	b_7	2.1394×10^{-1}
a_5	-1.1748	b_8	-1.99108×10^4
a_6	0.01169	b_9	2.77846×10^4
a_7	-2.6185×10^{-5}	b_{10}	9.72801×10
a_8	7.0661×10^{-8}	c_1	141.355
b_1	-2.34825×10^4	c_2	4202.07
b_2	3.15183×10^5	c_3	0.535
b_3	2.80269×10^6	c_4	0.004

is also reported in IAPWS-2008 (Cooper et al. 2008) report. The correlation for seawater is in terms of temperature (T) in °C, pressure (P) in MPa, and salinity (S) in kg/kg, as shown in Eq. (4.6). All the relevant constants are provided in Table 4.1.

$$\begin{aligned} h = {} & h_l + (P - P_o) \\ & \times \left(a_1 + a_2T + a_3T^2 + a_4T^3 + \frac{S}{1000}\left(a_5 + a_6T + a_7T^2 + a_8T^3\right) \right) \end{aligned} \tag{4.6}$$

h_l can be estimated from Eq. (4.7).

$$\begin{aligned} h_l = h_w - S\, \big(& b_1 + b_2S + b_3S^2 + b_4S^3 + b_5T \\ & + b_6T^2 + b_7T^3 + b_8ST + b_9S^2T + b_{10}ST^2\big) \end{aligned} \tag{4.7}$$

h_W can be calculated as in Eq. (4.8).

$$h_W = c_1 + c_2T - c_3T^2 + c_4T^3 \tag{4.8}$$

The correlation is valid in the salinity range of 0–120 g/kg, temperature range of 10–80 °C, and pressure of 0.1 MPa with an accuracy of ± 1.36%. However, it is valid too for the salinity range of 42–120 g/kg, temperature range of 10–120 °C, and pressure of 0–12 MPa with an accuracy of ± 1.47%.

Connors (1970) proposed a correlation for specific enthalpy in terms of temperature (T) in °C, and salinity (S) in kg/kg as in Eq. (4.9). The required constants are provided in Table 4.2.

$$h = aT - bT^2 - S\left(cT - dT^2\right) - S^2\left(e - fT + gT^2\right) \tag{4.9}$$

Table 4.2 Constant for Connors correlation (Connors 1970) for specific enthalpy

Constants	Values
a	4.2044
b	0.00057
c	6.99
d	0.0343
e	464
f	19.6
g	0.3

This was based on available heat capacity data as well as the heat of mixing experiments (see Appendix 2). Connors (1970) correlation is valid in the salinity range of 0–40 g/kg, temperature range of 0–30 °C, and pressure of 0.1 MPa with an accuracy of ± 1.5 J/kg.

Generous et al. (2020) proposed a novel correlation in terms of temperature (T), in °C and molality (m), in mol/kg,

$$h = aT + b \tag{4.10}$$

where,

$$a = 0.00019\,m^4 - 0.00454\,m^3 + 0.04718\,m^2 - 0.31195\,m + 4.18119$$

$$b = 0.00049\,m^5 - 0.01337\,m^4 + 0.21743\,m^3 - 1.57257\,m^2 - 0.51754\,m + 0.21693$$

This correlation is obtained from the Pitzer model to be easily used under a temperature range of 15–35 °C, and salinity range of 0–260 g/kg (0–6 mol) within accuracy of $R^2 = 1$. It is important to note that 'h' is a weak function of pressure.

IAPWS (1992) applies Eq. (4.3) to calculate specific enthalpy in J/kg K. It uses an equation of state in the form of specific Gibbs free energy that is dependent upon temperature, salinity, and pressure. The equation has two parts: namely, a water part and a saline part, as shown in Eq. (5.14) below:

$$g(S, T, p) = g^W(T, p) + g^S(S, T, p) \tag{4.11}$$

where p is the absolute pressure, in Pascal, T is the absolute temperature (ITS-90), and S is the salinity based on the Reference-Composition Salinity Scale, in kg/kg. The first term is calculated from IAPWS-95, while the second term is computed based on the reduced polynomial given below:

$$g^S(S, T, p)/g^* = \sum_{k=0}^{5}\sum_{j=0}^{6}\left(g_{1jk}\xi^2 \ln \xi + \sum_{i=2}^{7} g_{ijk}\xi^i\right)\tau^j \pi^k \tag{4.12}$$

Table 4.3 Value of constants used in Eq. (4.12)

Constants	Values
g^*	1 J/kg
S^*	0.040188617 kg/kg
T^*	40 K
p^*	10^8 Pa
T_0	273.15 K
p_0	101,325 Pa

ξ is the reduced salinity $\sqrt{S/S^*}$, τ is the reduced temperature $(T - T_0)/T^*$ and π is the reduced pressure $(p - p_0)/p^*$. The values for the constants appearing in these reduced quantities are given in Table 4.3.

For the coefficients of the saline part in Eq. (4.12), see Appendix 3.

4.5 Case Study

The specific enthalpy of saline water is a function of salinity, temperature, and pressure. The specific enthalpy usually decreases with an increase in salinity. The results obtained from Connors (1970), Bromley et al. (1970) (data in Appendix 4), Pitzer et al. (1984), Pitzer (1987), Nayar et al. (2016), and the proposed correlations along with IAPWS (1992) and Chou (1968) data are plotted in the salinity range of 0–6 molal (~ 260 g/kg) at 25 °C and 0.1 MPa, as shown in Fig. 4.1.

The model of Nayar et al. (2016) performs better compared to Bromley et al. (1970) with respect to IAPWS (1992) data. At higher salinities (> 120 g/kg), only the Pitzer model is compared with the experimental data of Chou (1968) (shown in Appendix 5) up to 6 molal (~ 260 g/kg). The maximum difference is less than 1.2%. While fitting the data, it was found that, for a fixed salinity, the variation of enthalpy is exactly linear for the temperature range investigated. The data are plotted for a temperature range of 15–35 °C, while a fitted polynomial equation ($R^2 = 1$) is provided in Eq. (4.1). After plotting the proposed correlation, the average difference compared to IAPWS (1992) and data of Chou (1968) for salinities up to and above 120 g/kg, respectively, was 0.214%. The maximum difference is found to be within 1.1%, and this occurs at 6 molal.

The enthalpy values estimated by these correlations/models have been plotted in Fig. 4.1b against IAPWS (1992) and data of Chou (1968) for salinities up to and above 120 g/kg, respectively, to show their accuracy. It is clear that the present correlation is the most accurate model over a wider salinity range.

The enthalpy results as a function of temperature (15–35 °C) obtained from the correlations of Connors (1970), Nayar et al. (2016), Pitzer model (Pitzer et al. 1984; Pitzer 1987), IAPWS (1992), and the current work at 40 g/kg and 0.1 MPa are plotted in Fig. 4.2a. It can be seen from the figure that the estimated specific enthalpy

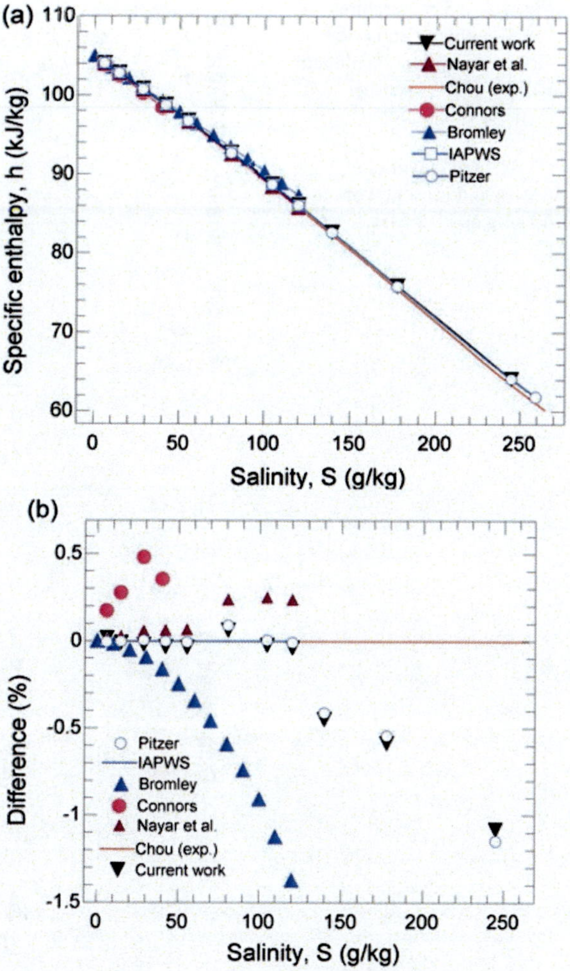

Fig. 4.1 **a** Salinity versus specific enthalpy and **b** the difference against the data of IAPWS (1992) and Chou (1968) for the methods of Pitzer (Pitzer et al. 1984; Pitzer 1987) for Connors (1970), Bromley et al. (1970), Nayar et al. (2016), and present work at 25 °C and 0.1 MPa

values largely overlap. The difference from the experimental data of Chou (1968) and IAPWS (1992) is shown in Fig. 4.2b. The Pitzer model shows the most accurate results (the difference is almost zero), while the present work correlation and Nayar et al. (2016) lies within ± 0.2% difference leaving Connors (1970) which is within ± 1%, as shown in Fig. 4.2b.

Based on the above discussion, some methods and correlations could be recommended to be used, as summarized in Table 4.4.

Precaution should be exercised while choosing a particular correlation. Correct selection may be dependent upon the type of application. For instance, thermal-based desalination may require correlation or a model that be used at high temperatures. On

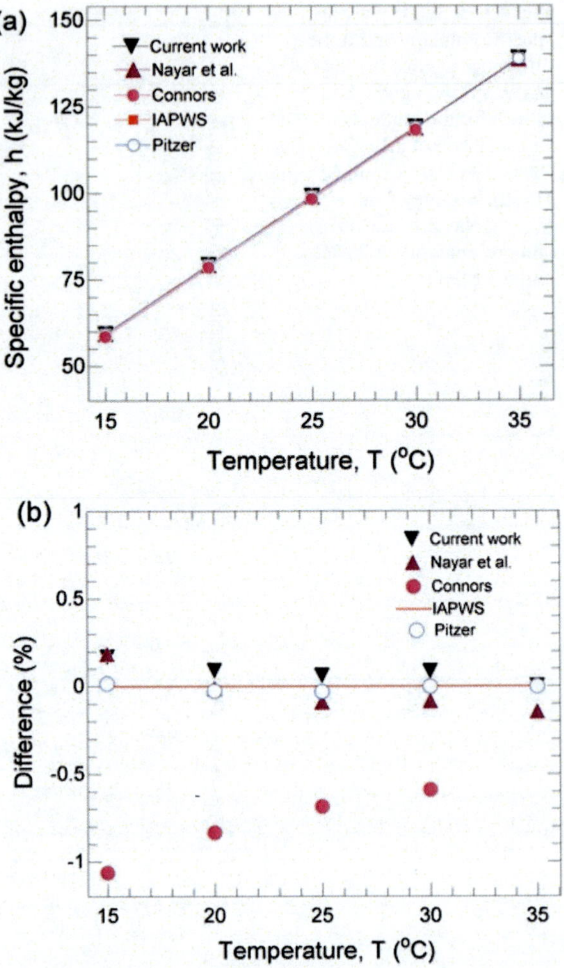

Fig. 4.2 **a** Temperature versus enthalpy and **b** the difference against the data of IAPWS (1992) for the methods of Connors (1970), Pitzer et al. (1984), Pitzer (1987), and Nayar et al. (2016) and present work at 40 g/kg and 0.1 MPa

the other hand, pressure-driven desalination technologies such as reverse osmosis and forward osmosis may require correlations that are more accurate at higher pressures. Table 4.4 proposes the recommended correlations and methods to be used to calculate the specific enthalpy of saline water.

It is also noted that nearly all correlations are for NaCl properties, while the multi-component correlations are inadequate. It is intensely suggested to perform large-scale experimental investigations to evaluate multi-component saline water properties at high values of salinity, temperature, and pressure.

Table 4.4 Summary of the recommended correlations and models for the enthalpy of saline water (Generous et al. 2020)

Correlation-model	Advantages	Applications	References
Pitzer et al. (Eq. 4.4)	• General for multi-component saline solution • The most accurate model (refer to Fig. 4.1)	• All membrane-based desalination technologies • All thermal-based desalination technologies	Pitzer et al. (1984), Pitzer (1987)
The current correlation (Eq. 4.10)	• Obtained from the Pitzer model to be easily used at ambient conditions, i.e., 15–35 °C (refer to Fig. 4.2) • Simplest and most accurate correlation for seawater (refer to Fig. 4.1)	• All membrane- and thermal-based desalination technologies	The present work
Connors (Eq. 4.9)	• One of the most accurate correlations to study the impact of salinity up to 40 g/kg within a difference of ± 0.5% (refer to Fig. 4.1) • Can be used up to 35 °C with a difference of around ± 1% (refer to Fig. 4.2)	• All thermal-based desalination technologies • Electrodialysis (ED) • Electrodialysis reversal (EDR)	Connors (1970)
Nayar et al. (Eq. 4.6)	• One of the most accurate correlations for a salinity range of 0–120 g/kg within a difference of ± 0.5% (refer to Fig. 4.1) • Valid in the temperature range of 0–35 °C with negligible difference (refer to Fig. 4.2)	• All membrane-based desalination technologies • All thermal-based desalination technologies • Seawater greenhouse • High-grade water recycling • Wave-powered desalination	Nayar et al. (2016)
Bromley (Eq. 4.5)	• Valid in a salinity range of 0–120 g/kg within a difference of ± 1.5% (refer to Fig. 4.1)	• All thermal-based desalination technologies • High-grade water recycling • Wave-powered desalination	Bromley et al. (1970)

4.6 EES Codes

The EES codes for the correlations discussed above:

Nayar et al. Correlation Based on IAPWS-2008

```
S = 0.035 "kg"
T = 25 "C"
P = 0.1 "MPa"
P_o = 0
h_tsp = h_ts + (P − P_o) * (a1 + a2 * T + a3 * T^2 + a4 * T^3 + (S * (a5 +
a6 * T + a7 * T^2 + a8 * T^3)))
h_ts = h_w − (S * (b1 + b2 * S + b3 * S^2 + b4 * S^3 + b5 * T + b6 * T^2 +
b7 * T^3 + b8 * T * S + b9 * S^2 * T + b10 * S * T^2))
h_w = 141.355 + 4202.07 * T − 0.535 * T^2 + 0.004 * T^3

a1 = 996.7767
a2 = − 3.2406
a3 = 0.0127
a4 = − 4.7723/10^5
a5 = − 1.1748
a6 = 0.01169
a7 = − 2.6185/10^5
a8 = 7.0661/10^8
b1 = − 2.34825 * 10^4
b2 = 3.15183 * 10^5
b3 = 2.80269 * 10^6
b4 = − 1.44606 * 10^7
b5 = 7.82607 * 10^3
b6 = − 4.41733 * 10^1
b7 = 2.1394/10^1
b8 = − 1.99108 * 10^4
b9 = 2.77846 * 10^4
b10 = 9.72801 * 10

"enthalpy in kJ/kg"

h_kJ = h_tsp/1000
```

Connors Correlations

```
"Validity 0–30 C and Salinity 10 to 40 g/kg"
S = 0.04 "kg"
T = 25 "C"
```

h = (4.20448 * T) – (0.00057 * T^2) – (S * ((6.99 * T) – (0.0343 * T^2))) – (S^2 * (464 – (19.6 * T) + (0.3 * T^2)))

This Work Correlation

T = 25 "C"
m = 0 "put proper value"
h = a * T + b
a = 0.00019 * m^4 – 0.00454 * m^3 + 0.04718 * m^2 – 0.31195 * m + 4.18119
b = 0.00049 * m^5 – 0.01337 * m^4 + 0.21743 * m^3 – 1.57257 * m^2 – 0.51754 * m + 0.21693

Appendix 1

See Tables 4.5, 4.6, 4.7, 4.8, 4.9 and 4.10.

Appendix 2

See Tables 4.11 and 4.12.

Appendix 3

See Table 4.13.

Appendix 4

See Table 4.14.

Appendix 5

See Table 4.15.

Table 4.5 The specific enthalpy of NaCl(aq), J/g K (P < 100 bar)

T	P	m = 0.1	m = 0.25	m = 0.5	m = 0.75	m = 1	m = 2	m = 3	m = 4	m = 5	m = 6
°C	bar	mol/kg									
0	1.0	– 1985.7	– 1968.2	– 1939.8	– 1912.4	– 1885.9	– 1788.4	– 1702.3	– 1625.2	– 1555.5	– 1491.6
10	1.0	– 1944.1	– 1927.1	– 1899.5	– 1872.8	– 1847.0	– 1751.6	– 1667.0	– 1591.1	– 1522.3	– 1459.2
20	1.0	– 1902.6	– 1886.0	– 1859.2	– 1833.2	– 1808.0	– 1714.6	– 1631.5	– 1556.7	– 1488.9	– 1426.6
25	1.0	– 1881.8	– 1865.5	– 1839.0	– 1813.3	– 1788.4	– 1696.1	– 1613.7	– 1539.5	– 1472.1	– 1410.3
30	1.0	– 1861.1	– 1845.0	– 1818.8	– 1793.4	– 1768.8	– 1677.5	– 1595.9	– 1522.3	– 1455.4	– 1393.9
40	1.0	– 1819.5	– 1803.9	– 1778.4	– 1753.6	– 1729.6	– 1640.2	– 1560.1	– 1487.8	– 1421.8	– 1361.2
50	1.0	– 1778.0	– 1762.8	– 1737.9	– 1713.8	– 1690.4	– 1602.9	– 1524.4	– 1453.3	– 1388.3	– 1328.5
60	1.0	– 1736.5	– 1721.6	– 1697.5	– 1673.9	– 1651.1	– 1565.6	– 1488.6	– 1418.7	– 1354.8	– 1295.8
70	1.0	– 1694.9	– 1680.5	– 1657.0	– 1634.0	– 1611.7	– 1528.2	– 1452.8	– 1384.2	– 1321.3	– 1263.3
80	1.0	– 1653.3	– 1639.2	– 1616.4	– 1594.1	– 1572.4	– 1490.8	– 1417.0	– 1349.7	– 1287.9	– 1230.7
90	1.0	– 1611.6	– 1598.0	– 1575.8	– 1554.1	– 1532.9	– 1453.4	– 1381.1	– 1315.2	– 1254.4	– 1198.2
100	1.0	– 1569.8	– 1556.7	– 1535.1	– 1514.0	– 1493.4	– 1415.9	– 1345.3	– 1280.6	– 1221.0	– 1165.7
110	1.4	– 1527.8	– 1515.1	– 1494.2	– 1473.8	– 1453.8	– 1378.3	– 1309.4	– 1246.0	– 1187.6	– 1133.2
120	2.0	– 1485.7	– 1473.4	– 1453.2	– 1433.4	– 1414.0	– 1340.6	– 1273.4	– 1211.5	– 1154.1	– 1100.7
130	2.7	– 1443.4	– 1431.6	– 1412.1	– 1392.9	– 2374.2	– 1302.9	– 1237.4	– 1176.8	– 1120.6	– 1068.2
140	3.6	– 1400.9	– 1389.6	– 1370.8	– 1352.3	– 1334.2	– 1265.1	– 1201.3	– 1142.2	– 1087.2	– 1035.7
150	4.8	– 1358.2	– 1347.3	– 1329.3	– 1311.5	– 1294.0	– 1227.2	– 1165.1	– 1107.5	– 1053.6	– 1003.2

(continued)

Table 4.5 (continued)

T	P	m = 0.1	m = 0.25	m = 0.5	m = 0.75	m = 1	m = 2	m = 3	m = 4	m = 5	m = 6
°C	bar	mol/kg									
160	6.2	− 1315.2	− 1304.8	− 1287.6	− 1270.5	− 1253.7	− 1189.1	− 1128.9	− 1072.7	− 1020.1	− 970.7
170	7.9	− 1272.0	− 1262.1	− 1245.7	− 1229.3	− 1213.2	− 1151.0	− 1092.6	− 1037.9	− 986.5	− 938.1
180	10.0	− 1228.4	− 1219.1	− 1203.5	− 1187.9	− 1172.5	− 1112.6	− 1056.2	− 1003.0	− 952.9	− 905.5
190	12.5	− 1184.4	− 1175.7	− 1161.0	− 1146.2	− 1131.5	− 1074.2	− 1019.6	− 968.0	− 919.2	− 872.9
200	15.5	− 1140.1	− 1132.0	− 1118.1	− 1104.2	− 1090.3	− 1035.5	− 983.0	− 933.0	− 885.4	− 840.2
210	19.1	− 1095.3	− 1087.8	− 1074.9	− 1061.8	− 1048.7	− 996.6	− 946.1	− 897.8	− 851.6	− 807.5
220	23.2	− 1049.9	− 1043.1	− 1031.2	− 1019.1	− 1006.8	− 957.4	− 909.1	− 862.4	− 817.6	− 774.7
230	28.0	− 1004.0	− 997.9	− 987.1	− 975.8	− 964.4	− 917.9	− 871.8	− 827.0	− 783.6	− 741.8
240	33.4	− 957.3	− 952.0	− 942.3	− 932.1	− 921.6	− 878.2	− 834.4	− 791.3	− 749.4	− 708.9
250	39.7	− 909.9	− 905.4	− 896.9	− 887.8	− 878.3	− 838.0	− 796.6	− 755.4	− 715.0	− 675.8
260	46.9	− 861.5	− 857.9	− 850.8	− 842.8	− 834.3	− 797.4	− 758.5	− 719.3	− 680.5	− 642.7
270	55.0	− 812.1	− 809.5	− 803.9	− 797.1	− 789.8	− 756.4	− 720.1	− 682.9	− 645.8	− 609.4
280	64.1	− 761.5	− 760.1	− 756.0	− 750.7	− 744.5	− 714.9	− 681.3	− 646.2	− 610.8	− 575.9
290	74.4	− 709.5	− 709.4	− 707.1	− 703.3	− 698.5	− 672.9	− 642.1	− 609.2	− 575.6	− 542.2
300	85.8	− 655.9	− 657.3	− 657.1	− 655.1	− 651.7	− 630.5	− 602.6	− 511.8	− 539.9	− 508.0

Table 4.6 The specific enthalpy of NaCl(aq), J/g K (P = 200 bar)

T	P	m = 0.1	m = 0.25	m = 0.5	m = 0.75	m = 1	m = 2	m = 3	m = 4	m = 5	m = 6
°C	bar	mol/kg									
0	200	– 1965.9	– 1948.6	– 1920.7	– 1893.7	– 1867.6	– 1771.3	– 1686.0	– 1609.6	– 1540.4	– 1477.0
10	200	– 1925.0	– 1908.3	– 1881.0	– 1854.7	– 1829.2	– 1734.8	– 1650.9	– 1675.6	– 1507.3	– 1444.7
20	200	– 1884.1	– 1867.4	– 1841.2	– 1815.5	– 1790.6	– 1698.1	– 1615.6	– 1541.4	– 1474.0	– 1412.2
25	200	– 1863.7	– 1847.5	– 1821.3	– 1795.9	– 1771.2	– 1679.6	– 1597.9	– 1524.3	– 1457.3	– 1395.9
30	200	– 1843.1	– 1827.2	– 1801.3	– 1776.2	– 1751.8	– 1661.2	– 1580.2	– 1507.1	– 1440.6	– 1379.6
40	200	– 1802.1	– 1786.5	– 1761.3	– 1736.7	– 1711.9	– 1624.2	– 1544.6	– 1472.7	– 1407.2	– 1346.9
50	200	– 1761.0	– 1745.0	– 1721.2	– 1697.2	– 1674.0	– 1587.1	– 1509.1	– 1438.4	– 1373.8	– 1314.4
60	200	– 1719.9	– 1705.1	– 1681.1	– 1657.7	– 1635.0	– 1550.0	– 1473.5	– 1404.0	– 1340.4	– 1281.8
70	200	– 1678.7	– 1664.4	– 1641.0	– 1618.2	– 1596.0	– 1512.9	– 1437.9	– 1369.6	– 1307.1	– 1249.3
80	200	– 1637.5	– 1623.6	– 1600.8	– 1578.6	– 1556.9	– 1475.7	– 1402.2	– 1335.3	– 1273.8	– 1216.8
90	200	– 1596.2	– 1582.7	– 1560.5	– 1538.9	– 1517.8	– 1438.5	– 1366.6	– 1300.9	– 1240.5	– 1184.4
100	200	– 1554.8	– 1541.7	– 1510.2	– 1499.1	– 1478.6	– 1401.3	– 1331.0	– 1266.5	– 1207.2	– 1152.1
110	200	– 1513.2	– 1500.6	– 1479.7	– 1459.3	– 1439.2	– 1364.0	– 1295.3	– 1232.2	– 1173.9	– 1119.7
120	200	– 1471.7	– 1459.4	– 1439.2	– 1419.4	– 1400.0	– 1326.7	– 1259.6	– 1197.8	– 1140.7	– 1087.4
130	200	– 1429.9	– 1418.1	– 1398.5	– 1379.2	– 1360.5	– 1289.3	– 1223.8	– 1163.4	– 1107.4	– 1055.2
140	200	– 1388.0	– 1376.5	– 1357.7	– 1339.2	– 1321.6	– 1251.9	– 1188.1	– 1129.1	– 1074.2	– 1022.9
150	200	– 1345.8	– 1334.9	– 1316.8	– 1298.9	– 1281.4	– 1214.4	– 1152.3	– 1094.7	– 1041.0	– 990.6

(continued)

Table 4.6 (continued)

T	P	m = 0.1	m = 0.25	m = 0.5	m = 0.75	m = 1	m = 2	m = 3	m = 4	m = 5	m = 6
°C	bar	mol/kg									
160	200	– 1303.5	– 1293.1	– 1275.7	– 1258.5	– 1241.6	– 1176.8	– 1116.5	– 1060.3	– 1007.7	– 958.4
170	200	– 1261.0	– 1251.1	– 1234.5	– 1218.0	– 1201.7	– 1139.1	– 1080.6	– 1025.8	– 974.5	– 926.2
180	200	– 1218.3	– 1208.8	– 1192.0	– 1177.3	– 1161.7	– 1101.4	– 1044.7	– 991.4	– 941.2	– 893.9
190	200	– 1175.3	– 1166.3	– 1161.3	– 1136.4	– 1121.5	– 1063.4	– 1008.7	– 956.9	– 907.9	– 861.6
200	200	– 1131.9	– 1123.6	– 1109.4	– 1095.2	– 1081.1	– 1025.6	– 972.7	– 922.3	– 874.6	– 829.3
210	200	– 1080.2	– 1080.5	– 1067.2	– 1053.8	– 1040.4	– 987.5	– 936.5	– 887.7	– 841.3	– 797.0
220	200	– 1044.1	– 1037.0	– 1024.7	– 1012.1	– 999.5	– 949.2	– 900.2	– 853.1	– 807.9	– 764.6
230	200	– 999.5	– 993.0	– 981.7	– 970.1	– 958.3	– 910.8	– 863.9	– 818.3	– 774.4	– 732.2
240	200	– 954.2	– 948.6	– 938.2	– 927.7	– 916.8	– 872.1	– 827.3	– 783.4	– 740.8	– 699.7
250	200	– 908.5	– 903.5	– 894.8	– 884.8	– 874.8	– 833.1	– 790.5	– 748.4	– 707.1	– 667.2
260	200	– 862.0	– 857.8	– 849.9	– 841.3	– 832.3	– 793.7	– 753.5	– 713.1	– 673.3	– 634.5
270	200	– 814.6	– 811.3	– 804.7	– 797.3	– 789.3	– 754.0	– 716.2	– 677.7	– 639.4	– 601.8
280	200	– 766.2	– 763.9	– 758.7	– 752.5	– 745.6	– 713.8	– 678.6	– 642.0	– 605.2	– 568.9
290	200	– 716.5	– 715.3	– 711.7	– 706.9	– 701.2	– 673.1	– 640.5	– 605.9	– 570.7	– 535.8
300	200	– 665.4	– 665.5	– 663.7	– 660.3	– 655.9	– 631.8	– 601.9	– 569.3	– 535.8	– 502.3

Table 4.7 The specific enthalpy of NaCl(aq), J/g K (P = 400 bar)

T	P	m = 0.1	m = 0.25	m = 0.5	m = 0.75	m = 1	m = 2	m = 3	m = 4	m = 5	m = 6
°C	bar	mol/kg									
0	400	– 1946.5	– 1929.5	– 1902.0	– 1875.3	– 1849.6	– 1754.4	– 1669.9	– 1594.1	– 1525.4	– 1462.4
10	400	– 1906.3	– 1889.7	– 1862.8	– 1836.8	– 1811.6	– 1718.1	– 1634.9	– 1560.2	– 1492.3	– 1430.2
20	400	– 1865.9	– 1849.3	– 1823.5	– 1798.0	– 1773.3	– 1681.6	– 1599.8	– 1526.1	– 1459.1	– 1397.7
25	400	– 1845.7	– 1829.7	– 1803.7	– 1778.5	– 1754.1	– 1663.3	– 1582.1	– 1509.0	– 1442.5	– 1381.5
30	400	– 1825.4	– 1809.5	– 1783.9	– 1759.0	– 1734.8	– 1644.9	– 1564.5	– 1491.9	– 1425.8	– 1365.2
40	400	– 1784.7	– 1769.3	– 1744.2	– 1719.9	– 1696.2	– 1608.1	– 1529.1	– 1457.7	– 1392.6	– 1332.7
50	400	– 1744.0	– 1729.0	– 1704.5	– 1680.7	– 1657.6	– 1571.3	– 1493.7	– 1423.5	– 1359.3	– 1300.2
60	400	– 1703.2	– 1688.6	– 1664.7	– 1641.5	– 1618.9	– 1534.4	– 1458.3	– 1389.2	– 1326.0	– 1267.7
70	400	– 1662.4	– 1648.2	– 1624.9	– 1602.2	– 1580.2	– 1497.5	– 1422.9	– 1355.0	– 1292.8	– 1235.3
80	400	– 1621.6	– 1607.7	– 1585.0	– 1562.9	– 1541.4	– 1460.6	– 1387.4	– 1320.8	– 1259.5	– 1202.9
90	400	– 1580.7	– 1567.2	– 1545.1	– 1523.5	– 1502.5	– 1423.6	– 1352.0	– 1286.6	– 1226.4	– 1170.6
100	400	– 1539.7	– 1526.5	– 1505.1	– 1484.1	– 1463.6	– 1386.6	– 1316.5	– 1252.4	– 1193.2	– 1138.4
110	400	– 1498.5	– 1485.8	– 1465.0	– 1444.6	– 1424.7	– 1349.6	– 1281.0	– 1218.2	– 1160.1	– 1106.2
120	400	– 1457.3	– 1445.0	– 1424.8	– 1405.0	– 1385.6	– 1312.5	– 1245.5	– 1184.0	– 1127.0	– 1074.0
130	400	– 1415.9	– 1404.1	– 1384.5	– 1365.3	– 1346.5	– 1275.3	– 1210.0	– 1149.8	– 1093.9	– 1041.9
140	400	– 1374.4	– 1363.0	– 1344.1	– 1325.5	– 1307.3	– 1238.2	– 1174.5	– 1115.6	– 1060.9	– 1009.8
150	400	– 1332.8	– 1321.8	– 1303.6	– 1285.7	– 1268.1	– 1201.0	– 1138.9	– 1081.4	– 1027.8	– 977.7

(continued)

Table 4.7 (continued)

T	P	m = 0.1	m = 0.25	m = 0.5	m = 0.75	m = 1	m = 2	m = 3	m = 4	m = 5	m = 6
°C	bar	mol/kg									
160	400	– 1291.0	– 1280.5	– 1263.0	– 1245.7	– 1228.7	– 1163.7	– 1103.4	– 1047.2	– 994.8	– 945.7
170	400	– 1249.1	– 1239.0	– 1222.2	– 1205.6	– 1189.3	– 1126.4	– 1067.8	– 1013.0	– 961.8	– 913.6
180	400	– 1206.9	– 1197.4	– 1181.3	– 1165.5	– 1149.7	– 1089.1	– 1032.2	– 978.8	– 928.7	– 881.5
190	400	– 1164.6	– 1155.5	– 1140.3	– 1125.1	– 1110.1	– 1051.7	– 996.6	– 944.6	– 895.7	– 849.5
200	400	– 1122.0	– 1113.5	– 1099.0	– 1084.6	– 1070.3	– 1014.2	– 960.9	– 910.4	– 862.6	– 817.4
210	400	– 1079.1	– 1071.2	– 1057.6	– 1043.9	– 1030.3	– 976.7	– 925.2	– 876.2	– 829.6	– 785.2
220	400	– 1036.0	– 1028.6	– 1015.9	– 1003.1	– 990.2	– 939.1	– 889.5	– 841.9	– 796.5	– 753.1
230	400	– 992.4	– 985.7	– 973.9	– 961.9	– 949.8	– 901.3	– 853.7	– 807.6	– 763.3	– 720.9
240	400	– 948.5	– 942.4	– 931.6	– 920.5	– 909.3	– 863.4	– 817.8	– 773.2	– 730.1	– 688.6
250	400	– 904.1	– 898.7	– 889.0	– 878.8	– 868.4	– 825.3	– 781.8	– 738.8	– 696.8	– 656.3
260	400	– 859.2	– 854.5	– 845.9	– 836.8	– 827.2	– 787.1	– 745.6	– 704.2	– 663.5	– 623.8
270	400	– 813.8	– 809.8	– 802.4	– 794.3	– 785.7	– 748.6	– 709.4	– 669.5	– 630.0	– 591.2
280	400	– 767.6	– 764.6	– 758.4	– 751.4	– 743.8	– 709.9	– 672.9	– 634.7	– 596.3	– 558.5
290	400	– 720.7	– 718.6	– 713.8	– 707.9	– 701.4	– 670.8	– 636.1	– 599.6	– 562.4	– 525.5
300	400	– 672.9	– 671.8	– 668.5	– 663.9	– 658.5	– 631.4	– 599.0	– 564.1	– 528.2	– 492.2

Table 4.8 The specific enthalpy of NaCl(aq), J/g K (P = 600 bar)

T	P	m = 0.1	m = 0.25	m = 0.5	m = 0.75	m = 1	m = 2	m = 3	m = 4	m = 5	m = 6
°C	bar	mol/kg									
0	600	– 1927.7	– 1910.9	– 1883.7	– 1857.4	– 1831.9	– 1737.7	– 1653.9	– 1578.6	– 1510.4	– 1447.9
10	600	– 1888.0	– 1871.6	– 1845.0	– 1819.2	– 1794.2	– 1701.6	– 1619.0	– 1544.8	– 1477.4	– 1415.7
20	600	– 1848.0	– 1832.0	– 1806.0	– 1780.7	– 1756.2	– 1665.3	– 1584.1	– 1510.8	– 1444.3	– 1383.4
25	600	– 1827.9	– 1812.1	– 1786.4	– 1761.4	– 1737.2	– 1647.1	– 1566.5	– 1493.8	– 1427.7	– 1367.1
30	600	– 1807.8	– 1792.2	– 1766.7	– 1742.0	– 1718.0	– 1628.8	– 1548.9	– 1476.8	– 1411.2	– 1350.9
40	600	– 1767.5	– 1752.2	– 1727.4	– 1703.2	– 1679.7	– 1592.2	– 1513.7	– 1442.7	– 1378.0	– 1318.5
50	600	– 1727.1	– 1712.2	– 1687.9	– 1664.3	– 1641.3	– 1555.6	– 1478.5	– 1408.6	– 1344.8	– 1286.1
60	600	– 1686.7	– 1672.2	– 1648.4	– 1625.3	– 1602.9	– 1518.9	– 1443.2	– 1374.5	– 1311.6	– 1253.7
70	600	– 1646.2	– 1632.1	– 1608.9	– 1586.3	– 1564.4	– 1482.2	– 1407.9	– 1340.4	– 1278.4	– 1221.2
80	600	– 1605.7	– 1591.9	– 1569.3	– 1547.3	– 1525.9	– 1445.4	– 1372.6	– 1306.3	– 1245.3	– 1189.0
90	600	– 1565.1	– 1551.7	– 1529.7	– 1508.2	– 1487.3	– 1408.7	– 1337.3	– 1272.2	– 1212.3	– 1156.8
100	600	– 1524.4	– 1511.4	– 1490.0	– 1469.0	– 1448.6	– 1371.9	– 1302.0	– 1238.1	– 1179.3	– 1124.7
110	600	– 1483.7	– 1471.0	– 1450.2	– 1429.8	– 1409.9	– 1335.0	– 1266.7	– 1204.1	– 1146.3	– 1092.6
120	600	– 1442.8	– 1430.5	– 1410.3	– 1390.5	– 1371.2	– 1298.1	– 1231.4	– 1170.0	– 1113.3	– 1060.6
130	600	– 1401.8	– 1389.9	– 1370.3	– 1351.1	– 1332.4	– 1261.2	– 1196.0	– 1136.0	– 1080.4	– 1028.6
140	600	– 1360.7	– 1349.2	– 1330.3	– 1311.7	– 1293.5	– 1224.3	– 1160.7	– 1102.0	– 1047.5	– 996.6
150	600	– 1319.5	– 1308.4	– 1290.2	– 1272.2	– 1254.5	– 1187.4	– 1125.4	– 1068.0	– 1014.6	– 964.7

(continued)

Table 4.8 (continued)

T	P	m = 0.1	m = 0.25	m = 0.5	m = 0.75	m = 1	m = 2	m = 3	m = 4	m = 5	m = 6
°C	bar	mol/kg									
160	600	– 1278.2	– 1267.5	– 1249.9	– 1232.6	– 1215.5	– 1150.4	– 1090.1	– 1034.0	– 981.7	– 932.8
170	600	– 1236.7	– 1226.5	– 1209.6	– 1192.9	– 1176.5	– 1113.4	– 1054.7	– 1000.0	– 943.9	– 901.0
180	600	– 1195.1	– 1185.4	– 1169.2	– 1153.2	– 1137.3	– 1076.4	– 1019.4	– 966.1	– 916.1	– 869.1
190	600	– 1153.3	– 1144.1	– 1128.6	– 1113.3	– 1098.1	– 1039.4	– 984.1	– 932.1	– 883.2	– 837.2
200	600	– 1111.3	– 1102.6	– 1088.0	– 1073.3	– 1058.8	– 1002.3	– 948.8	– 898.2	– 850.4	– 805.3
210	600	– 1069.1	– 1060.9	– 1047.1	– 1033.2	– 1019.4	– 965.2	– 913.5	– 864.2	– 817.6	– 773.4
220	600	– 1026.7	– 1019.1	– 1006.1	– 993.0	– 979.9	– 928.1	– 878.1	– 830.3	– 784.7	– 741.4
230	600	– 984.0	– 977.0	– 964.9	– 952.6	– 940.2	– 890.9	– 842.8	– 796.4	– 751.9	– 709.4
240	600	– 941.0	– 934.6	– 923.5	– 912.0	– 900.4	– 853.7	– 807.4	– 762.4	– 719.0	– 677.4
250	600	– 897.7	– 891.9	– 881.8	– 871.2	– 860.5	– 816.3	– 772.0	– 728.4	– 686.1	– 645.2
260	600	– 854.0	– 848.9	– 839.8	– 830.2	– 820.3	– 778.9	– 736.6	– 694.4	– 653.1	– 613.0
270	600	– 809.9	– 805.5	– 797.5	– 788.9	– 779.9	– 741.5	– 701.1	– 660.4	– 620.0	– 580.6
280	600	– 765.3	– 761.7	– 754.9	– 747.4	– 739.3	– 703.9	– 665.6	– 626.3	– 586.8	– 548.1
290	600	– 720.2	– 717.5	– 711.9	– 705.5	– 698.5	– 666.2	– 630.0	– 592.0	– 553.5	– 515.2
300	600	– 674.5	– 672.7	– 668.5	– 663.3	– 657.3	– 628.4	– 594.3	– 557.6	– 519.8	– 482.0

Table 4.9 The specific enthalpy of NaCl(aq), J/g K (P = 800 bar)

T	P	m = 0.1	m = 0.25	m = 0.5	m = 0.75	m = 1	m = 2	m = 3	m = 4	m = 5	m = 6
°C	bar	mol/kg									
0	800	– 1909.2	– 1892.7	– 1865.8	– 1839.7	– 1814.5	– 1721.1	– 1638.0	– 1563.2	– 1495.5	– 1433.5
10	800	– 1869.9	– 1853.7	– 1827.4	– 1801.8	– 1777.1	– 1685.2	– 1603.3	– 1529.5	– 1462.6	– 1401.4
20	800	– 1830.4	– 1814.5	– 1788.7	– 1763.7	– 1739.4	– 1649.1	– 1568.4	– 1495.7	– 1429.6	– 1369.1
25	800	– 1810.5	– 1794.8	– 1769.2	– 1744.5	– 1720.4	– 1631.0	– 1550.9	– 1478.7	– 1413.1	– 1352.9
30	800	– 1790.5	– 1775.0	– 1749.7	– 1725.2	– 1701.4	– 1612.8	– 1533.4	– 1461.8	– 1396.5	– 1336.7
40	800	– 1750.5	– 1735.3	– 1710.6	– 1686.6	– 1663.3	– 1576.4	– 1498.4	– 1427.8	– 1363.4	– 1304.3
50	800	– 1710.4	– 1695.6	– 1671.4	– 1648.0	– 1625.2	– 1539.9	– 1463.3	– 1393.8	– 1330.3	– 1272.0
60	800	– 1670.3	– 1655.8	– 1632.2	– 1609.3	– 1586.9	– 1503.4	– 1428.2	– 1359.8	– 1297.3	– 1239.7
70	800	– 1630.1	– 1616.0	– 1592.9	– 1570.5	– 1548.7	– 1466.9	– 1393.0	– 1325.8	– 1264.2	– 1207.2
80	800	– 1589.9	– 1576.1	– 1553.6	– 1531.7	– 1510.4	– 1430.3	– 1357.9	– 1291.8	– 1231.2	– 1175.1
90	800	– 1549.6	– 1536.2	– 1514.2	– 1492.9	– 1472.0	– 1393.7	– 1322.7	– 1257.9	– 1198.2	– 1143.0
100	800	– 1509.2	– 1496.2	– 1474.8	– 1454.0	– 1433.6	– 1357.1	– 1287.5	– 1223.9	– 1165.3	– 1110.9
110	800	– 1468.7	– 1456.1	– 1435.3	– 1415.0	– 1395.2	– 1320.4	– 1252.4	– 1190.0	– 1132.4	– 1079.0
120	800	– 1428.2	– 1415.9	– 1395.7	– 1376.0	– 1356.7	– 1283.8	– 1217.2	– 1156.1	– 1099.6	– 1047.1
130	800	– 1387.5	– 1375.6	– 1356.0	– 1336.9	– 1318.1	– 1247.1	– 1182.0	– 1122.2	– 1066.8	– 1015.2
140	800	– 1346.8	– 1335.3	– 1316.3	– 1297.7	– 1279.5	– 1210.4	– 1146.9	– 1088.3	– 1034.0	– 983.4
150	800	– 1305.9	– 1294.8	– 1276.5	– 1258.5	– 1240.8	– 1173.7	– 1111.7	– 1054.5	– 1001.3	– 951.7

(continued)

Table 4.9 (continued)

T	P	m = 0.1	m = 0.25	m = 0.5	m = 0.75	m = 1	m = 2	m = 3	m = 4	m = 5	m = 6
°C	bar	mol/kg									
160	800	– 1265.0	– 1254.3	– 1236.6	– 1219.2	– 1202.2	– 1137.0	– 1076.6	– 1020.7	– 968.6	– 919.9
170	800	– 1223.9	– 1213.7	– 1196.7	– 1179.9	– 1163.4	– 1100.2	– 1041.5	– 986.9	– 935.9	– 888.2
180	800	– 1182.8	– 1173.0	– 1156.7	– 1140.5	– 1124.6	– 1063.5	– 1006.4	– 953.1	– 903.3	– 856.5
190	800	– 1141.4	– 1132.1	– 1116.6	– 1101.1	– 1085.8	– 1026.8	– 971.4	– 919.4	– 870.6	– 824.8
200	800	– 1100.0	– 1091.2	– 1076.3	– 1061.6	– 1046.9	– 990.1	– 936.3	– 885.7	– 838.0	– 793.1
210	800	– 1058.4	– 1050.1	– 1036.0	– 1022.0	– 1008.0	– 953.4	– 901.3	– 852.0	– 805.4	– 761.4
220	800	– 1016.6	– 1008.8	– 995.6	– 982.3	– 969.0	– 916.6	– 866.3	– 818.3	– 772.8	– 729.6
230	800	– 974.6	– 967.3	– 955.0	– 942.4	– 929.9	– 879.9	– 831.4	– 784.7	– 740.2	– 697.9
240	800	– 932.3	– 925.7	– 914.2	– 902.5	– 890.7	– 843.2	– 796.4	– 751.1	– 707.6	– 666.0
250	800	– 889.8	– 883.8	– 873.3	– 862.4	– 851.4	– 806.4	– 761.5	– 717.5	– 674.9	– 634.1
260	800	– 847.1	– 841.7	– 832.2	– 822.2	– 812.0	– 769.7	– 726.6	– 684.0	– 642.3	– 602.1
270	800	– 804.0	– 799.3	– 790.8	– 781.9	– 772.5	– 733.0	– 691.8	– 650.4	– 609.6	– 569.9
280	800	– 760.5	– 756.6	– 749.3	– 741.4	– 732.9	– 696.3	– 657.1	– 616.9	– 576.8	– 537.5
290	800	– 716.7	– 713.6	– 707.6	– 700.7	– 693.3	– 659.7	– 622.4	– 583.4	– 543.9	– 504.9
300	800	– 672.5	– 670.3	– 665.6	– 659.9	– 653.5	– 623.1	– 587.8	– 549.8	– 510.8	– 471.7

Table 4.10 The specific enthalpy of NaCl(aq), J/g K (P = 1000 bar)

T	P	m = 0.1	m = 0.25	m = 0.5	m = 0.75	m = 1	m = 2	m = 3	m = 4	m = 5	m = 6
°C	bar	mol/kg									
0	1000	– 1891.2	– 1874.8	– 1848.2	– 1822.4	– 1797.4	– 1704.8	– 1622.1	– 1547.9	– 1480.6	– 1419.2
10	1000	– 1852.2	– 1836.2	– 1810.0	– 1784.7	– 1760.2	– 1669.0	– 1587.6	– 1514.3	– 1447.9	– 1387.1
20	1000	– 1812.9	– 1797.2	– 1771.6	– 1746.8	– 1722.7	– 1633.1	– 1552.9	– 1480.6	– 1414.9	– 1354.9
25	1000	– 1793.2	– 1777.6	– 1752.3	– 1727.7	– 1703.9	– 1615.0	– 1535.5	– 1463.7	– 1398.5	– 1338.7
30	1000	– 1773.5	– 1758.0	– 1732.9	– 1708.6	– 1685.0	– 1597.0	– 1518.1	– 1446.8	– 1382.0	– 1322.5
40	1000	– 1733.6	– 1718.6	– 1694.0	– 1670.2	– 1647.1	– 1560.7	– 1483.1	– 1412.9	– 1348.9	– 1290.2
50	1000	– 1693.3	– 1679.1	– 1655.1	– 1631.8	– 1609.1	– 1524.4	– 1448.2	– 1379.1	– 1315.9	– 1257.9
60	1000	– 1654.0	– 1639.6	– 1616.1	– 1593.3	– 1571.1	– 1488.0	– 1413.2	– 1345.2	– 1283.0	– 1225.6
70	1000	– 1614.0	– 1600.0	– 1577.1	– 1554.7	– 1533.0	– 1451.7	– 1378.2	– 1311.3	– 1249.9	– 1193.3
80	1000	– 1574.1	– 1560.4	– 1538.0	– 1516.2	– 1494.9	– 1415.2	– 1343.1	– 1277.4	– 1217.0	– 1161.2
90	1000	– 1534.0	– 1520.7	– 1498.8	– 1477.5	– 1456.8	– 1378.8	– 1308.1	– 1243.5	– 1184.2	– 1129.1
100	1000	– 1493.9	– 1480.9	– 1459.6	– 1438.9	– 1418.6	– 1342.4	– 1273.1	– 1209.7	– 1151.3	– 1097.2
110	1000	– 1453.3	– 1441.1	– 1420.4	– 1400.1	– 1380.4	– 1305.9	– 1238.1	– 1175.9	– 1118.6	– 1065.3
120	1000	– 1413.5	– 1401.2	– 1381.1	– 1361.3	– 1342.1	– 1269.0	– 1203.0	– 1142.1	– 1085.8	– 1033.5
130	1000	– 1373.1	– 1361.2	– 1341.7	– 1322.5	– 1303.8	– 1232.9	– 1168.0	– 1108.4	– 1053.2	– 1001.8
140	1000	– 1332.7	– 1321.2	– 1302.2	– 1283.6	– 1265.4	– 1196.4	– 1133.0	– 1074.6	– 1020.5	– 970.1
150	1000	– 1292.2	– 1281.1	– 1262.7	– 1244.7	– 1227.0	– 1159.9	– 1098.1	– 1040.9	– 987.9	– 938.5

(continued)

Table 4.10 (continued)

T	P	m = 0.1	m = 0.25	m = 0.5	m = 0.75	m = 1	m = 2	m = 3	m = 4	m = 5	m = 6
°C	bar	mol/kg									
160	1000	– 1251.6	– 1240.9	– 1223.2	– 1205.7	– 1188.6	– 1123.4	– 1063.1	– 1007.3	– 955.4	– 906.9
170	1000	– 1210.9	– 1200.6	– 1183.5	– 1166.7	– 1150.2	– 1086.9	– 1028.2	– 973.7	– 922.8	– 875.3
180	1000	– 1170.1	– 1160.3	– 1143.9	– 1127.7	– 1111.7	– 1050.4	– 993.3	– 940.1	– 890.4	– 843.8
190	1000	– 1129.2	– 1119.8	– 1104.1	– 1088.6	– 1073.2	– 1014.0	– 958.5	– 906.5	– 857.9	– 812.3
200	1000	– 1088.2	– 1079.3	– 1064.3	– 1049.4	– 1034.7	– 977.6	– 923.7	– 873.0	– 825.5	– 780.7
210	1000	– 1047.1	– 1038.6	– 1024.4	– 1010.2	– 996.1	– 941.2	– 888.9	– 839.6	– 793.0	– 749.2
220	1000	– 1005.8	– 997.9	– 984.5	– 971.0	– 957.5	– 904.8	– 854.2	– 806.2	– 760.6	– 717.7
230	1000	– 964.4	– 957.0	– 944.4	– 931.7	– 918.9	– 868.4	– 819.6	– 772.8	– 728.3	– 686.1
240	1000	– 922.3	– 915.9	– 904.2	– 892.3	– 880.2	– 832.1	– 785.0	– 739.5	– 695.9	– 654.4
250	1000	– 881.0	– 874.7	– 863.9	– 852.8	– 841.5	– 795.9	– 750.5	– 706.2	– 663.5	– 622.7
260	1000	– 838.9	– 833.3	– 823.5	– 813.2	– 802.8	– 759.7	– 716.0	– 673.0	– 631.1	– 590.9
270	1000	– 796.7	– 791.7	– 782.9	– 773.6	– 764.0	– 723.6	– 681.7	– 639.8	– 598.7	– 559.0
280	1000	– 754.1	– 749.9	– 742.2	– 733.9	– 725.2	– 687.6	– 647.5	– 606.8	– 566.3	– 526.9
290	1000	– 711.3	– 707.9	– 701.4	– 694.2	– 686.4	– 651.7	– 613.5	– 573.8	– 533.7	– 494.4
300	1000	– 668.2	– 665.7	– 660.5	– 654.4	– 647.7	– 616.2	– 579.8	– 540.8	– 501.0	– 461.5

Table 4.11 Partial specific enthalpy of sea salt at 2 °C as a function of concentration (Connors 1970)

S_1 (‰)	S_2 (‰)	T (°C)	ΔT (°C)	h_2* (J/g)[a]
10	60	1.93	– 0.067	– 860
10	60	1.9	– 0.063	– 820
15	54.6	2.05	– 0.041	– 840
15	54.6	2.02	– 0.039	– 820
15	54.6	2.86	– 0.04	– 830
20	49.6	2.00	– 0.021	– 780
20	49.6	1.99	– 0.022	– 810
24.8	45.4	2.08	– 0.01	– 820
24.8	45.4	2.03	– 0.01	– 790

[a] This is an empirically determined constant used to calculate the partial specific enthalpy of sea salt

Table 4.12 Temperature dependency of the heat of mixing of seawaters (Connors 1970)

T (°C)	S_1 (‰)	S_2 (‰)	ΔT (°C)
25.3	10.24	60.5	– 0.025
24.6	10.24	60.5	– 0.025
21.4	10.24	60.5	– 0.028
20	10.24	60.5	– 0.029
15	10.24	60.5	– 0.037
14.2	10.24	60.5	– 0.038
9.15	10.24	60.5	– 0.049
3.1	10.24	60.5	– 0.066
2.75	10.24	60.5	– 0.068
2	10.24	60.5	– 0.067

Table 4.13 Coefficients for the saline part of the Gibbs function used by IAPWS (1992)

i	j	k	g_{ijk}	i	j	k	g_{ijk}
1	0	0	0.581 281 456 626 732 × 10^4	2	2	1	– 0.860 764 303 783 977 × 10^3
2	0	0	0.141 627 648 484 197 × 10^4	3	2	1	0.383 058 066 002 476 × 10^3
3	0	0	– 0.243 214 662 381 794 × 10^4	2	3	1	0.694 244 814 133 268 × 10^3
4	0	0	0.202 580 115 603 697 × 10^4	3	3	1	– 0.460 319 931 801 257 × 10^3
5	0	0	– 0.109 166 841 042 967 × 10^4	2	4	1	– 0.297 728 741 987 187 × 10^3
6	0	0	0.374 601 237 877 840 × 10^3	3	4	1	0.234 565 187 611 355 × 10^3
7	0	0	– 0.485 891 069 025 409 × 10^2	2	0	2	0.384 794 152 978 599 × 10^3
1	1	0	0.851 226 734 946 706 × 10^3	3	0	2	– 0.522 940 909 281 335 × 10^2
2	1	0	0.168 072 408 311 545 × 10^3	4	0	2	– 0.408 193 978 912 261 × 10
3	1	0	– 0.493 407 510 141 682 × 10^3	2	1	2	– 0.343 956 902 961 561 × 10^3
4	1	0	0.543 835 333 000 098 × 10^3	3	1	2	0.831 923 927 801 819 × 10^2
5	1	0	– 0.196 028 306 689 776 × 10^3	2	2	2	0.337 409 530 269 367 × 10^3
6	1	0	0.367 571 622 995 805 × 10^2	3	2	2	– 0.541 917 262 517 112 × 10^2
2	2	0	0.880 031 352 997 204 × 10^3	2	3	2	– 0.204 889 641 964 903 × 10^3
3	2	0	– 0.430 664 675 978 042 × 10^2	2	4	2	0.747 261 411 387 560 × 10^2
4	2	0	– 0.685 572 509 204 491 × 10^2	2	0	3	– 0.965 324 320 107 458 × 10^2
2	3	0	– 0.225 267 649 263 401 × 10^3	3	0	3	0.680 444 942 726 459 × 10^2
3	3	0	– 0.100 227 370 861 875 × 10^2	4	0	3	– 0.301 755 111 971 161 × 10^2
4	3	0	0.493 667 694 856 254 × 10^2	2	1	3	0.124 687 671 116 248 × 10^3
2	4	0	0.914 260 447 751 259 × 10^2	3	1	3	– 0.294 830 643 494 290 × 10^2
3	4	0	0.875 600 661 808 945	2	2	3	– 0.178 314 556 207 638 × 10^3
4	4	0	– 0.171 397 577 419 788 × 10^2	3	2	3	0.256 398 487 389 914 × 10^2
2	5	0	– 0.216 603 240 875 311 × 10^2	2	3	3	0.113 561 697 840 594 × 10^3
4	5	0	0.249 697 009 569 508 × 10	2	4	3	– 0.364 872 919 001 588 × 10^2
2	6	0	0.213 016 970 847 183 × 10	2	0	4	0.158 408 172 766 824 × 10^2
2	0	1	– 0.331 049 154 044 839 × 10^4	3	0	4	– 0.341 251 932 441 282 × 10
3	0	1	0.199 459 603 073 901 × 10^3	2	1	4	– 0.316 569 643 860 730 × 10^2
4	0	1	– 0.547 919 133 532 887 × 10^2	2	2	4	0.442 040 358 308 000 × 10^2
5	0	1	0.360 284 195 611 086 × 10^2	2	3	4	– 0.111 282 734 326 413 × 10^2
2	1	1	0.729 116 529 735 046 × 10^3	2	0	5	– 0.262 480 156 590 992 × 10
3	1	1	– 0.175 292 041 186 547 × 10^3	2	1	5	0.704 658 803 315 449 × 10
4	1	1	– 0.226 683 558 512 829 × 10^2	2	2	5	– 0.792 001 547 211 682 × 10

Table 4.14 Enthalpy of sea salt solutions at thermal saturation (Bromley et al. 1970)

T (°C)	Salinity, wt. (%)												
	0	1	2	3	4	5	6	7	8	9	10	11	12
0	0	0.007	0.001	− 0.029	− 0.076	− 0.145	− 0.235	− 0.345	− 0.477	− 0.63	− 0.805	− 0.996	− 1.209
5	5.032	4.958	4.874	4.771	4.654	4.52	4.368	4.2	4.014	3.811	3.591	3.358	3.109
10	10.05	9.9	9.744	9.572	9.389	9.191	8.979	8.754	8.515	8.261	7.994	7.718	7.429
15	15.056	14.834	14.61	14.371	14.125	13.866	13.596	13.315	13.023	12.719	12.404	12.084	11.753
20	20.056	19.765	19.475	19.173	18.866	18.548	18.221	17.885	17.541	17.187	16.824	16.458	16.084
25	25.055	24.698	24.344	23.981	23.614	23.239	22.856	22.466	22.07	21.666	21.255	20.843	20.425
30	30.051	29.629	29.214	28.791	28.365	27.934	27.496	27.053	26.605	26.151	25.693	25.235	24.773
35	35.044	34.559	34.083	33.601	33.118	32.631	32.139	31.644	31.145	30.642	30.135	29.631	29.124
40	40.038	39.491	38.955	38.415	37.876	37.333	36.788	36.24	35.691	35.138	34.583	34.032	33.48
45	45.034	44.425	43.83	43.232	42.637	42.04	41.441	40.842	40.241	39.639	39.037	38.439	37.842
50	50.033	49.363	48.709	48.054	47.403	46.752	46.1	45.448	44.797	44.146	43.495	42.851	42.208
55	55.03	54.3	53.59	52.87	52.17	51.46	50.76	50.06	49.35	48.65	47.95	47.26	46.58
60	60.03	59.24	58.47	57.7	56.94	56.18	55.42	54.67	53.91	53.16	52.42	51.68	50.94
65	65.03	64.18	63.35	62.53	61.71	60.9	60.09	59.28	58.48	57.68	56.88	56.1	55.32
70	70.04	69.13	68.24	67.36	66.49	65.62	64.76	63.9	63.04	62.19	61.35	60.52	59.69
75	75.05	74.08	73.13	72.19	71.26	70.34	69.43	68.52	67.61	66.71	65.82	64.94	64.07
80	80.07	79.03	78.03	77.03	76.05	75.07	74.1	73.14	72.19	71.24	70.3	69.37	68.45
85	85.09	83.99	82.93	81.87	80.84	79.8	78.78	77.77	76.76	75.77	74.78	73.8	72.84
90	90.12	88.95	87.83	86.72	85.63	84.54	83.47	82.4	81.35	80.3	79.26	78.24	77.23
95	95.15	93.92	92.74	91.57	90.43	89.29	88.16	87.04	85.93	84.84	83.75	82.68	81.62
100	100.23	98.94	97.7	96.48	95.27	94.08	92.89	91.72	90.57	89.42	88.28	87.16	86.06

(continued)

Table 4.14 (continued)

T (°C)	Salinity, wt. (%)												
	0	1	2	3	4	5	6	7	8	9	10	11	12
105	105.23	103.88	102.58	101.3	100.03	98.79	97.55	96.33	95.12	93.92	92.74	91.57	90.42
110	110.29	108.87	107.51	106.17	104.85	103.55	102.26	100.98	99.72	98.47	97.24	96.02	94.82
115	115.35	113.87	112.45	111.05	109.67	108.31	106.97	105.64	104.33	103.03	101.74	100.48	99.23
120	120.42	118.87	117.39	115.93	114.5	113.08	111.68	110.3	108.93	107.58	106.25	104.94	103.64
125	125.51	123.9	122.35	120.83	119.34	117.87	116.42	114.98	113.56	112.16	110.77	109.41	108.07
130	130.6	128.92	127.31	125.74	124.18	122.66	121.14	119.66	118.18	116.73	115.29	113.88	112.49
135	135.72	133.97	132.3	130.66	129.05	127.46	125.89	124.35	122.82	121.31	119.83	118.36	116.92
140	140.83	139.02	137.28	135.58	133.91	132.26	130.64	129.04	127.46	125.89	124.35	122.84	121.35
145	145.97	144.09	142.28	140.52	138.79	137.08	135.4	133.74	132.11	130.49	128.9	127.33	125.79
150	151.13	149.18	147.31	145.48	143.69	141.92	140.18	138.46	136.77	135.1	133.45	131.83	130.24
155	156.3	154.27	152.33	150.44	148.58	146.76	144.95	143.18	141.43	139.7	138	136.33	134.68
160	161.48	159.38	157.37	155.41	153.49	151.61	149.75	147.91	146.11	144.32	142.57	140.84	139.14
165	166.7	164.52	162.44	160.41	158.43	156.48	154.56	152.66	150.8	148.96	147.14	145.36	143.61
170	171.91	169.65	167.5	165.41	163.36	161.34	159.36	157.4	155.48	153.58	151.71	149.88	148.07
175	177.17	174.83	172.61	170.44	168.33	166.25	164.2	162.19	160.2	158.24	156.32	154.43	152.56
180	182.43	180.01	177.71	175.48	173.29	171.15	169.04	166.96	164.91	162.9	160.91	158.97	157.05
185	187.73	185.23	182.86	180.56	178.3	176.09	173.91	171.77	169.67	167.59	165.55	163.54	161.57
190	193.04	190.45	188.01	185.63	183.3	181.02	178.78	176.58	174.41	172.28	170.17	168.11	166.08
195	193.39	195.72	193.2	190.74	188.35	186	183.69	181.43	179.19	177	174.83	172.71	170.62
200	203.77	201.01	198.41	195.88	193.41	190.99	188.62	186.28	183.99	181.73	179.51	177.32	175.18

Table 4.15 Specific enthalpy of NaCl(aq) at 25 °C and 1 atmospheric pressure from Chou (1968)

Salt	Enthalpy, h
%	kJ/kg K
26.496	68.0388
24.4954	70.8859
17.7822	81.5209
13.9571	87.4246
11.4863	91.0254
9.7588	93.4538
6.0931	98.1852
3.1423	101.5766
1.5962	103.1677
0.8045	103.9632
0.3234	104.4238
0.1619	104.5913
0.0648	104.6750
0.0324	104.7169
0.0065	104.7169
0.0032	104.7169
0.0006	104.7169

References

Bromley LA (1973) Thermodynamic properties of strong electrolytes in aqueous solutions. AIChE J 19(2):313–320. https://doi.org/10.1002/aic.690190216

Bromley LA, Diamond AE, Salami E, Wilkins DG (1970) Heat capacities and enthalpies of sea salt solutions to 200 °C. J Chem Eng Data 15(2):246–253. https://doi.org/10.1021/je60045a038

Chou JCS (1968) Thermodynamic properties of aqueous sodium chloride solutions from 32 to 350 °F. Georgia Institute of Technology

Chou JCS, Rowe AM (1969) Enthalpies of aqueous sodium chloride solutions from 32 to 350 °F. Desalination 6(1):105–115. https://doi.org/10.1016/S0011-9164(00)80014-4

Connors DN (1970) On the enthalpy of seawater. Limnol Oceanogr 15(4):587–594. https://doi.org/10.4319/lo.1970.15.4.0587

Cooper JR, Mary Q, Road ME, Associates SI (2008) The international association for the properties of water and steam release (IAPWS-2008) on the formulation for the the thermodynamic properties of water and steam, Sept 2008, pp 1–19

Feistel R (1993) Equilibrium thermodynamics of seawater revisited. Prog Oceanogr 31(2):101–179. https://doi.org/10.1016/0079-6611(93)90024-8

Feistel R (2008) A Gibbs function for seawater thermodynamics for −6 to 80 °C and salinity up to 120 g kg^{-1}. Deep Sea Res Part 1 Oceanogr Res Pap 55(12):1639–1671. https://doi.org/10.1016/j.dsr.2008.07.004

Generous MM, Qasem NAA, Qureshi BA, Zubair SM (2020) A comprehensive review of saline water correlations and data-part I: thermodynamic properties. Arab J Sci Eng 45(11):8817–8876. https://doi.org/10.1007/s13369-020-05019-y

IAPWS (1992) The international association for the properties of water and steam: revised (June 2014). Supplementary release on saturation properties of ordinary water substance, vol 10, issue Aug 1992

Kukulka DJ, Gebhart B, Mollendorf JC (1987) Thermodynamic and transport properties of pure and saline water. Adv Heat Transfer 18(C):325–363. https://doi.org/10.1016/S0065-2717(08)70121-7

Millero FJ (1983) The thermodynamics of seawater part II: thermochemical properties. Ocean Sci Eng 8(1):1–40

Nayar KG, Sharqawy MH, Banchik LD, Lienhard JH (2016) Thermophysical properties of seawater: a review and new correlations that include pressure dependence. Desalination 390:1–24. https://doi.org/10.1016/j.desal.2016.02.024

Pitzer KS (1987) A thermodynamic model for aqueous solutions of liquid-like density. Rev Mineral 17:97–142

Pitzer KS, Peiper JC, Busey RH (1984) Thermodynamic properties of aqueous sodium chloride solutions. J Phys Chem Ref Data 13(1):1–102. https://doi.org/10.1063/1.555709

Qasem NAA, Generous MM, Qureshi BA, Zubair SM (2019) Thermophysical properties of saline water: a comprehensive review of existing correlations and data under different conditions—part II. Int J Refrig

Singh D, Bromley LRA (1973) Relative enthalpies of sea salt solutions at 0° to 75 °C. J Chem Eng Data 18(2):174–181. https://doi.org/10.1021/je60057a028

Thiel GP, Lienhard JH (2014) Treating produced water from hydraulic fracturing: composition effects on scale formation and desalination system selection. Desalination 346:54–69. https://doi.org/10.1016/j.desal.2014.05.001

Chapter 5
Specific Entropy

Abstract The specific entropy of saline water is important for designing desalination plants and examining performance and irreversibility. It is a thermodynamic property computed based on the differentiation of the Gibbs free energy with respect to temperature at constant pressure. This chapter explains the thermodynamic approaches, multicomponent models, correlation, and data used to estimate specific entropy at various salinity and temperature values. Besides, the calculated specific entropy values from the correlations and methods are compared and assessed against the experimental data of seawater. A new correlation for NaCl-specific entropy with high accuracy is suggested for a salinity of 0–260 g/kg and ambient conditions. Some correlations and methods are recommended based on their accuracy and operating conditions. Furthermore, the codes used for the investigated methods are provided.

Keywords Specific entropy · Thermodynamics · Correlations · Methods · Irreversibility · Saline water

5.1 Introduction

The entropy of saline water is a valuable property to be considered for designing desalination plants and examining performance and irreversibility. It is a thermodynamic property that is calculated based on the differentiation of the Gibbs free energy with respect to temperature at constant pressure (Sharqawy et al. 2011).

Pitzer et al. (1984) proposed a comprehensive model in 1987 that can be conveniently used to calculate the entropy of multicomponent saline solutions for a high range of salinity.

A limited number of implicit correlations for entropy was also proposed. Millero (1983) proposed an implicit correlation valid in salinity and temperature up to 40 g/kg and 40 °C, respectively. Similar implicit correlations were proposed by Sun et al. (2008) and Nayar et al. (2016). All these correlations cover a salinity range of 0–120 g/kg. A correlation proposed by Sun et al. (2008) covers wider temperature and pressure ranges of 0–374 °C and 0–100 MPa.

N. A. A. Qasem et al., *Thermodynamic and Thermophysical Properties of Saline Water*, Springer Water, https://doi.org/10.1007/978-3-031-35193-8_5

5.2 Thermodynamic Approach

It is a thermodynamic property that cannot be directly measured. Instead, it is calculated based on the differentiation of the Gibbs free energy with respect to temperature at constant pressure (Sharqawy et al. 2011).

$$s = -\left(\frac{dg}{dT}\right)_P \tag{5.1}$$

In 1989, Slesarenko and Shtim (1989) used characteristic functions based on thermodynamic differential equations to calculate entropy. The total specific entropy is related to the apparent and partial entropies, like enthalpy, as shown below:

$$s = x^{\phi}s = x\bar{s}_s + (1 - x)\bar{s}_w \tag{5.2}$$

5.3 Multiple Components Models

The Extended Pitzer model (Pitzer et al. 1984; Pitzer 1987; Thiel and Lienhard 2014) can be used for specific entropy of multiple components of aqueous solutions.

$$\frac{G^{ex}}{n_w RT} = \sum_i m_i(\ln\gamma_i + (1 - \phi)) \tag{5.3}$$

$$S^{ex} = \left(L - G^{ex}\right)/T \tag{5.4}$$

$$^{\phi}s = S/n_2 \tag{5.5}$$

The multicomponent activity coefficient and osmotic coefficient are described in Part II (Qasem et al. 2021) based on Thiel and Lienhard (2014). L is determined from the expression in Chap. 4 (specific enthalpy). n_2 is the moles of solute. The model was initially tested for aqueous NaCl, and values of specific entropy are given in Appendix 1, taken from Pitzer et al. (1984), for a considerable variation of temperature (0–300 °C), pressure (100 kPa–100 MPa), and molality (0.1–6 mol/kg). It is important to note that specific entropy is a weak function of pressure and temperature limitation depends upon measured parameters. The complete data set was provided for the convenience of checking all correlations provided in the next section against the work of Pitzer et al. (1984).

5.4 Correlations and Data

Correlations are the easiest way to calculate the saltwater entropy as they are fitted from the experimental data. Most of the available specific entropy correlations in the literature are for seawater. Sun et al. (2008) calculated specific entropy values at vapor pressures using the following familiar equation that relates it to specific heat capacity:

$$s = s_0 + \int_0^t \frac{C_p}{t + 273.15} dt \tag{5.6}$$

The specific heat data was fitted and then applied in Eq. (5.6). s_0 is the value of s, which corresponds to the temperature-independent terms and coefficients (1, p, p^2, p^3, p^4, S, S^2, S^3, and S^4) shown in Table 5.1. Based upon this, Sun et al. (2008) proposed a correlation for seawater and brackish water entropy in terms of temperature (T) in °C, pressure (P) in MPa, and salinity (S) in g/kg. All the required constants necessary for this correlation are provided in Table 5.1.

$$s = \sum_{i=1}^{24} m_i n_i - \sum_{i=1}^{10} X_i Y_i \tag{5.7}$$

A sample of the data used is provided in Appendix 2. The correlation is valid in the salinity range of 0–40 g/kg, temperature range of 0–374 °C, and pressure of 0.1–100 MPa with an accuracy of ± 1%.

Nayar et al. (2016) correlation, also reported in IAPWS-2008 (Cooper et al. 2008), is expressed in terms of temperature (T) in °C, pressure (P) in MPa, and salinity (S) in kg/kg. The main correlation is reported in Eq. (5.8), and all the required constants are provided in Table 5.2.

$$s = s_1 + (P - P_o)\left(a_1 + a_2 T + a_3 T^2 + a_4 T^3\right) + S\left(a_5 + a_6 T + a_7 T^2 + a_8 T^3\right) \tag{5.8}$$

s_1 can be calculated from Eq. (5.9).

$$s_1 = s_2 - S\left(b_1 + b_2 S + b_3 S^2 + b_4 S^3 + b_5 T + b_6 T^2 + b_7 T^3 + b_8 ST + b_9 S^2 T + b_{10} ST^2\right) \tag{5.9}$$

s_2 is estimated from Eq. (5.10).

$$s_2 = c_1 + c_2 T + c_3 T^2 + c_4 T^3 + c_5 T^4 \tag{5.10}$$

The correlation is valid in the salinity range of 0–120 g/kg, temperature range of 10–120 °C, and pressure of 0.1–12 MPa with an accuracy of ± 1.2%.

Table 5.1 Constant for Sun et al. (2008) correlations for calculation of specific entropy

m_i	n_i	X_i	Y_i
7.7711828×10^{-3}	1	-4.679909×10^{4}	S
1.5011735×10^{-2}	T	2.845857×10^{-5}	S^2
$-2.3741029 \times 10^{-5}$	T^2	-3.505039×10^{-7}	S^3
3.7544485×10^{-8}	T^3	1.3550718×10^{-9}	S^4
$-1.5222796 \times 10^{-11}$	T^4	1.8388961×10^{-5}	ST
$6.0718455 \times 10^{-15}$	T^5	-8.138308×10^{-8}	ST^2
$-1.4393952 \times 10^{-4}$	P	2.547241×10^{-10}	ST^3
$-5.01934525 \times 10^{-6}$	PT	-3.648801×10^{-8}	S^2T
$2.54031415 \times 10^{-9}$	PT^2	$2.4962954 \times 10^{-10}$	S^3T
$-8.249480 \times 10^{-11}$	PT^3	4.3458579×10^{-9}	PST
$-5.0169318 \times 10^{-15}$	PT^4		
$-1.2192506 \times 10^{-6}$	P^2		
$1.07139472 \times 10^{-8}$	P^2T		
7.972446×10^{-11}	P^2T^2		
$1.3348922 \times 10^{-13}$	P^2T^3		
$4.38423518 \times 10^{-9}$	P^3		
$4.7087670 \times 10^{-11}$	P^3T		
$-1.6847016 \times 10^{-13}$	P^3T^2		
$-6.1607962 \times 10^{-17}$	P^3T^3		
$-6.2806718 \times 10^{-12}$	P^4		
$8.1815113 \times 10^{-14}$	P^4T^2		
$8.8902636 \times 10^{-17}$	P^4T^3		
$2.9409367 \times 10^{-15}$	P^5		
$-4.20129614 \times 10^{-17}$	P^5T		

Table 5.2 Constants for the correlations of Nayar et al. (2016) for specific entropy

Constants	Values	Constants	Values
a_1	-4.4786×10^{-3}	b_5	2.562×10^{1}
a_2	-1.1654×10^{-2}	b_6	-1.443×10^{-1}
a_3	6.1154×10^{-5}	b_7	5.879×10^{-4}
a_4	-2.0696×10^{-7}	b_8	-6.111×10^{1}
a_5	-1.5531×10^{-3}	b_9	8.041×10^{1}
a_6	4.0054×10^{-5}	b_{10}	3.035×10^{-1}
a_7	-1.4193×10^{-7}	c_1	0.1543
a_8	3.3142×10^{-10}	c_2	15.383
b_1	-4.231×10^{2}	c_3	-2.996×10^{-2}
b_2	1.463×10^{4}	c_4	8.193×10^{-5}
b_3	-9.880×10^{4}	c_5	-1.370×10^{-7}
b_4	3.095×10^{5}		

Table 5.3 Constants for Millero (1983) correlation for specific entropy calculations

Constants	Values	Constants	Values
a_1	1.4218×10^{-3}	C_2	1.5903×10^{-8}
a_2	-3.1137×10^{-7}	C_3	-2.3525×10^{-10}
a_3	4.2446×10^{-9}	d_1	0.1543
b_1	-2.1762×10^{-4}	d_2	15.383
b_2	4.1426×10^{-7}	d_3	-2.996×10^{-2}
b_3	-1.6285×10^{-9}	d_4	8.193×10^{-5}
C_1	1.0201×10^{-5}	d_5	-1.370×10^{-7}

Millero (1983) determined the specific entropy of seawater using enthalpy and free energy data (see Appendix 3 for sample). He then fitted the data and proposed a correlation in terms of temperature (T), in °C, and salinity (S), in g/kg. The main correlation is Eq. (5.11), and the required constants are reported in Table 5.3.

$$s = s_1 + \left(AS + BS^{1.5} + CS^2\right) \tag{5.11}$$

s_1 can be calculated by Eq. (5.12).

$$s_1 = (d_1 + d_2T + d_3T^2 + d_4T^3 + d_5T^4)/1000 \tag{5.12}$$

where,

$$A = a_1 + a_2T + a_3T^2$$
$$B = b_1 + b_2T + b_3T^2$$
$$C = C_1 + C_2T + C_3T^2$$

The correlation is valid in salinity range of 0–40 g/kg, temperature range of 0–40 °C, and pressure of 0.1 MPa with accuracy of ± 8% (from IAPWS 2008) (Cooper and Dooley 2008).

El-Dessouky and Ettouney (2002) enlisted a correlation for the specific entropy of pure water. The correlation is a function of temperature (T) in °C, as in Eq. (5.13). and the required constants are provided in Table 5.4.

$$s = a_1 + a_2\,T + a_3\,T^2 + a_4\,T^3 \tag{5.13}$$

The correlation of El-Dessouky and Ettouney (2002) is valid in the temperature range of 5–200 °C and pressure of 0.1 MPa with an accuracy of ± 0.4%.

Generous et al. (2020) proposed a correlation for the calculation of the specific entropy of seawater in terms of molality (m) in mol/kg (molal).

$$s_{sw} = s_w - \left(a_1m + a_2m^2 + a_3m^3\right) \tag{5.14}$$

Table 5.4 Constants for correlation of El-Dessouky and Ettouney (2002) for specific entropy of pure water

Constants	Values
a_1	-5.7846×10^{-4}
a_2	0.015297
a_3	-2.63129×10^{-5}
a_4	4.11959×10^{-8}

Table 5.5 Constants for correlation of Generous et al. (2020) for specific entropy of seawater

Constants	Values
a_1	0.0138507
a_2	0.00117115
a_3	9.80648×10^{-5}

where s_w is the pure water specific entropy at 25 °C (= 0.3672). The constants for Eq. (5.14) are provided in Table 5.5.

This correlation is obtained from the Pitzer model to be easily used at 25 °C. It covers a wider salinity range of 0–260 g/kg (0–6 mol) with a maximum and average accuracy of ± 0.13% and ± 0.08%, respectively.

It should be noted that IAPWS (1992) applies Eq. (5.1) to calculate specific entropy, in J/kg K, by using an equation of state in the form of specific Gibbs free energy that is dependent upon temperature, salinity, and pressure. The rest of the details are the same as explained in Chap. 4.

5.5 Case Study

Figure 5.1 shows the specific entropy of saline water as a function of salinity at 25 °C and 0.1 MPa. The methods and correlations of Pitzer et al. (1984), Nayar et al. (2016), IAPWS (1992), Millero (1983), Sun et al. (2008), and present work are plotted in the salinity range of 0–6 molal (0 to ~ 260 g/kg) at 25 °C and 0.1 MPa, as shown in Fig. 5.1a. The entropy values calculated by these correlations/models have been tested against IAPWS (1992) and data of Chou (1968) for salinities up to and above 120 g/kg, respectively, to show their accuracy, as shown in Fig. 5.1b. All correlations/methods are reasonably close to the data of Chou (1968) (shown in Appendix 4) except for Millero (1983), which was found to have a maximum difference of 9%. This difference is similar to the one reported by Sharqawy et al. (2011).

It is also noted that only the Pitzer model (1984) has an application range beyond a salinity of 120 g/kg (~ 2.335 molal). The average difference is found to be 1.1%. Based on Fig. 5.1a, b, it can be seen that the Pitzer model and present correlation can accurately predict the specific entropy under the tabulated conditions up to 6 molal. It should be noted that, near 6 molal, the difference of the Pitzer model (1984) shows

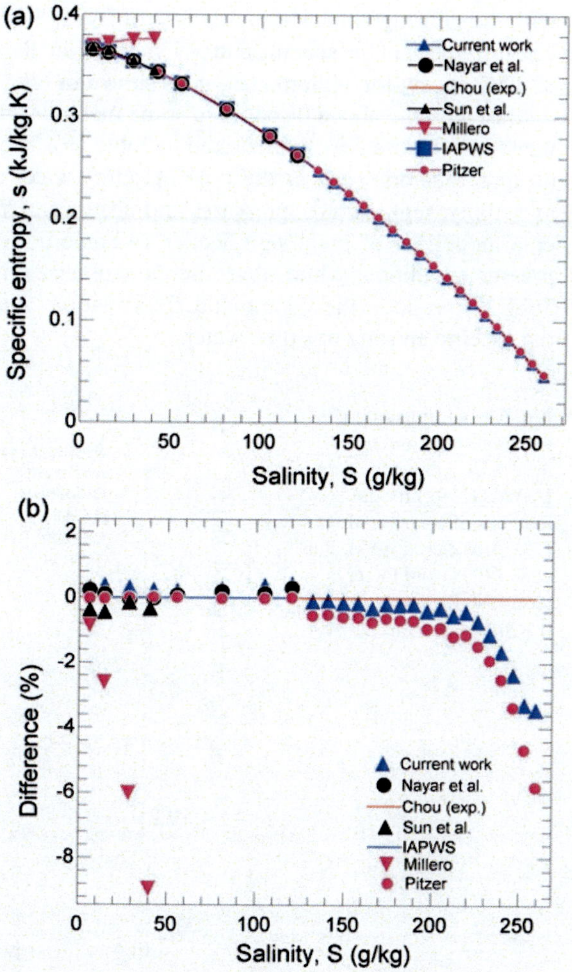

Fig. 5.1 **a** Salinity versus entropy and **b** the difference against the data of Chou (1968) and IAPWS (1992) for the methods of Pitzer et al. (1984), present work, Nayar et al. (2016), Millero (1983) and Sun et al. (2008) at 25 °C and 0.1 MPa

to be around 5.8%. This is only due to shifting the original data of the Pitzer model (1984), which uses a reference of 0 °C and 0 g/kg salinity to the reference of IAPWS (1992), which uses a reference of 0 °C and 35 g/kg salinity. It results in the entropy value being close to zero near 6 molal and, therefore, even a small difference in the prediction of the entropy results in a significant percentage difference. It should also be noted here that the maximum difference in the entropy value encountered was 0.00266 kJ/kg K. Therefore, the Pitzer model (1984) is found to be very reliable at high salinities.

The dependence of the specific entropy on temperature is shown in Fig. 5.2. The results of Nayar et al. (2016), Millero (1983), Pitzer model (1984), IAPWS (1992), and Sun et al. (2008) methods in a temperature range of 0–35 °C at 40 g/kg and 0.1 MPa, are plotted in Fig. 5.2a. All the correlations seem to have a similar trend with the same agreement with the data of IAPWS (1992) except for Millero (1983). The difference of the specific entropy values from the data of IAPWS (1992) is within ± 0.5% except for Millero (1983), as shown in Fig. 5.2b.

Based on the above discussion, some methods and correlations could be recommended to be used, as summarized in Table 5.6. The estimation of entropy depends on many factors, such as the type of saline water and operating conditions, such as salinity, temperature, pressure, and accuracy. Criteria of a selected correlation or a model are of great significance as some correlations are excellent for high-pressure conditions while others are more precise for high-temperature applications. Table 5.6 proposes the recommended correlations and methods to be used to calculate the specific entropy of saline water.

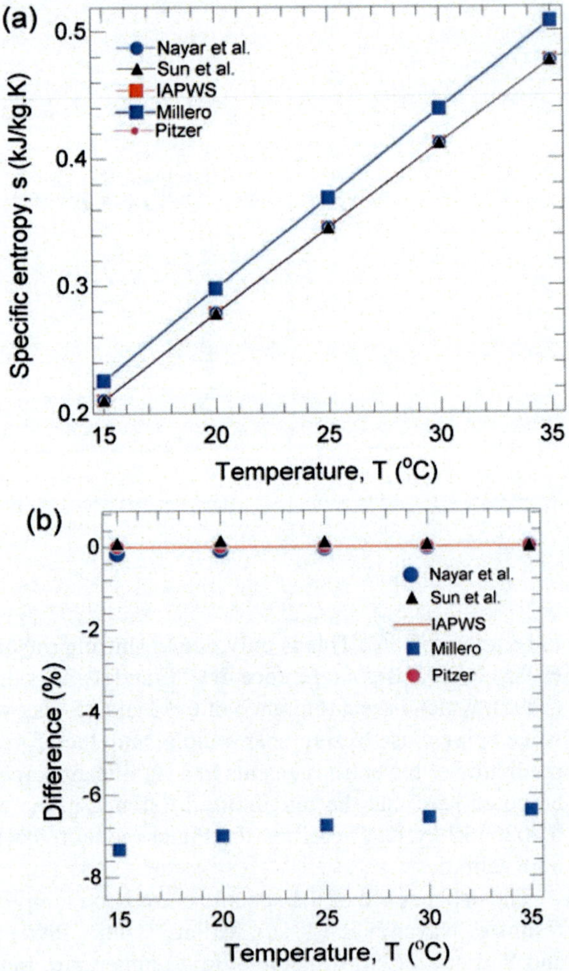

Fig. 5.2 **a** Temperature versus entropy and **b** the difference against the data of IAPWS (1992) for the methods of Nayar et al. (2016), Millero (1983), Sun et al. (2008) and Pitzer et al. (1984) at 40 g/kg and 0.1 MPa

Table 5.6 Summary of the recommended correlations and models for the specific entropy of saline water (Generous et al. 2020)

Correlation-model	Advantages	Applications	References
Pitzer et al. (Eq. 5.5)	• General for multicomponent saline solution • Accurate model (refer to Fig. 5.1)	• All membrane-based desalination technologies • All thermal-based desalination technologies	Pitzer et al. (1984) and Pitzer (1987)
The current model (Eq. 5.10)	• Most accurate and simple model at ambient conditions for seawater • It is obtained from the Pitzer model with high accuracy	• All membrane-based desalination technologies • All thermal-based desalination technologies	The present work
Sun et al. (Eq. 5.7)	• Valid in a salinity range of 0–40 g/kg within a difference of ± 1% (refer to Fig. 5.1) • It generates accurate results up to a temperature of 40 °C (refer to Fig. 5.2)	• All membrane-based desalination technologies • All thermal-based desalination technologies • Seawater greenhouse • High-grade water recycling • Wave-powered desalination	Sun et al. (2008)
Nayar et al. (Eq. 5.8)	• Valid in a salinity range of 0–120 g/kg with negligible difference (refer to Fig. 5.1) • Predicts more accurate results (within ± 8% difference (refer to Fig. 5.2)) at a temperature range 0–35 °C	• All membrane-based desalination technologies • All thermal-based desalination technologies • Seawater greenhouse • High-grade water recycling • Wave-powered desalination	Nayar et al. (2016)

Like the rest of the properties, we need more work to develop a model and correlation for saline water with multicomponent composition and extensive experimental work to for the purpose of comparison.

5.6 EES Codes

The EES codes for the correlations discussed above are:

Sun et al. Correlation

```
"Correlation for entropy as a Function of Temperature, Pressure, and Salinity"
"Please note that here in the model, m, n, x, and y are notations to represent terms."

"Variable to be defined by the user"
P = 0.1 "MPA"
T = 20 "C"
S = 35 "g/kg"

m[1] = 7.771182883/10^3
m[2] = 1.50117356/10^2
m[3] = – 2.37410293/10^5
m[4] = 3.75444856/10^8
m[5] = – 1.52227969/10^11
m[6] = 6.07184558/10^15
m[7] = – 1.43939529/10^4
m[8] = – 5.01934525/10^6
m[9] = 2.54031415/10^9
m[10] = – 8.2494804/10^11
m[11] = – 5.01693186/10^15
m[12] = – 1.21925066/10^6
m[13] = 1.07139472/10^8
m[14] = 7.9724463/10^11
m[15] = 1.3348922/10^13
m[16] = 4.38423518/10^9
m[17] = 4.7087670/10^11
m[18] = – 1.68470164/10^13
m[19] = – 6.16079622/10^17
m[20] = – 6.2806718/10^12
m[21] = 8.18151132/10^14
m[22] = 8.89026362/10^17
m[23] = 2.94093673/10^15
m[24] = – 4.20129614/10^17

n[1] = 1
n[2] = t
n[3] = t^2
```

```
n[4] = t^3
n[5] = t^4
n[6] = t^5
n[7] = p
n[8] = p * t
n[9] = p * t^2
n[10] = p * t^3
n[11] = p * t^4
n[12] = p^2
n[13] = p^2 * t
n[14] = p^2 * t^2
n[15] = p^2 * t^3
n[16] = p^3
n[17] = p^3 * t
n[18] = p^3 * t^2
n[19] = p^3 * t^3
n[20] = p^4
n[21] = p^4 * t^2
n[22] = p^4 * t^3
n[23] = p^5
n[24] = p^5 * t

x[1] = S
x[2] = S^2
x[3] = S^3
x[4] = S^4
x[5] = S * t
x[6] = S * t^2
x[7] = S * t^3
x[8] = S^2 * t
x[9] = S^3 * t
x[10] = p * S * t

y[1] = – 4.67990975/10^4
y[2] = 2.84585789/10^5
y[3] = – 3.50503953/10^7
y[4] = 1.35507185/10^9
y[5] = 1.83889613/10^5
y[6] = – 8.13830857/10^8
y[7] = 2.54724177/10^10
```

y[8] = − 3.64880194/10^8
y[9] = 2.49629547/10^10
y[10] = 4.34585798/10^9

"Relation for Saline Water entropy as a Function of Temperature, Pressure, and Salinity"

S_entropy = **sum**(m[i] * n[i], i = 1, 24) − **sum**(x[i] * y[i], i = 1, 10)

Nayar et al. Correlation Based on IAPWS-2008 (Cooper et al. 2008)

S = 0.034986 "kg/kg"
T = 25 "C"
P = 0
P_o = 0.1

S_t_s_p = S_t_s + (P − P_o) * (a_1 + a_2 * T + a_3 * T^2 + a_4 * T^3) + S * (a_5 + a_6 * T + a_7 * T^2 + a_8 * T^3)

"Entropy as a function of temperature and salinity is given as"

S_t_s = S_t − S * (b_1 + b_2 * S + b_3 * S^2 + b_4 * S^3 + b_5 * T + b_6 * T^2 + b_7 * T^3 + b_8 * S * T + b_9 * S^2 * T + b_10 * S * T^2)

"Entropy as a function of temperature is given as"

S_t = C_1 + C_2 * T + C_3 * T^2 + C_4 * T^3 + C_5 * T^4

"Constants"
a_1 = − 4.4786/10^3
a_2 = − 1.1654/10^2
a_3 = 6.1154/10^5
a_4 = − 2.0696/10^7
a_5 = − 1.5531/10^3
a_6 = 4.0054/10^5
a_7 = − 1.4193/10^7
a_8 = 3.3142/10^10

b_1 = − 4.231 * 10^2
b_2 = 1.463 * 10^4
b_3 = − 9.880 * 10^4

```
b_4 = 3.095 * 10^5
b_5 = 2.562 * 10^1
b_6 = – 1.443/10^1
b_7 = 5.879/10^4
b_8 = – 6.111 * 10^1
b_9 = 8.041 * 10^1
b_10 = 3.035/10^1

C_1 = 0.1543
C_2 = 15.383
C_3 = – 2.996/10^2
C_4 = 8.193/10^5
C_5 = – 1.370/10^7

S_e = S_t_s_p/1000
```

Millero Correlation

```
S = 0.030 "kg/kg"
T = 20 "C"
S_t_s = S_t + (A * S + B * S^1.5 + C * S^2)

"Where is S is practical salinity of the saline water and A, B, and C are constants."

A = a_1 + a_2 * T + a_3 * T^2

a_1 = 1.4218/10^3
a_2 = – 3.1137/10^7
a_3 = 4.2446/10^9

B = b_1 + b_2 * T + b_3 * T^2

b_1 = – 2.1762/10^4
b_2 = 4.1426/10^7
b_3 = – 1.6285/10^9

C = C_1 + C_2 * T + C_3 * T^2

C_1 = 1.0201/10^5
C_2 = 1.5903/10^8
```

C_3 = − 2.3525/10^10

"Entropy as a function of temperature is given as"

S_t = d_1 + d_2 * T + d_3 * T^2 + d_4 * T^3 + d_5 * T^4

"Constants"

d_1 = 0.1543
d_2 = 15.383
d_3 = − 2.996/10^2
d_4 = 8.193/10^5
d_5 = − 1.370/10^7

S_e = S_t_s/1000

This Work Correlation

m = 0.5 "mol/kg"
s_sw = s_w − (a_1 * m + a_2 * m^2 + a_3 * m^3)

s_w = 0.3672
a_1 = 0.0138507
a_2 = 0.00117115
a_3 = 9.80648/10^5

Appendix 1

Tables 5.7, 5.8, 5.9, 5.10, 5.11 and 5.12.

Appendix 2

See Table 5.13.

Appendix 3

See Table 5.14.

Table 5.7 The specific entropy of NaCl(aq), J/g K (P < 100 bar)

T	P	m = 0.1	m = 0.25	m = 0.5	m = 0.75	m = 1	m = 2	m = 3	m = 4	m = 5	m = 6
°C	bar	mol/kg									
0	1.0	3.513	3.507	3.492	3.475	3.457	3.376	3.289	3.202	3.117	3.036
10	1.0	3.663	3.654	3.637	3.617	3.597	3.508	3.416	3.325	3.236	3.152
20	1.0	3.807	3.797	3.777	3.755	3.732	3.637	3.539	3.444	3.352	3.265
25	1.0	3.878	3.866	3.845	3.822	3.798	3.699	3.600	3.502	3.409	3.321
30	1.0	3.947	3.935	3.912	3.888	3.864	3.761	3.659	3.559	3.465	3.375
40	1.0	4.081	4.048	4.043	4.018	3.991	3.882	3.775	3.671	3.573	3.481
50	1.0	4.212	4.197	4.171	4.143	4.114	3.999	3.887	3.780	3.679	3.584
60	1.0	4.338	4.323	4.294	4.264	4.234	4.113	3.996	3.885	3.781	3.683
70	1.0	4.461	4.444	4.414	4.382	4.350	4.224	4.102	3.987	3.880	3.780
80	1.0	4.581	4.563	4.530	4.497	4.463	4.331	4.205	4.086	3.976	3.873
90	1.0	4.697	4.678	4.644	4.609	4.574	4.436	4.305	4.183	4.069	3.964
100	1.0	4.811	4.790	4.754	4.718	4.681	4.538	4.402	4.277	4.160	4.052
110	1.4	4.922	4.900	4.862	4.824	4.786	4.637	4.497	4.368	4.248	4.138
120	2.0	5.030	5.007	4.968	4.928	4.888	4.734	4.590	4.457	4.335	4.222
130	2.7	5.136	5.112	5.071	5.029	4.988	4.828	4.680	4.544	4.418	4.303
140	3.6	5.240	5.215	5.172	5.128	5.086	4.921	4.768	4.629	4.500	4.383
150	4.8	5.342	5.316	5.271	5.226	5.181	5.011	4.855	4.711	4.580	4.460
160	6.2	5.442	5.414	5.368	5.321	5.275	5.100	4.939	4.792	4.658	4.536
170	7.9	5.540	5.512	5.463	5.415	5.367	5.186	5.021	4.871	4.734	4.610
180	10.0	5.637	5.607	5.557	5.507	5.458	5.271	5.102	4.949	4.809	4.682
190	12.5	5.732	5.761	5.649	5.597	5.546	5.355	5.181	5.024	4.882	4.753
200	15.5	5.826	5.794	5.740	5.686	5.634	5.437	5.259	5.099	4.954	4.822
210	19.1	5.919	5.885	5.829	5.774	5.720	5.517	5.335	5.172	5.024	4.890
220	23.2	6.011	5.976	5.918	5.861	5.805	5.597	5.410	5.243	5.092	4.956
230	28.0	6.102	6.066	6.005	5.946	5.889	5.675	5.484	5.313	5.160	5.021
240	33.4	6.193	6.155	6.092	6.031	5.972	5.752	5.557	5.383	5.226	5.085
250	39.7	6.283	6.243	6.178	6.150	6.054	5.828	5.628	5.451	5.291	5.148
260	46.9	6.373	6.331	6.264	6.199	6.136	5.903	5.699	5.517	5.355	5.209
270	55.0	6.462	6.419	6.349	5.282	6.217	5.978	5.769	5.584	5.418	5.270
280	64.1	6.553	6.507	6.434	6.364	6.297	6.052	5.828	5.649	5.480	5.329
290	74.4	6.643	6.596	6.520	6.447	6.377	6.125	5.906	5.713	5.542	5.388
300	85.8	6.735	6.685	6.605	6.529	6.457	6.197	5.973	5.777	5.602	5.446

Table 5.8 The specific entropy of NaCl(aq), J/g K (P = 200 bar)

T	P	m = 0.1	m = 0.25	m = 0.5	m = 0.75	m = 1	m = 2	m = 3	m = 4	m = 5	m = 6
°C	bar	mol/kg									
0	200	3.511	3.506	3.491	3.473	3.454	3.371	3.284	3.196	3.110	3.029
10	200	3.661	3.652	3.634	3.614	3.593	3.503	3.410	3.318	3.229	3.145
20	200	3.803	3.792	3.772	3.750	3.727	3.630	3.532	3.437	3.345	3.258
25	200	3.872	3.861	3.839	3.816	3.792	3.692	3.592	3.495	3.401	3.313
30	200	3.940	3.928	3.906	3.882	3.857	3.754	3.651	3.552	3.457	3.367
40	200	4.074	4.060	4.036	4.010	3.983	3.874	3.767	3.663	3.565	3.473
50	200	4.203	4.188	4.162	4.134	4.105	3.991	3.878	3.771	3.670	3.576
60	200	4.328	4.312	4.284	4.254	4.224	4.104	3.987	3.876	3.772	3.675
70	200	4.459	4.433	4.402	4.371	4.339	4.213	4.092	3.978	3.871	3.771
80	200	4.568	4.550	4.518	4.485	4.452	4.320	4.195	4.077	3.966	3.864
90	200	4.683	4.664	4.630	4.596	4.561	4.424	4.294	4.172	4.059	3.955
100	200	4.796	4.776	4.740	4.704	4.667	4.525	4.391	4.266	4.150	4.043
110	200	4.906	4.884	4.847	4.809	4.771	4.624	4.485	4.357	4.238	4.128
120	200	5.013	4.990	4.951	4.912	4.872	4.720	4.577	4.445	4.323	4.211
130	200	5.118	5.094	5.053	5.012	4.972	4.814	4.667	4.532	4.407	4.292
140	200	5.221	5.196	5.153	5.111	5.068	4.906	4.755	4.616	4.488	4.371
150	200	5.321	5.295	5.251	5.207	5.563	4.995	4.840	4.698	4.568	4.449
160	200	5.420	5.393	5.347	5.301	5.256	5.083	4.924	4.778	4.645	4.524
170	200	5.517	5.489	5.441	5.394	5.347	5.169	5.096	4.857	4.721	4.598
180	200	5.613	5.583	5.534	5.485	5.436	5.253	5.086	4.934	4.796	4.670
190	200	5.706	5.676	5.625	5.574	5.524	5.336	5.164	5.009	4.868	4.740
200	200	5.799	5.767	5.714	5.662	5.611	5.417	5.241	5.083	4.939	4.809
210	200	5.890	5.858	5.803	5.749	5.696	5.496	5.317	5.155	5.009	4.877
220	200	5.981	5.947	5.890	5.834	5.779	5.575	5.391	5.226	5.078	4.943
230	200	6.070	6.035	5.976	5.918	5.892	5.652	5.454	5.296	5.145	5.008
240	200	6.159	6.122	6.061	6.002	5.944	5.728	5.536	5.395	5.211	5.072
250	200	6.248	6.209	6.146	6.085	6.025	5.804	5.607	5.432	5.276	5.135
260	200	6.336	6.296	6.230	6.167	6.105	5.878	5.677	5.499	5.340	5.197
270	200	6.424	6.382	6.314	6.249	6.185	5.952	5.747	5.565	5.403	5.257
280	200	6.512	6.469	6.398	6.330	6.265	6.025	5.815	5.630	5.465	5.317
290	200	6.601	6.556	6.482	6.412	6.345	6.098	5.844	5.695	5.527	5.377
300	200	6.691	6.644	6.567	6.494	6.424	6.171	5.952	5.759	5.589	5.436

Table 5.9 The specific entropy of NaCl(aq), J/g K (P = 400 bar)

T	P	m = 0.1	m = 0.25	m = 0.5	m = 0.75	m = 1	m = 2	m = 3	m = 4	m = 5	m = 6
°C	bar	mol/kg									
0	400	3.513	3.505	3.489	3.471	3.451	3.366	3.278	3.190	3.104	3.022
10	400	3.658	3.648	3.629	3.609	3.588	3.497	3.403	3.311	3.223	3.138
20	400	3.798	3.787	3.766	3.744	3.720	3.623	3.525	3.430	3.338	3.251
25	400	3.866	3.855	3.833	3.810	3.785	3.685	3.585	3.488	3.394	3.306
30	400	3.934	3.922	3.899	3.875	3.850	3.746	3.644	3.544	3.449	3.360
40	400	4.066	4.052	4.028	4.002	3.975	3.866	3.759	3.655	3.557	3.465
50	400	4.194	4.179	4.153	4.125	4.096	3.982	3.870	3.163	3.662	3.567
60	400	4.318	4.302	4.274	4.244	4.214	4.094	3.978	3.763	3.763	3.666
70	400	4.439	4.422	4.391	4.360	4.329	4.203	4.083	3.969	3.862	3.762
80	400	4.556	4.538	4.506	4.473	4.440	4.309	4.184	4.067	3.957	3.855
90	400	4.670	4.651	4.618	4.583	4.549	4.413	4.283	4.162	4.050	3.946
100	400	4.782	4.761	4.726	4.690	4.654	4.513	4.380	4.255	4.140	4.033
110	400	4.890	4.869	4.832	4.795	4.757	4.611	4.474	4.346	4.227	4.128
120	400	4.997	4.974	4.936	4.897	4.858	4.707	4.565	4.434	4.313	4.201
130	400	5.100	5.077	5.037	4.996	4.956	4.800	4.654	4.520	4.396	4.282
140	400	5.202	5.178	5.136	5.094	5.052	4.891	4.741	4.604	4.477	4.360
150	400	5.302	5.276	5.233	5.189	5.146	4.980	4.826	4.685	4.556	4.437
160	400	5.399	5.373	5.328	5.283	5.238	5.067	4.909	4.765	4.633	4.512
170	400	5.495	5.468	5.421	5.374	5.328	5.152	4.991	4.843	4.708	4.585
180	400	5.589	5.560	5.512	5.464	5.416	5.235	5.070	4.919	4.782	4.650
190	400	5.681	5.652	5.602	5.552	5.503	5.317	5.148	4.994	4.854	4.727
200	400	5.772	5.742	5.690	5.638	5.588	5.397	5.224	5.067	4.925	4.795
210	400	5.862	5.830	5.776	5.723	5.671	5.476	5.299	5.139	4.994	4.863
220	400	5.951	5.917	5.862	5.807	5.754	5.553	5.372	5.209	5.062	4.928
230	400	6.038	6.003	5.946	5.890	5.835	5.628	5.444	5.278	5.328	4.993
240	400	6.124	6.089	6.029	5.971	5.914	5.703	5.514	5.346	5.194	5.057
250	400	6.210	6.173	6.112	6.052	5.993	5.776	5.584	5.412	5.258	5.119
260	400	6.295	6.257	6.193	6.131	6.071	5.849	5.652	5.477	5.321	5.180
270	400	6.380	6.340	6.274	6.210	6.148	5.920	5.720	5.542	5.383	5.241
280	400	6.464	6.422	6.354	6.288	6.225	5.991	5.785	5.606	5.445	5.301
290	400	6.548	6.505	6.434	6.366	6.301	6.061	5.852	5.668	5.505	5.360
300	400	6.632	6.587	6.514	6.444	6.376	6.130	5.917	5.031	5.566	5.419

Table 5.10 The specific entropy of NaCl(aq), J/g K (P = 600 bar)

T	P	m = 0.1	m = 0.25	m = 0.5	m = 0.75	m = 1	m = 2	m = 3	m = 4	m = 5	m = 6
°C	bar	mol/kg									
0	600	3.511	3.502	3.486	3.467	3.447	3.361	3.272	3.183	3.097	3.016
10	600	3.654	3.644	3.625	3.604	3.582	3.491	3.397	3.305	3.216	3.131
20	600	3.792	3.781	3.76	3.738	3.714	3.617	3.519	3.423	3.331	3.243
25	600	3.86	3.848	3.826	3.803	3.779	3.678	3.578	3.48	3.387	3.298
30	600	3.927	3.915	3.892	3.867	3.842	3.739	3.636	3.537	3.442	3.352
40	600	4.058	4.044	4.02	3.993	3.967	3.858	3.751	3.648	3.55	3.458
50	600	4.185	4.17	4.144	4.116	4.087	3.973	3.861	3.755	3.654	3.559
60	600	4.308	4.292	4.264	4.234	4.205	4.085	3.969	3.859	3.755	3.658
70	600	4.428	4.411	4.381	4.35	4.318	4.193	4.073	3.96	3.853	3.754
80	600	4.544	4.526	4.495	4.162	4.429	4.299	4.175	4.058	3.948	3.847
90	600	4.657	4.639	4.605	4.571	4.527	4.402	4.273	4.153	4.041	3.937
100	600	4.768	4.748	4.713	4.677	4.642	4.502	4.369	4.245	4.13	4.024
110	600	4.876	4.815	4.818	4.781	4.744	4.599	4.462	4.335	4.218	4.109
120	600	4.981	4.959	4.921	4.882	4.844	4.694	4.553	4.423	4.302	4.191
130	600	5.084	5.061	5.021	4.981	4.941	4.787	4.642	4.509	4.385	4.272
140	600	5.185	5.161	5.12	5.078	5.037	4.877	4.739	4.592	4.466	4.35
150	600	5.283	5.258	5.216	5.173	5.13	4.966	4.813	4.73	4.544	4.426
160	600	5.38	5.354	5.309	5.265	5.221	5.052	4.896	4.753	4.621	4.501
170	600	5.474	5.447	5.401	5.356	5.31	5.136	4.976	4.83	4.696	4.573
180	600	5.567	5.539	5.492	5.444	5.397	5.215	5.055	4.906	4.769	4.645
190	600	5.659	5.629	5.58	5.531	5.483	5.3	5.132	4.98	4.841	4.714
200	600	5.748	5.718	5.667	5.617	5.567	5.379	5.208	5.052	4.911	4.782
210	600	5.836	5.805	5.753	5.7	5.649	5.456	5.282	5.123	4.98	4.849
220	600	5.923	5.891	5.837	5.783	5.73	5.533	5.354	5.193	5.047	4.914
230	600	6.009	5.975	5.919	5.864	5.81	5.607	5.425	5.261	5.113	4.979
240	600	6.094	6.059	6.001	5.944	5.888	5.68	5.495	5.328	5.178	5.042
250	600	6.177	6.141	6.081	6.023	5.965	5.753	5.563	5.394	5.241	5.104
260	600	6.26	6.223	6.161	6.1	6.042	5.813	5.63	5.458	5.304	5.165
270	600	6.342	6.303	6.239	6.177	6.117	5.893	5.696	5.521	5.365	5.225
280	600	6.423	6.383	6.317	6.253	6.191	5.962	5.761	5.583	5.426	5.284
290	600	6.504	6.462	6.394	6.328	6.264	6.029	5.824	5.645	5.486	5.343
300	600	6.585	6.541	6.47	6.402	6.336	6.096	5.887	5.705	5.545	5.402

Table 5.11 The specific entropy of NaCl(aq), J/g K (P = 800 bar)

T	P	m = 0.1	m = 0.25	m = 0.5	m = 0.75	m = 1	m = 2	m = 3	m = 4	m = 5	m = 6
°C	bar	mol/kg									
0	800	3.508	3.499	3.482	3.463	3.442	3.356	3.266	3.177	3.091	3.009
10	800	3.649	3.639	3.620	3.599	3.577	3.485	3.391	3.298	3.209	3.124
20	800	3.786	3.775	3.754	3.731	3.708	3.610	3.512	3.416	3.324	3.236
25	800	3.854	3.842	3.820	3.796	3.772	3.671	3.571	3.473	3.380	3.291
30	800	3.920	3.908	3.885	3.860	3.835	3.732	3.629	3.530	3.435	3.345
40	800	4.050	4.036	4.012	3.985	3.959	3.850	3.743	3.640	3.542	3.450
50	800	4.176	4.161	4.135	4.107	4.079	3.964	3.853	3.747	3.646	3.552
60	800	4.298	4.282	4.254	4.225	4.195	4.076	3.960	3.850	3.747	3.650
70	800	4.417	4.400	4.370	4.340	4.308	4.184	4.064	3.951	3.845	3.746
80	800	4.532	4.515	4.483	4.451	4.418	4.289	4.165	4.448	3.940	3.838
90	800	4.645	4.626	4.593	4.559	4.525	4.391	4.263	4.143	4.032	3.928
100	800	4.755	4.735	4.700	4.665	4.630	4.495	4.359	4.235	4.121	4.015
110	800	4.862	4.841	4.805	4.768	4.731	4.587	4.452	4.325	4.208	4.100
120	800	4.966	4.945	4.907	4.869	4.831	4.682	4.542	4.413	4.293	4.182
130	800	5.068	5.046	5.007	4.967	4.927	4.774	4.631	4.498	4.375	4.262
140	800	5.168	5.145	5.104	5.063	5.022	4.864	4.717	4.581	4.455	4.340
150	800	5.266	5.241	5.199	5.157	5.114	4.952	4.801	4.662	4.533	4.416
160	800	5.361	5.336	5.292	5.248	5.205	5.038	4.883	4.741	4.610	4.490
170	800	5.455	5.429	5.383	5.338	5.293	5.121	4.963	4.818	4.684	4.562
180	800	5.547	5.520	5.473	5.426	5.380	5.203	5.041	4.893	4.757	4.633
190	800	5.637	5.609	5.560	5.512	5.465	5.283	5.118	4.967	4.829	4.702
200	800	5.726	5.696	5.646	5.597	5.548	5.362	5.193	5.039	4.898	4.770
210	800	5.813	5.782	5.731	5.679	5.629	5.439	5.266	5.109	4.966	4.836
220	800	5.898	5.867	5.813	5.761	5.709	5.514	5.338	5.178	5.033	4.901
230	800	5.983	5.950	5.895	5.841	5.787	5.588	5.408	5.246	5.099	4.965
240	800	6.066	6.032	5.975	5.919	5.865	5.660	5.477	5.312	5.163	5.028
250	800	6.148	6.113	6.054	5.997	5.940	5.731	5.544	5.377	5.226	5.089
260	800	6.229	6.192	6.132	6.073	6.015	5.800	5.610	5.440	5.288	5.150
270	800	6.309	6.271	6.209	6.148	6.088	5.869	5.675	5.502	5.348	5.210
280	800	6.388	6.349	6.285	6.222	6.161	5.936	5.738	5.564	5.408	5.269
290	800	6.467	6.426	6.359	6.294	6.232	6.001	5.800	5.624	5.467	5.327
300	800	6.545	6.502	6.433	6.366	6.302	6.066	5.861	5.683	5.526	5.386

Table 5.12 The specific entropy of NaCl(aq), J/g K (P = 1000 bar)

T	P	m = 0.1	m = 0.25	m = 0.5	m = 0.75	m = 1	m = 2	m = 3	m = 4	m = 5	m = 6
°C	bar	mol/kg									
0	1000	3.504	3.495	3.477	3.458	3.437	3.350	3.260	3.171	3.085	3.002
10	1000	3.644	3.634	3.614	3.593	3.571	3.478	3.384	3.292	3.202	3.117
20	1000	3.780	3.769	3.748	3.725	3.701	3.603	3.505	3.409	3.317	3.229
25	1000	3.847	3.835	3.813	3.789	3.765	3.664	3.564	3.466	3.373	3.284
30	1000	3.913	3.900	3.877	3.853	3.828	3.724	3.622	3.522	3.427	3.337
40	1000	4.042	4.028	4.004	3.977	3.951	3.842	3.735	3.632	3.535	3.442
50	1000	4.167	4.152	4.126	4.098	4.070	3.956	3.845	3.739	3.638	3.544
60	1000	4.288	4.273	4.245	4.216	4.186	4.067	3.951	3.842	3.739	3.642
70	1000	4.406	4.390	4.360	4.329	4.298	4.174	4.055	3.942	3.836	3.738
80	1000	4.521	4.504	4.472	4.440	4.408	4.279	4.156	4.040	3.931	3.830
90	1000	4.633	4.615	4.582	4.548	4.514	4.381	4.253	4.134	4.023	3.920
100	1000	4.742	4.723	4.688	4.653	4.618	4.480	4.349	4.226	4.112	4.006
110	1000	4.848	4.828	4.792	4.759	4.719	4.576	4.441	4.315	4.199	4.091
120	1000	4.952	4.931	4.893	4.856	4.818	4.670	4.531	4.402	4.283	4.173
130	1000	5.053	5.031	4.992	4.953	4.914	4.762	4.619	4.487	4.365	4.252
140	1000	5.152	5.129	5.089	5.048	5.008	4.851	4.705	4.570	4.445	4.330
150	1000	5.249	5.225	5.183	5.141	5.100	4.939	4.789	4.650	4.523	4.406
160	1000	5.344	5.319	5.276	5.233	5.189	5.024	4.870	4.729	4.599	4.479
170	1000	5.437	5.411	5.366	5.322	5.277	5.107	4.950	4.806	4.673	4.551
180	1000	5.528	5.501	5.455	5.409	5.363	5.188	5.028	4.881	4.746	4.622
190	1000	5.617	5.589	5.542	5.494	5.447	5.268	5.104	4.954	4.817	4.691
200	1000	5.705	5.676	5.627	5.578	5.529	5.346	5.178	5.025	4.886	4.758
210	1000	5.791	5.761	5.710	5.660	5.610	5.422	5.251	5.095	4.954	4.824
220	1000	5.875	5.844	5.792	5.740	5.689	5.497	5.322	5.164	5.020	4.889
230	1000	5.959	5.926	5.872	5.819	5.767	5.569	5.392	5.231	5.081	4.952
240	1000	6.041	6.007	5.951	5.897	5.843	5.641	5.460	5.296	5.149	5.014
250	1000	6.121	6.087	6.029	5.973	5.917	5.711	5.526	5.361	5.211	5.075
260	1000	6.201	6.165	6.106	6.048	5.991	5.779	5.591	5.424	5.273	5.136
270	1000	6.279	6.242	6.181	6.121	6.063	5.847	5.655	5.485	5.333	5.195
280	1000	6.357	6.319	6.255	6.194	6.134	5.912	5.718	5.546	5.392	5.254
290	1000	6.434	6.394	6.329	6.265	6.203	5.976	5.779	5.605	5.450	5.312
300	1000	6.510	6.468	6.401	6.335	6.271	6.039	5.838	5.663	5.508	5.370

Table 5.13 Specific entropy values (in kJ/kg K) used for the correlation of Sun et al. (2008)

Pressure (MPa)	Temperature (°C)	S (g/kg)								
		0	5	10	15	20	25	30	35	40
0	0	– 0.148	2.484	3.344	3.492	3.149	2.416	1.353	0	– 1.616
0	5	76.252	78.234	78.478	78.032	77.111	75.817	74.207	72.318	70.178
0	10	151.077	152.466	152.145	151.153	149.702	147.89	145.773	143.388	140.762
0	15	224.448	225.29	224.447	222.947	221.001	218.706	216.115	213.265	210.183
0	20	296.463	296.794	295.461	293.485	291.073	288.321	285.282	281.992	278.476
0	25	367.2	367.049	365.252	362.823	359.968	356.781	353.314	349.602	345.67
0	30	436.725	436.116	433.875	431.013	427.732	424.126	420.247	416.129	411.796
0	35	505.097	504.046	501.378	498.098	494.407	490.397	486.119	481.608	476.886
0	40	572.365	570.888	567.808	564.124	560.036	555.634	550.971	546.078	540.981
10	0	0.341	2.774	3.453	3.432	2.929	2.046	0.84	– 0.648	– 2.392
10	5	75.943	77.774	77.878	77.298	76.251	74.835	73.108	71.108	68.862
10	10	150.081	151.354	150.925	149.828	148.277	146.368	144.159	141.685	138.973
10	15	222.848	223.601	222.673	221.093	219.069	216.698	214.035	211.116	207.967
10	20	294.323	294.585	293.187	291.149	288.678	285.868	282.775	279.432	275.866
10	25	364.57	364.368	362.523	360.048	357.149	353.919	350.411	346.661	342.693
10	30	433.645	433.003	430.729	427.835	424.523	420.887	416.979	412.833	408.473
10	35	501.598	500.536	497.852	494.554	490.844	486.813	482.515	477.982	473.239
10	40	568.474	567.013	563.937	560.25	556.154	551.741	547.061	542.15	537.031
20	0	0.469	2.722	3.237	3.062	2.414	1.394	0.059	– 1.552	– 3.412
20	5	75.35	77.044	77.02	76.319	75.156	73.63	71.797	69.696	67.353
20	10	148.862	150.03	149.502	148.311	146.668	144.671	142.377	139.822	137.031
20	15	221.075	221.747	220.743	219.088	216.994	214.554	211.826	208.844	205.634

(continued)

Table 5.13 (continued)

Pressure (MPa)	Temperature (°C)	S (g/kg)								
		0	5	10	15	20	25	30	35	40
20	20	292.051	292.252	290.795	288.701	286.175	283.314	280.171	276.781	273.17
20	25	361.844	361.597	359.708	357.191	354.252	350.984	347.44	343.656	339.655
20	30	430.5	429.828	427.525	424.603	421.263	417.6	413.666	409.496	405.113
20	35	498.062	496.99	494.293	490.979	487.252	483.203	478.886	474.334	469.572
20	40	564.57	563.125	560.054	556.365	552.262	547.837	543.143	538.214	533.074
40	0	− 0.246	1.692	1.919	1.477	0.578	− 0.679	− 2.237	− 4.058	− 6.116
40	5	73.389	74.846	74.6	73.688	72.325	70.606	68.59	66.313	63.801
40	10	145.807	146.797	146.097	144.742	142.941	140.792	138.351	135.654	132.728
40	15	217.045	217.58	216.445	214.664	212.447	209.89	207.048	203.957	200.642
40	20	287.137	287.234	285.677	283.487	280.87	277.921	274.693	271.223	267.535
40	25	356.118	355.796	353.833	351.247	348.241	344.91	341.306	337.465	333.411
40	30	424.022	423.303	420.952	417.983	414.598	410.891	406.915	402.705	398.284
40	35	490.878	489.793	487.075	483.737	479.982	475.905	471.557	466.974	462.181
40	40	556.718	555.307	552.246	548.556	544.441	539.997	535.276	530.315	525.138
100	0	− 8.535	− 7.277	− 7.661	− 8.667	− 10.088	− 11.83	− 13.839	− 16.08	− 18.528
100	5	62.499	63.47	62.771	61.43	59.658	57.551	55.166	52.538	49.694
100	10	132.6	133.242	132.208	130.529	128.417	125.967	123.237	120.262	117.069
100	15	201.735	202.02	200.638	198.618	196.17	193.389	190.331	187.033	183.519
100	20	269.891	269.807	268.072	265.708	262.925	259.816	256.437	252.822	248.998
100	25	337.067	336.62	334.535	331.829	328.709	325.269	321.563	317.627	313.485
100	30	403.269	402.487	400.064	397.02	393.561	389.782	385.737	381.461	376.979
100	35	468.508	467.436	464.701	461.33	457.532	453.402	448.998	444.353	439.494
100	40	532.797	531.507	528.499	524.818	520.682	516.189	511.399	506.348	501.065

Table 5.14 Specific entropy of seawater at 35 g/kg determined by Millero (1983)

Temperature	Entropy, s
°C	10^2 kJ/kg K
0	1.718
5	1.735
10	1.815
15	1.858
20	1.897
25	1.933
30	1.968
35	2.001
40	2.035

Appendix 4

See Table 5.15.

Table 5.15 Specific entropy of NaCl(aq) at 25 °C and 1 atmospheric pressure from Chou (1968)

Molality, m	Entropy, s
mol/kg	kJ/kg K
0.1	0.3705
0.2	0.3722
0.3	0.3731
0.4	0.3731
0.5	0.3731
0.6	0.3726
0.7	0.3718
0.8	0.3705
0.9	0.3693
1	0.3676
1.2	0.3639
1.4	0.3597
1.6	0.3551
1.8	0.3500
2	0.3433
2.2	0.3387
2.4	0.3329

(continued)

Table 5.15 (continued)

Molality, m	Entropy, s
mol/kg	kJ/kg K
2.6	0.3266
2.8	0.3203
3	0.3136
3.2	0.3069
3.4	0.2998
3.6	0.2931
3.8	0.2860
4	0.2789
4.2	0.2713
4.4	0.2642
4.6	0.2567
4.8	0.2495
5	0.2420
5.2	0.2345
5.4	0.2269
5.6	0.2194
5.8	0.2119
6	0.2047

References

Chou JCS (1968) Thermodynamic properties of aqueous sodium chloride solutions from 32 to 350 °F. Georgia Institute of Technology

Cooper JR, Dooley RB (2008) The international association for the properties of water and steam for the viscosity of ordinary water substance. International Organization

Cooper JR, Mary Q, Road ME, Associates SI (2008) The international association for the properties of water and steam release (IAPWS-2008) on the formulation for the the thermodynamic properties of water and steam, Sept 2008, pp 1–19

El-Dessouky HT, Ettouney HM (2002) Fundamentals of salt water desalination. https://doi.org/10.1016/b978-0-444-50810-2.x5000-3

Generous MM, Qasem NAA, Qureshi BA, Zubair SM (2020) A comprehensive review of saline water correlations and data-part I: thermodynamic properties. Arab J Sci Eng 45(11):8817–8876. https://doi.org/10.1007/s13369-020-05019-y

IAPWS (1992) The international association for the properties of water and steam: revised (June 2014). Supplementary release on saturation properties of ordinary water substance, vol 10, issue Aug 1992

Millero FJ (1983) The thermodynamics of seawater part II: thermochemical properties. Ocean Sci Eng 8(1):1–40

Nayar KG, Sharqawy MH, Banchik LD, Lienhard JH (2016) Thermophysical properties of seawater: a review and new correlations that include pressure dependence. Desalination 390:1–24. https://doi.org/10.1016/j.desal.2016.02.024

Pitzer KS (1987) A thermodynamic model for aqueous solutions of liquid-like density. Rev Mineral 17:97–142

Pitzer KS, Peiper JC, Busey RH (1984) Thermodynamic properties of aqueous sodium chloride solutions. J Phys Chem Ref Data 13(1):1–102. https://doi.org/10.1063/1.555709

Qasem NAA, Generous MM, Qureshi BA, Zubair SM (2021) A comprehensive review of saline water correlations and data: part II—thermophysical properties. Arab J Sci Eng 46(3):1941–1979. https://doi.org/10.1007/s13369-020-05020-5

Sharqawy MH, Lienhard VJH, Zubair SM (2011) Thermophysical properties of seawater: a review of existing correlations and data. Desalin Water Treat 29(1–3):355–355. https://doi.org/10.5004/dwt.2011.2947

Slesarenko V, Shtim A (1989) Determination of seawater enthalpy and entropy during the calculation of thermal desalination plants. Desalination 71(2):203–210. https://doi.org/10.1016/0011-9164(89)80009-8

Sun H, Feistel R, Koch M, Markoe A (2008) New equations for density, entropy, heat capacity, and potential temperature of a saline thermal fluid. Deep Sea Res Part 1 Oceanogr Res Pap 55(10):1304–1310. https://doi.org/10.1016/j.dsr.2008.05.011

Thiel GP, Lienhard JH (2014) Treating produced water from hydraulic fracturing: composition effects on scale formation and desalination system selection. Desalination 346:54–69. https://doi.org/10.1016/j.desal.2014.05.001

Chapter 6
Latent Heat of Vaporization

Abstract The latent heat of vaporization/condensation of saline water represents the energy needed for phase change from liquid to vapor or reverse. It is an important thermodynamic property for thermal desalination and water treatment processes, especially when water is separated by vaporization. This chapter introduced the thermodynamic approaches, correlation, and data used to estimate latent heat of vaporization at different salinity and temperature values. Furthermore, the estimated latent heat of vaporization values from the correlations and methods are compared and assessed against the experimental data of seawater. Some correlations and methods are recommended based on their accuracy and operating conditions. Additionally, the codes used in this chapter are supplied.

Keywords Latent heat of vaporization · Thermodynamics · Correlations · Methods · Phase change · Saline water

6.1 Introduction

The latent heat of vaporization is the amount of energy required in the process of phase change at saturation temperature and pressure. All thermal desalination technologies are based on the idea of phase change from liquid to vapor to get rid of impurities in saline water and then condense the vapor back into the liquid state (pure water). Therefore, it is one of the most significant thermodynamic properties for the design and operation of industrial processes that involve the phase change phenomenon.

The latent heat of vaporization can be described as the energy required by the unit mass of a substance in the process of phase transition. The idea can be summarized in the form of the following well-known relation (Hunter and Bliss 2005)

$$h_{fg} = h_g - h_f \tag{6.1}$$

N. A. A. Qasem et al., *Thermodynamic and Thermophysical Properties of Saline Water*, Springer Water, https://doi.org/10.1007/978-3-031-35193-8_6

As an ionic compound, upon dissociation in water, produces ionic species with a positive and negative charge, extra forces are established between solute ions and solvent molecules due to the dipole nature of water molecules. The presence of these forces costs more energy for the phase transition of saline water, and it is expected to increase with the increase of salinity (Han et al. 2006). Another important factor in the determination of the latent heat of vaporization is the external pressure to which saline water is subjected. The phenomenon of evaporation declines due to higher pressure, and more energy will be required to completely evaporate the unit mass of saline water.

Therefore, the purpose of this section is to briefly summarize the conventional correlations reported in the literature. The following sections summarize the correlations for the calculation of the latent heat of vaporization of saline water.

6.2 Thermodynamic Approach

The latent heat of evaporation can be directly calculated by using correlations and models reported for enthalpy and entropy of saline water by using the Clausius–Clapeyron relation as given below (Sahu et al. 2018) in Eq. (6.2).

$$\int_{f}^{g} \mathrm{Tds} = \int_{f}^{g} \mathrm{dh} + \mathrm{vdp} \tag{6.2}$$

For a constant pressure and temperature during evaporation, Eq. (6.2) can be written as

$$\int_{f}^{g} \mathrm{Tds} = \int_{f}^{g} \mathrm{dh} \tag{6.3}$$

$$\mathrm{h_g} - \mathrm{h_f} = \mathrm{T}(\mathrm{s_g} - \mathrm{s_f}) \tag{6.4}$$

$$\mathrm{h_{fg}} = \mathrm{T}(\mathrm{s_g} - \mathrm{s_f}) \tag{6.5}$$

Then, the latent heat of vaporization can readily be calculated from correlations and models summarized for entropy and enthalpy. Generally, the thermodynamic approach is cumbersome and needs a lot of details, so researchers proposed some methods and correlations to simplify the latent heat of vaporization estimation for multicomponent saline water and NaCl seawater.

6.3 Correlations and Data

Correlations are the easiest way to calculate the saltwater heat capacity as they are fitted from the experimental data. Most of the available latent heat of vaporization correlations in the literature are for seawater. Valderrama et al. (2015) proposed a correlation for the calculation of latent heat of vaporization (L_s) in J/kg of seawater in terms of temperature (T) in K and salinity (S) in g/kg is as Eq. (6.6). The constants are provided in Table 6.1.

$$L_s = \left(a_1 + a_2(T - 273.15) + a_3(T - 373.15)^2 + a_4(T - 273.15)^3 + a_5(T - 273.15)^4\right)\left(1 - \frac{S}{1000}\right) \quad (6.6)$$

The correlation is valid in the salinity range of 0–120 g/kg and temperature range of 0–200 °C with an accuracy of ± 1%.

Chou (1971) suggested a correlation for the calculation of latent heat of vaporization of seawater (L_s) in cal/g in terms of temperature (T) in K and salinity (S) in g/kg are given by Eq. (6.7), and related constants are summarized in Table 6.2.

$$L_s - L_w = T\left(a_1S + a_2S^{1.5} + a_3S^2\right) - \left(a_4S + a_5S^{1.5} + a_6S^2\right) \quad (6.7)$$

The latent heat of vaporization of pure water (L_w) can be estimated as in Eq. (6.8).

$$L_w = b_1 + b_2T + b_3T^2 + b_4T^3 \quad (6.8)$$

The correlation is valid in the salinity range of 0–120 g/kg and temperature range of 0–200 °C with an accuracy of ± 5%.

Table 6.1 Constants for Eq. (6.6)

Constants	Values	Constants	Values
a_1	2.501×10^6	a_4	-8.103×10^{-3}
a_2	-2.369×10^3	a_5	-2.079×10^{-5}
a_3	2.678×10^{-1}		

Table 6.2 Constants for Chou (1971) correlation for the latent heat of vaporization of seawater

Constants	Values	Constants	Values
a_1	$0.92968267 \times 10^{-4}$	a_6	$0.38753341 \times 10^{-3}$
a_2	$-0.1282345 \times 10^{-4}$	b_1	854.52547
a_3	0.195061×10^{-5}	b_2	-1.5939341
a_4	$0.77803108 \times 10^{-2}$	b_3	$0.34368756 \times 10^{-2}$
a_5	$0.75123967 \times 10^{-3}$	b_4	$-0.38341991 \times 10^{-5}$

Table 6.3 Constants for the correlation proposed by El-Dessouky and Ettouney (2002) for the latent heat of vaporization of pure water

Constants	Values	Constants	Values
b_1	2501.897149	b_3	1.992217×10^{-3}
b_2	– 2.407064037	b_4	-1.5863×10^{-5}

Table 6.4 Constants for the correlation of latent heat of pure water reported in the IAPWS-1995 (2002) report

Constants	Values	Constants	Values
a_1	2.501×10^6	a_4	-8.103×10^{-3}
a_2	-2.369×10^3	a_5	-2.079×10^{-5}
a_3	2.678×10^{-1}		

El-Dessouky and Ettouney (2002) proposed a correlation for the latent heat of vaporization of pure water (L_w) in kJ/kg in terms of temperature (T) in °C is given in Eq. (6.9), and related constants are provided in Table 6.3.

$$L_w = b_1 + b_2T + b_3T^2 + b_4T^3 \tag{6.9}$$

The correlation is valid in the temperature range of 5–200 °C and pressure with an accuracy of ± 0.026% (from steam table values).

Another correlation for the latent heat of vaporization of pure water (L_w) is taken from the IAPWS-1995 report (2002). Latent heat of vaporization of pure water (L_w) in J/kg in terms of temperature (T) in °C is provided in Eq. (6.10). The associated constants are provided in Table 6.4.

$$L_w = b_1 + b_2T + b_3T^2 + b_4T^3 \tag{6.10}$$

The correlation is valid in the temperature range of 0–200 °C with an accuracy of ±0.01% (according to IAPWS-1995 (2002) data).

6.4 Case Study

Like the above-mentioned thermodynamic properties, the latent heat of the vaporization of saline water is sensitive to salinity and temperature. Therefore, two figures are plotted to study the effect of these two variables. Figure 6.1a shows the effect of salinity on the values of L_s obtained by Valderrama et al. (2015) and Chou (1971) correlations and data of Nayar et al. (2016) at 20 °C. It can be seen from the figure that Valderrama et al. (2015) method predicts values close to those of Nayar et al. (2016). Chou (1971) correlation shows almost independent values with respect to salinity. The difference is significant when using Chou (1971) correlation at high salinity, as shown in Fig. 6.1b.

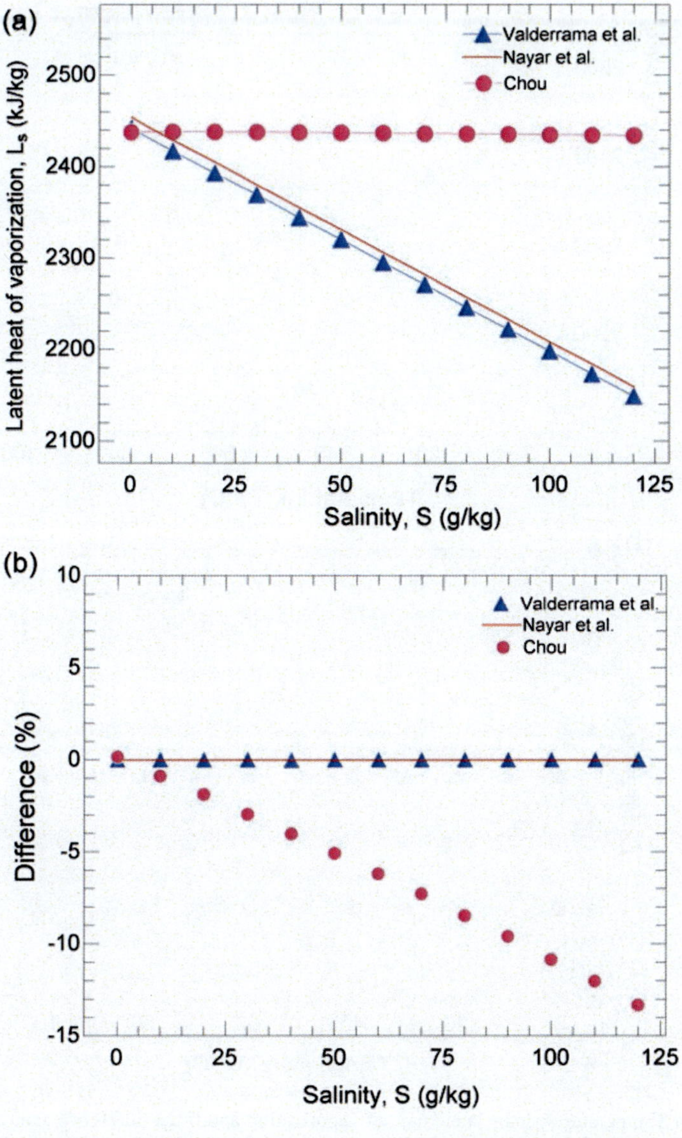

Fig. 6.1 **a** Salinity versus latent heat of vaporization and **b** the difference against the data of Nayar et al. (2016) for the correlations of Valderrama et al. (2015) and Chou (1971) at 20 °C

The effect of temperature (see Fig. 6.2) on L_s is illustrated at 30 g/kg. Similar to the effect of salinity, Valderrama et al. (2015) correlation provide overlapped values with those of Nayar et al. (2016). The difference using Chou (1971) is about ± 3%.

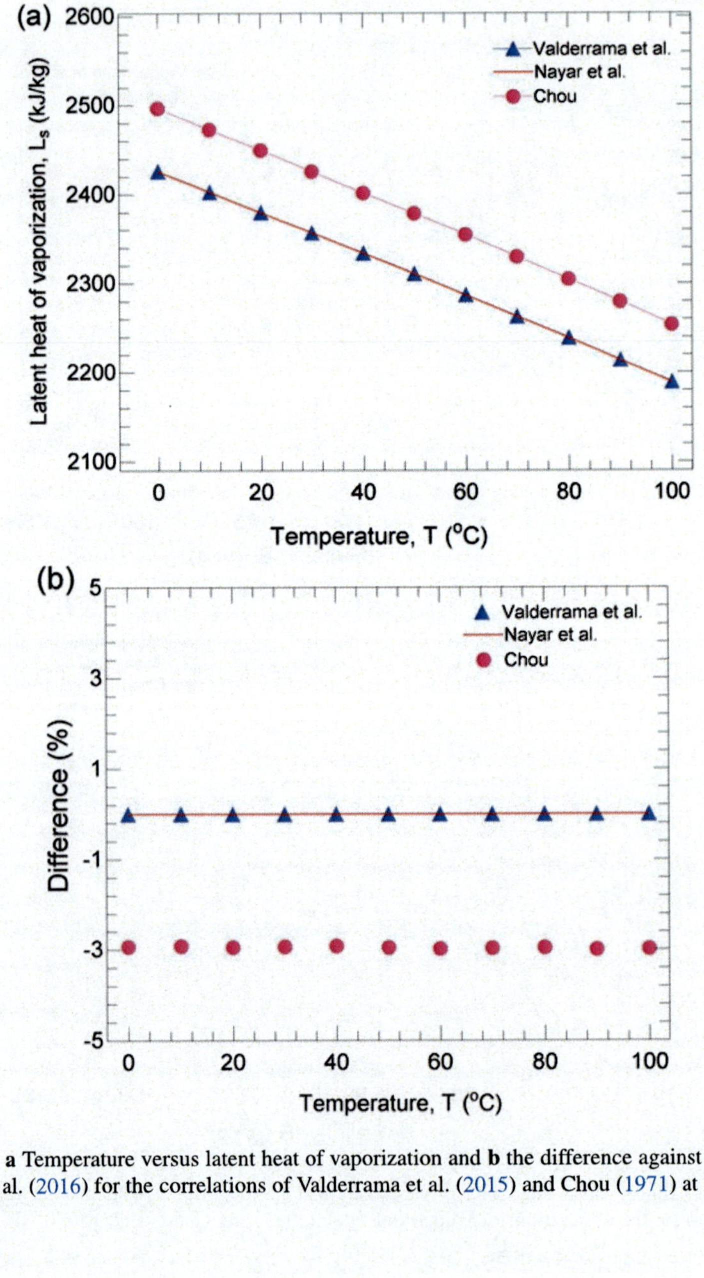

Fig. 6.2 **a** Temperature versus latent heat of vaporization and **b** the difference against the data of Nayar et al. (2016) for the correlations of Valderrama et al. (2015) and Chou (1971) at 30 g/kg

Table 6.5 Summary of the recommended correlations and models for the latent heat of vaporization of saline water (Generous et al. 2020)

Correlation-model	Advantages	Applications	References
Valderrama et al. (Eq. 6.6)	• The most accurate in terms of salinity within a difference of ± 0.5% (refer to Fig. 6.1) for a salinity range of 0–125 g/kg • The best correlations involve the temperature effect within a difference of ± 0.5% (refer to Fig. 6.1)	• All thermal desalination technologies • Seawater greenhouse • High-grade water recycling • Wave-powered desalination	Valderrama et al. (2015)
Chou (Eq. 6.7)	• Temperature effects can be studied in the range of (0–100 °C) with a difference of up to ± 3% (refer to Fig. 6.2)	• All thermal desalination technologies • Seawater greenhouse • High-grade water recycling • Wave-powered desalination	Chou (1971)

Based on the above discussion, some methods and correlations could be recommended to be used, as summarized in Table 6.5.

The latent heat of vaporization is important for the thermal process, especially thermal desalination. Therefore, it is of great significance for thermal technologies to have near-perfect correlations and models for the estimation of latent heat for the proper design of these technologies.

Studies showed that tremendous data had been reported for latent heat calculation of seawater and NaCl solution, but we have not found a comprehensive model for the estimation of latent that can be used for multicomponent composition. Therefore, we strongly recommend a study in the development of a model that can calculate the latent heat of all saline solutions with any given composition.

6.5 EES Codes

The EES codes for the correlations discussed above are:

Valderrama et al. Correlation

"L_s and L_w are lant heat of vaporization of salt water and pure water respectively."

T = 298.5 "K"

S = 35 "g/kg"

L_s = (a1 + (a2 * (T – 273.15)) + (a3 * (T – 273.15)^2) + (a4 * (T – 273.15)^3) + (a5 * (T – 273.15)^4)) * (1 – S/1000)

"Constants"
a1 = 2.501 * 10^6
a2 = – 2.369 * 10^3
a3 = 2.678/10^1
a4 = – 8.103/10^3
a5 = – 2.079/10^5

"L_s in kJ/kg"

L_kJ = L_s/1000

Chou Correlation

"L_s and L_w are lant heat of vaporization of salt water and pure water respectively"

T = 298.5 "K"
S = 35 "g/kg"

L_s – L_w = T * (a1 * S + a2 * S^1.5 + a3 * S^2) – (a4 * S + a5 * S^1.5 + a6 * S^2)

L_w = b1 + b2 * T + b3 * T^2 + b4 * T^3

"Constants"

a1 = 0.92968267/10^4
a2 = – 0.1282345/10^4
a3 = 0.195061/10^5
a4 = 0.77803108/10^2
a5 = 0.75123967/10^3
a6 = 0.38753341/10^3

b1 = 854.52547
b2 = – 1.5939341
b3 = 0.34368756/10^2

b4 = − 0.38341991/10^5

"L_s in kJ/kg"

L_kJ = L_s * 4.18

El Dessouky and Ettouney Correlation

"L_s and L_w are lant heat of vaporization of salt water and pure water respectively"

T = 25 "in C, this value is put as an example"

L_w = b1 + b2 * T + b3 * T^2 + b4 * T^3

"Constants"

b1 = 2501.897149
b2 = − 2.40706403
b3 = 1.192217/10^2
b4 = − 1.5863/10^5

IAPWS 1995

T = 20 "in C, this value is put as an example"
L = a1 + a2 * T + a3 * T^2 + a4 * T^3+ a5 * T^4

a1 = 2.501 * 10^6
a2 = − 2.369 * 10^3
a3 = 2.678/10
a4 = − 8.103/10^3
a5 = − 2.079/10^5
L_SI = L/1000

References

Chou JCS (1971) Thermal properties of sea water. Technical report 49 (issue 49). Hawaii University, Water Resources Research Center

El-Dessouky HT, Ettouney HM (2002) Fundamentals of salt water desalination. https://doi.org/10.1016/b978-0-444-50810-2.x5000-3

Generous MM, Qasem NAA, Qureshi BA, Zubair SM (2020) A comprehensive review of saline water correlations and data-part I: thermodynamic properties. Arab J Sci Eng 45(11):8817–8876. https://doi.org/10.1007/s13369-020-05019-y

Han B, Choi JH, Dantzig JA, Bischof JC (2006) A quantitative analysis on latent heat of an aqueous binary mixture. Cryobiology 52(1):146–151. https://doi.org/10.1016/j.cryobiol.2005.09.007

Hunter JB, Bliss H (2005) Thermodynamic properties of aqueous salt solutions. Ind Eng Chem 36(10):945–953. https://doi.org/10.1021/ie50418a019

Nayar KG, Sharqawy MH, Banchik LD, Lienhard JH (2016) Thermophysical properties of seawater: a review and new correlations that include pressure dependence. Desalination 390:1–24. https://doi.org/10.1016/j.desal.2016.02.024

Sahu P, Krishnaswamy S, Ponnani K, Pande NK (2018) A thermodynamic approach to selection of suitable hydrate formers for seawater desalination. Desalination 436:144–151. https://doi.org/10.1016/j.desal.2018.02.001

Valderrama JO, Campusano RA, Toro AS (2015) Correlation and prediction of saline solution properties for their use in mineral processing using artificial neural networks. J Water Reuse Desalin 5(4):454–464. https://doi.org/10.2166/wrd.2015.132

Wagner W, Pruß A (2002) Revised release on the (IAPWS) formulation 1995 for the thermodynamic properties of ordinary water substance for general and scientific use. J Phys Chem Ref Data 31(2):387–535. https://doi.org/10.1158/0008-5472.CAN-09-0634

Chapter 7
Boiling Point Elevation

Abstract Boiling point elevation is a thermodynamic property that measures the difference in boiling temperature of a pure solvent and that of a solution at the same pressure. It is important to determine the correct boiling point of solutions for heat exchanger design and different applications involving thermal-based desalination technologies. This chapter explains multicomponent models, correlation, and data used to estimate boiling point elevation at different salinity and temperature values. Besides, the calculated boiling point elevation values from the correlations and methods are compared and assessed against the experimental data of seawater. This chapter suggests some correlations and methods based on their accuracy and operating conditions. Also, the codes used for the investigated methods are supplied.

Keywords Boiling point elevation · Thermodynamics · Correlations · Method · Saline water

7.1 Introduction

Boiling point elevation (BPE) is defined as the difference in boiling temperature of a pure solvent and that of a solution at the same pressure. It is a key parameter in the estimation of the right boiling point of aqueous solutions and heat exchanger design for various applications, including thermal-based desalination technologies (MSF, ME, VC). It is also a prime factor to be considered in the evaporation processes of multicomponent liquids in the chemical and pharmaceutical industries (Al-Shammiri 2002; Aminian 2010; Yue et al. 2018).

The BPE is a colligative property that depends upon the number of solute particles. New forces between solute and solvent particles are established when a non-volatile solute is dissolved in water. These forces reduce the rate of evaporation. Due to a reduction in the rate of evaporation, an additional temperature is required for the vapor pressure to be equal to the external pressure. Therefore, the boiling point of the saline solution increases. As BPE is a colligative property, therefore, it is strongly affected

N. A. A. Qasem et al., *Thermodynamic and Thermophysical Properties of Saline Water*, Springer Water, https://doi.org/10.1007/978-3-031-35193-8_7

by salinity. Temperature and external pressure also affect boiling point elevation. It is obvious that increasing non-volatile solute concentration in aqueous solutions results in higher BPE due to a strong solute–solvent interaction that adversely reduces the vapor pressure (Yue et al. 2018).

Thermal-based desalination technologies require an accurate prediction of BPE of saline water as corrosion is also a temperature-sensitive process. Therefore, it emphasizes the fact that the BPE of the solution must be known. It will help to precisely design desalination technology to avoid the destructive phenomenon of corrosion (Brogioli et al. 2018).

This chapter summarizes various correlations to calculate the BPE of saline water as a function of temperature and salinity. This chapter summarizes various correlations for the calculation of BPE.

7.2 Multiple Components Models

Multicomponent models were proposed to be general for all types of saltwater under different salinity values. Major constituents of the aqueous ionic solutions are cations, anions, and water. The boiling point elevation (BPE) for single electrolyte solutions can be calculated using Van't Hoff method (Obille 2017). The following formula can be used for a solution of any salt dissolved in any concentration:

$$\mathrm{BPE} = \mathrm{K_b\ m\ i} \tag{7.1}$$

where K_b is ebullioscopic constant, *m* is the molality of the solution in mol/kg, and *i* is Van't Hoff factor. The ebullioscopic constants for some common solvents are enlisted in Table 7.1.

Van't Hoff factor for covalent compounds such as sugar is 1. For ionic compounds, it can be approximated as:

$$\mathrm{i} = \frac{\text{moles of particles}}{\text{moles of solute}}$$

The method is valid at the temperature of 25 °C with an accuracy of ±5%. However, salinity is not specified.

Table 7.1 The ebullioscopic constant for some common solvents

Solvent	Ebullioscopic constant (K_b)
Water	0.512
Acetic acid	3.07
Phenol	3.04
Naphthalene	5.8

Xinlei and Wang's (2009) model is also a multicomponent model that can be used for the BPE calculation of a single aqueous electrolyte solution using Eq. (7.2).

$$-\mathrm{RT}\ln(\mathrm{a_i}) = \Delta\mathrm{H}_{\mathrm{T_b}}^{\mathrm{vap}}\left(\frac{1}{\mathrm{T_b}} - \frac{1}{\mathrm{T_b}+\mathrm{BPE}}\right) + \Delta\mathrm{C}_{\mathrm{p}}^{\mathrm{Vap}}\left(\left(\frac{\ln(\mathrm{T_b}+\mathrm{BPE})}{\mathrm{T_b}}\right) - \frac{\mathrm{BPE}}{\mathrm{T_b}+\mathrm{BPE}}\right) \tag{7.2}$$

where $\Delta\mathrm{H}_{\mathrm{T_b}}^{\mathrm{vap}}$ is the enthalpy of vaporization of the pure solvent (usually water) at boiling temperature $\left(40.657\frac{\mathrm{kJ}}{\mathrm{mol}}\right)$ at the boiling temperature $\mathrm{T_b}$ and $\Delta\mathrm{C}_{\mathrm{p}}^{\mathrm{Vap}}$ is the difference of isobaric heat capacity for liquid and vapors at the normal boiling temperature of the solvent.

$$\Delta\mathrm{C}_{\mathrm{p}}^{\mathrm{Vap}} = \mathrm{C}_{\mathrm{p_{liquid}}} - \mathrm{C}_{\mathrm{p_{vapors}}} \tag{7.3}$$

For water,

$$\mathrm{C}_{\mathrm{p_{liquid}}} = 4.2176\frac{\mathrm{kJ}}{\mathrm{kg\,K}}$$

$$\mathrm{C}_{\mathrm{p_{vapors}}} = 2.0784\frac{\mathrm{kJ}}{\mathrm{kg\,K}}$$

$\mathrm{a_i}$ is the solution activity (refer to Chap. 11 for details). A Taylor series expansion that kept only the first two terms resulted in an explicit equation for the BPE as shown below:

$$\mathrm{BPE} = \frac{\Delta\mathrm{H}_{\mathrm{T_b}}^{\mathrm{vap}} - 2R\mathrm{T_b}\ln(\mathrm{a_i}) + \sqrt{\left(\Delta\mathrm{H}_{\mathrm{T_b}}^{\mathrm{vap}}\right)^2 - 2\Delta\mathrm{C}_{\mathrm{p}}^{\mathrm{Vap}}T_b^2\ln(\mathrm{a_i})}}{2\left(\Delta\mathrm{H}_{\mathrm{T_b}}^{\mathrm{vap}}/\mathrm{T_b} + 0.5\Delta\mathrm{C}_{\mathrm{p}}^{\mathrm{Vap}} - R\ln(\mathrm{a_i})\right)} \tag{7.4}$$

The solution requires the properties of the pure solvent. The Antoine equation was suggested to be used for this, and the constants for it are provided in Appendix 1.

The Xinlei and Wang model (2009) is valid in the temperature range of 10–350 °C. However, the salinity and accuracy are not specified by the source.

7.3 Correlations and Data

Correlations are the easiest way to calculate the saltwater boiling point elevation as they are fitted from the experimental data. Most of the available specific entropy correlations in the literature are for seawater. El-Dessouky and Ettouney (2002) proposed a correlation for the BPE calculation of seawater (see Appendix 2 for data) in terms of temperature (T), in °C, and salinity (S), in wt%, as expressed in Eq. (7.5). The associated constants are provided in Table 7.2.

Table 7.2 The constant for El-Dessouky and Ettouney (2002) correlation for BPE calculation of seawater

Constants	Values	Constants	Values
a_1	8.325×10^{-2}	a_6	-5.2×10^{-7}
a_2	1.883×10^{-4}	a_7	1.522×10^{-4}
a_3	4.02×10^{-6}	a_8	-3×10^{-6}
a_4	-7.625×10^{-4}	a_9	-3×10^{-8}
a_5	9.02×10^{-5}		

$$\mathrm{BPE} = \mathrm{AS} + \mathrm{BS}^2 + \mathrm{CS}^3 \tag{7.5}$$

where,

$$\mathrm{A} = a_1 + a_2\mathrm{T} + a_3\mathrm{T}^2$$
$$\mathrm{B} = a_4 + a_5\mathrm{T} + a_6\mathrm{T}^2$$
$$\mathrm{C} = a_7 + a_8\mathrm{T} + a_9\mathrm{T}^2$$

The correlation is valid in the salinity range of 0–160 g/kg, temperature range of 10–80 °C, and corresponding saturation pressures. The accuracy is not reported by the source. However, the maximum difference is ± 4.8% from the date of Fabuss and Korosi (1966) (see Appendix 3).

Bromley (1973) proposed a correlation (see Appendix 4 for experimental data) for the BPE calculation of seawater in terms of temperature (T), in K, and practical salinity (S_p), in kg/kg, as expressed in Eq. (7.6). The associated constants are provided in Table 7.3.

$$\mathrm{BPE} = \frac{\mathrm{S_pT}}{a_1}\left(1 + a_2\mathrm{T} + a_3\mathrm{S_p} + a_4\mathrm{TS}^{0.5} + \frac{a_5\mathrm{S_p}(1 - \mathrm{S_p})}{\mathrm{T}} + a_6\,\mathrm{S_pT}\frac{\mathrm{T} + a_7}{\mathrm{T} + a_8}\right) \tag{7.6}$$

The relation between salinity and practical salinity is provided in Eq. (7.7).

$$\mathrm{S} = 1.00472\,\mathrm{S_p} \tag{7.7}$$

Table 7.3 Constants for correlation proposed by Bromley (1973) to the calculation of BPE

Constants	Values	Constants	Values
a_1	13,832	a_5	– 2.583
a_2	0.00137	a_6	– 0.0152
a_3	17.86	a_7	– 225.9
a_4	– 0.00272	a_8	– 236

Table 7.4 Constants for the correlation of Sharqawy et al. (2011) for BPE calculation

Constants	Values	Constants	Values
a_1	17.95	a_4	6.56
a_2	2.823×10^{-1}	a_5	5.267×10^{-2}
a_3	-4.584×10^{-4}	a_6	1.536×10^{-4}

The correlation is valid in the salinity range of 0–120 g/kg, temperature range of 0–200 °C, and corresponding saturation pressures with an accuracy of ± 0.1%.

Sharqawy et al. (2011) proposed a correlation for BPE calculation of seawater in terms of temperature (T), in °C, and salinity (S), in kg/kg, as expressed in Eq. (7.8). The associated constants are provided in Table 7.4.

$$BPE = AS + BS^2 \tag{7.8}$$

A and B are expressed in terms of temperature as,

$$A = a_1 + a_2T + a_3T^2$$
$$B = a_4 + a_5T + a_6T^2$$

The Sharqawy et al. (2011) correlation is valid in a salinity range of 0–120 g/kg, temperature range of 0–200 °C, and corresponding saturation pressures with an accuracy of ± 0.018 K.

Hakuta et al. (1974) proposed a correlation for BPE calculation of seawater in terms of temperature (T), in °C and salinity (S), in wt%, as expressed in Eq. (7.9). The associated constants are provided in Table 7.5 (see Appendix 5 for experimental data).

$$BPE = AS + BS^2 \tag{7.9}$$

A and B are expressed in terms of temperature as,

$$A = a_1 + a_2T + a_3T^2$$
$$B = a_4 + a_5T + a_6T^2$$

Table 7.5 Constants for the correlation of Hakuta et al. (1974) for BPE calculation

Constants	Values	Constants	Values
a_1	0.528764×10^{-1}	a_4	0.320553×10^{-2}
a_2	0.826030×10^{-3}	a_5	-0.144367×10^{-4}
a_3	-0.315082×10^{-7}	a_6	0.184416×10^{-6}

Table 7.6 Constants for BPE calculation using Fabuss and Korosi (1966) correlation

Constants	Values	Constants	Values
a_1	34.64	a_5	0.0257
a_2	0.2009	a_6	0.0193×10^{-2}
a_3	0.2867×10^{-2}	a_7	0.0001×10^{-4}
a_4	0.002×10^{-4}		

The Hakuta et al. (1974) correlation is valid in a salinity range of 17–110 g/kg, temperature range of 60–130 °C, and corresponding saturation pressures with an accuracy of ± 0.020 °C.

Fabuss and Korosi (1966) proposed a correlation for BPE calculation of seawater in terms of temperature (T), in °C and practical salinity (S_p), in g/kg, as expressed in Eq. (7.10). The associated constants are provided in Table 7.6.

$$\mathrm{BPE} = \mathrm{A}\left(\frac{\mathrm{S_p}}{\mathrm{a_1}}\right) + \mathrm{B}\left(\frac{\mathrm{S_p}}{\mathrm{a_1}}\right) \tag{7.10}$$

A and B are expressed in terms of temperature as,

$$\mathrm{A} = \mathrm{a_2} + \mathrm{a_3T} + \mathrm{a_4T^2}$$
$$\mathrm{B} = \mathrm{a_5} + \mathrm{a_6T} + \mathrm{a_7T^2}$$

The Fabuss and Korosi (1966) correlation is valid in a salinity range of 35–100 g/kg, temperature range of 20–180 °C, and corresponding saturation pressures with an accuracy of ± 0.7%.

7.4 Case Study

The boiling point elevation (BPE) of saline water is a function of salinity and temperature. Figure 7.1 shows the effect of salinity on BPE and the accuracy of the investigated correlations and models (i.e., Xinlei and Wang (2009), Hakuta et al. (1974), El-Dessouky and Ettouney (2002), Fabuss and Korosi (1966), Sharqawy et al. (2011), and Bromley (1973)) against the experimental data of Fabuss and Korosi (1966) at 25 °C. The increase in salinity results in an increase in BPE. Figure 7.1a shows that Hakuta et al. (1974), El-Dessouky and Ettouney (2002), and Sharqawy et al. (2011) methods provide a trend close to that of the experimental data (Fabuss and Korosi 1966). The difference values from the experimental data are shown in Fig. 7.1b. The correlation of Sharqawy et al. (2011) seems to be the most accurate one within a ± 3% difference from the experimental data. Moreover, the correlation of Fabuss and Korosi (1966) also exhibits a good agreement within ± 4% error. Xinlei and Wang's method (2009) has a significant difference at low salinities, as shown in Fig. 7.1b.

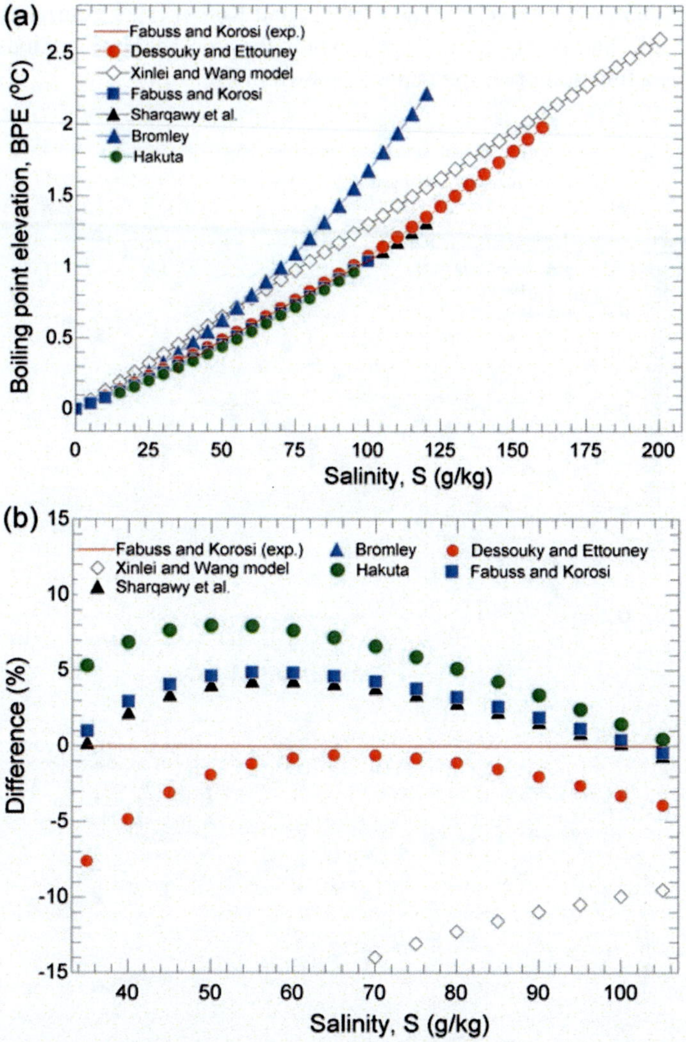

Fig. 7.1 **a** Salinity versus boiling point elevation and **b** the difference against the experimental data of Fabuss and Korosi (1966) for the methods of Xinlei and Wang (2009), Hakuta et al. (1974), El-Dessouky and Ettouney (2002), Sharqawy et al. (2011), Fabuss and Korosi (1966), and Bromley (1973) at 25 °C and saturated pressure

Figure 7.2 is dedicated to showing the independence of BPE on the temperature from 0 to 100 °C and comparing the methods of Xinlei and Wang (2009), Sharqawy et al. (2011), Bromley (1973), El-Dessouky and Ettouney (2002), Fabuss and Korosi (1966) and Hakuta et al. (1974) with the experimental data of Fabuss and Korosi (1966) at 35 g/kg and corresponding saturation pressures. All the correlations provide overlapping results except those of Bromley (1973), as shown in Fig. 7.2a. Within ±

4% error, Sharqawy et al. (2011) and Fabuss and Korosi (1966) correlations, along with the Xinlei and Wang model (2009), exhibit the most accurate methods to predict the BPE as a function of temperature, as shown in Fig. 7.2b.

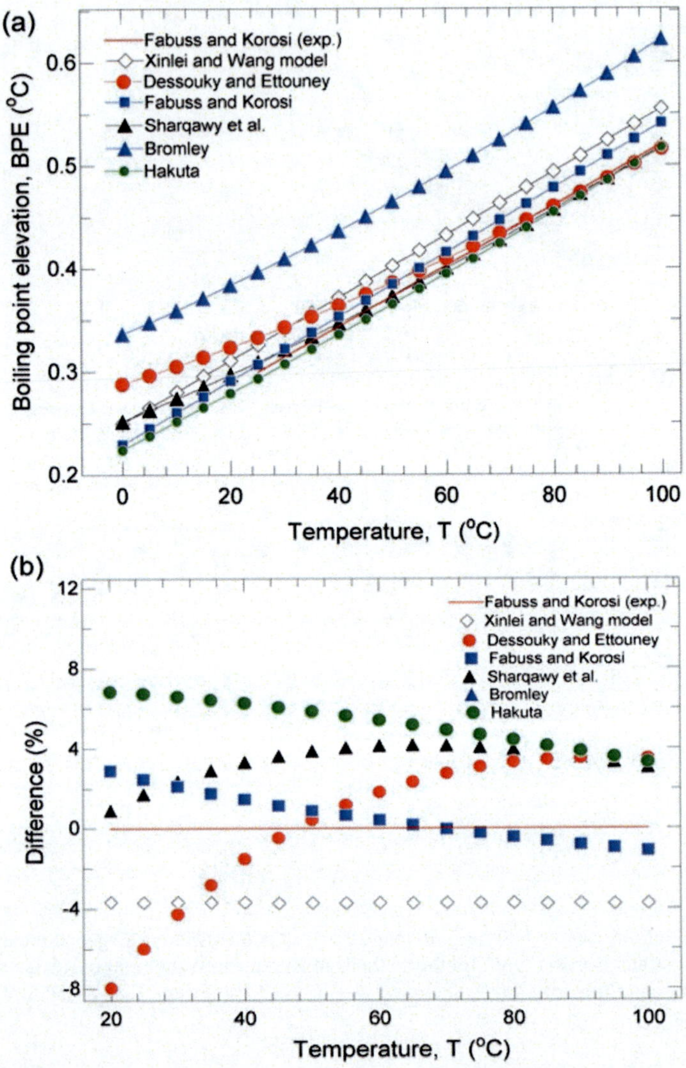

Fig. 7.2 **a** Temperature versus boiling point elevation (BPE) and **b** the difference against the experimental data for the methods of Fabuss and Korosi (1966) for Xinlei and Wang model (2009), Bromley (1973), Sharqawy et al. (2011), El-Dessouky and Ettouney (2002), Fabuss and Korosi (1966), and Hakuta et al. (1974) at saturated pressures and 35 g/kg

Similar to the latent heat and heat capacity, boiling point elevation is more important for thermal-driven technologies like multi-effect evaporation (MEE). The boiling point elevation plays a very important role in the design of stages of MEE plants.

As the measurement of BPE is linked with the design of various thermal-based systems, so proper selection is of great significance. Table 7.7 proposes some correlations for this purpose with associated applications and advantages.

Table 7.7 Summary of the recommended correlations and models for the BPE of saline water (Generous et al. 2020)

Correlation—model	Advantages	Applications	References
Xinlei and Wang model [Eq. (7.2)]	• General for multicomponent saline solution • Valid in a temperature range of 0–100 °C within a difference of ± 3.5% (refer to Fig. 7.2)	• All thermal-based desalination technologies • Seawater greenhouse • High-grade water recycling • Wave-powered desalination	(Xinlei and Wang 2009)
Sharqawy et al. [Eq. (7.8)]	• Valid in the salinity range of 0–120 g/kg within a difference of less than ± 4% (refer to Fig. 7.1) • Cover a wider range of temperatures (0–100 °C) with considerable accuracy (difference within ± 4% (refer to Fig. 7.2)	• All thermal-based desalination technologies • Seawater greenhouse • High-grade water recycling • Wave-powered desalination	(Sharqawy et al. 2011)
Hakuta [Eq. (7.9)]	• Valid in a salinity range of 0–110 g/kg within a difference of ± 5% (refer to Fig. 7.1) • More precise at higher temperature values (within a difference of ± 7% (refer to Fig. 7.2) up to a temperature range of 35–100 °C)	• All thermal-based desalination technologies • Seawater greenhouse • High-grade water recycling • Wave-powered desalination	(Hakuta et al. 1974)
El-Dessouky and Ettouney [Eq. (7.5)]	• It generates more accurate values at a higher salinity range (40–160 g/kg) within a difference of ± 5% (refer to Fig. 7.1) • Covers a wider range of temperatures (40–100 °C) with considerable accuracy (difference within ± 4%, refer to Fig. 7.2)	• All thermal-based desalination technologies	(El-Dessouky and Ettouney 2002)

(continued)

Table 7.7 (continued)

Correlation—model	Advantages	Applications	References
Fabuss and Korosi [Eq. (7.10)]	• Valid in the salinity range of 0–100 g/kg within a difference of ± 4% (refer to Fig. 7.1) • Covers a wide range of temperatures (0–100 °C) with considerable accuracy (difference within ± 3%, refer to Fig. 7.2)	• All thermal-based desalination technologies	(Fabuss and Korosi 1966)

7.5 EES Codes

The EES codes of the previously discussed correlations are:

BPE Usisng Van't Hoff Factor

T_b = i * k_b * m

"ebullioscopic constant for water"

k_b = 0.512

"molality (m) of salt"

{m = 0.6} "approx 0.5 mol/kg"

"Vant Hoff Factor (moles of particles/moles of solute)"

i = 2/1

"similarly BPE with for any salt solution can be calculated."

Xinlei and Wang Model

"Normal Boiling Temperature, T_b"

T = 25 "C"
P = 0.1 "MPa"
T_b = 373.15 "K"

"Enthalpy of vaporization, kJ mol^−1, of pure water at T_b =373.15 K"

dh_vap = 40,657 "J/mol"

"C_p of liquid water at 373.15 K"

```
    C_p_l = 4217.6 "J/kg K"
"C_p of vapors at 373.15 K"
    C_p_v = 2078.4 "J/kg K"
"Heat capacity difference for water at the normal boiling point of solvent"
    dC_p = C_p_l - C_p_v
"Gas Constant"
    R = 8.314 "J/mol K"
"Molecular Weight of solvent in g/mol"
    W = 18.015/1000 "g/mol"
"BPE equation"
    - R * ln(a_i) = dh_vap * ((1/T_b) - (1/(T_b + BPE))) + (dC_p * (ln((T_b +
    BPE)/T_b)) - (BPE/(T_b + BPE)))
"Activity of solvent a_i"
"m is molality in mol/kg, phi is osmotic coefficient"
    ln(a_i) = - nu * m_w * W * phi/1000
"satiochiometric ionization coefficients for cations and anions"
    nu = v_c + v_a
"osmotic coefficient"
    phi = 1 - ((z[1] * z[2]) * A_phi * (I^0.5/(I + b * I^0.5))) + ((S/(T * (v_c + v_
    a))) * ((2 * n)/(2 * n - 1)) * I^(2 * n))

    A_phi = 3.36901532/10^1 - (6.32100430/10^4 * T) + (9.14252359/T) -
    (1.35143986/10^2 * ln(T)) + (2.26089488/10^3/(T - 263)) + (1.92118597/10^6
    * T^2) + (45.2586464/(680 - T))

    I = 0.5 * sum(m[i] * Z[i]^2, i = 1, 2)
"let, for example, have a solution with NaCl dissolved in 1 kg of water."
"note 1. Na 2. Cl"
    m[1] = 0.469 "mol/kg"
    m[2] = 0.546 "mol/kg"
    m_w = 53.6 "mol/kg"
"Stichiometric coefficents"
    v_c = 1
```

v_a = 1

"magnitude of charge"

z[1] = 1
Z[2] = 1

"b is the distance between ions."

"S is solvation parameters defining interaction between ions and solvent."

" n is related to the distance between ion and solvent molecules."

"for NaCl"

b = 9.2452
S = − 295.2606
n = 0.22025

"Avogardro number L"

L = 6.022 * 10^23
e = 1.6021/10^19

"boltzman constant"

K = 1.3805/10^23

"Density of water"

d_s = 1000 "kg/m^3"

"Static dielectric Constant of the solvent."

D = 78.382 * ((1 − (4.5230/10^3) * (T − 25) + (9.5592/10^6) * (T − 25)^2)) * (1 + (4.8618/10^4) * P)

El-Dessouky and Ettouney Correlation

T = 25 "C"
S = 3.5 "%"

BPE = A * S + B * S^2 + C * S^3

A = 8.325/10^2 + 1.883/10^4 * T + 4.02/10^6 * T^2
B = − 7.625/10^4 + 9.02/10^5 * T − 5.2/10^7 * T^2
C = 1.522/10^4 − 3/10^6 * T − 3/10^8 * T^2

Bromley Correlation

"Temperature"

T = 298.15 "K"

```
S = 0.035 "g/kg"
S_p = S/1.00472

BPE = ((S_p * T^2)/(13832)) * (1 + (0.00137 * T) + (17.86 * S_p) – (0.00272
* T * S_p^0.5) – (2.583 * S_p * (1 – S_p)/T) – (0.0152 * S_p * T * (T – 255.9)/
(T – 236)))
```

Sharqawy et al. Correlation

```
T = 0 "in C, value here is an example"
S = 0.035 "g/kg"

BPE = A * S^2 + B * S

A = 17.95 + (2.823/10^1 * T) – (4.584/10^4 * T^2)
B = 6.56 + (5.267/10^2 * T) + (1.536/10^4 * T^2)
```

Hakuta et al. Correlation

```
T = 25 "C"
S = 3.5 "%"

BPE = A * S + B * S^2

A = a1 + a2 * T + a3 * T^2
B = a4 + a5 * T + a6 * T^2

a1 = 0.528764/10^1
a2 = 0.826030/10^3
a3 = – 0.315082/10^7
a4 = 0.320553/10^2
a5 = – 0.144367/10^4
a6=0.184416/10^6
```

Fabuss and Korosi correlation:

```
T = 25 "C"
S = 35 "g/kg"
S_p = S/1.00472

BPE = A * (S_p/34.46) + B * (S_p/34.46)^2

A = 0.2009 + (0.2867/10^2 * T) + (0.002/10^4 * T^2)
B = 0.0257 + (0.0193/10^2 * T) + (0.0001/10^4 * T^2)
```

Appendix 1

See Table 7.8.

Appendix 2

See Table 7.9.

Appendix 3

See Tables 7.10 and 7.11.

Appendix 4

See Table 7.12.

Appendix 5

See Table 7.13.

Table 7.8 Constants needed for the Antoine equation used for pure solvents

Solvents	T (K)	A	B	C
Water[a]	273–303	5.40221	1838.68	− 31.737
	304–333	5.20389	1733.93	− 39.485
	334–363	5.0768	1659.79	− 45.854
	379–573	3.55959	643.748	− 198.043
Methanol[a]	288.0–356.83	5.20409	1581.34	− 33.50
Ethanol[a]	292.77–366.63	5.24677	1598.67	− 33.50
Acetonitrile[b]	298.15–343.15	14.973	3410	− 25.415

[a] $\log_{10}(p/bar) = A - \{B/(T/K + C)\}$
[b] $\log_{10}(p/\text{kPa}) = A - \{B/(T/K + C)\}$

Table 7.9 Sea water variation data for BPE, in °C, used by El-Dessouky and Ettouney (2002)

T (°C)	Salinity (wt%)						
	1	2	3	4	5	6	7
10	0.085	0.171	0.258	0.348	0.441	0.538	0.639
15	0.087	0.175	0.266	0.359	0.456	0.556	0.662
20	0.089	0.18	0.273	0.37	0.47	0.575	0.684
25	0.091	0.185	0.281	0.381	0.485	0.593	0.706
30	0.093	0.19	0.29	0.393	0.5	0.612	0.727
35	0.096	0.195	0.298	0.405	0.516	0.63	0.748
40	0.099	0.201	0.307	0.417	0.531	0.648	0.769
45	0.101	0.207	0.316	0.43	0.546	0.666	0.789
50	0.104	0.213	0.326	0.443	0.562	0.684	0.809
55	0.108	0.22	0.336	0.456	0.578	0.703	0.829
60	0.111	0.227	0.346	0.469	0.594	0.721	0.848
65	0.115	0.234	0.357	0.483	0.61	0.739	0.866
70	0.118	0.241	0.368	0.497	0.627	0.756	0.885
75	0.122	0.249	0.379	0.511	0.643	0.774	0.903
80	0.126	0.257	0.391	0.525	0.66	0.792	0.921
85	0.13	0.265	0.402	0.54	0.677	0.81	0.938
90	0.135	0.274	0.415	0.555	0.694	0.828	0.955
95	0.139	0.283	0.427	0.571	0.711	0.845	0.971
100	0.144	0.292	0.44	0.587	0.728	0.863	0.987
105	0.149	0.301	0.453	0.603	0.746	0.88	1.003
110	0.154	0.311	0.467	0.619	0.764	0.898	1.018

Table 7.10 BPE of sea water and its concentrates, °C

f_{Cl}[a]	Cl (‰)[b]	S (‰)[c]	Temperature (°C)								
			20	40	60	80	100	120	140	160	180
1.0	19.00	34.46	0.30	0.34	0.40	0.47	0.53	0.59	0.66	0.71	0.77
1.2	22.80	41.35	0.36	0.42	0.49	0.57	0.67	0.72	0.80	0.87	0.93
1.4	26.61	48.24	0.43	0.50	0.58	0.68	0.77	0.86	0.95	1.03	1.10
1.6	30.41	55.13	0.51	0.58	0.68	0.79	0.89	1.00	1.10	1.19	1.28
1.8	34.21	62.03	0.57	0.67	0.79	0.90	1.01	1.14	1.26	1.36	1.46
2.0	38.01	68.92	0.65	0.75	0.88	1.02	1.16	1.29	1.42	1.54	1.65
2.2	41.81	75.81	0.72	0.85	0.98	1.13	1.27	1.44	1.58	1.72	1.84
2.4	45.61	82.70	0.81	0.94	1.07	1.26	1.43	1.59	1.75	1.90	2.04
2.6	49.41	89.59	0.89	1.04	1.21	1.39	1.59	1.75	1.92	2.09	2.24
2.8	53.21	96.48	0.98	1.13	1.32	1.52	1.70	1.91	2.10	2.28	2.45
3.0	57.01	103.38	1.06	1.23	1.44	1.66	1.86	2.08	2.28	2.48	2.66

[a]Chlorinity factor (multiplying factor of chlorinity)
[b]Chlorinity (parts of equivalent chloride per thousand parts of sea water)
[c]Salinity (parts of salt per thousand parts of sea water). For ionic composition, *see* Table 7.9

Table 7.11 Ionic composition of standard sea water and its concentrates (mol/kg$_w$)

	Concentration factor, $\frac{f_{Cl}{}^{a}}{f_s{}^{b}}$										
	$\frac{1}{1.00000}$	$\frac{1.2}{1.20900}$	$\frac{1.4}{1.44210}$	$\frac{1.6}{1.63600}$	$\frac{1.8}{1.85500}$	$\frac{2.0}{2.07700}$	$\frac{2.2}{2.30200}$	$\frac{2.4}{2.53100}$	$\frac{2.6}{2.76300}$	$\frac{2.8}{2.99900}$	$\frac{3.0}{3.24000}$
Na^+	0.47631	0.57582	0.07683	0.77938	0.88347	0.98918	1.09651	1.20553	1.31625	1.42873	1.543
Mg^{-2}	0.05423	0.06556	0.07706	0.08873	0.10059	0.11262	0.12484	0.13725	0.14986	0.16267	0.17568
Ca^{-2d}	0.01034	0.0125	0.01469	0.01692	0.01918	0.02147	0.0238	0.02617	0.02857	0.03102	0.0335
K^-	0.01007	0.01217	0.0143	0.01648	0.01868	0.02091	0.02318	0.02,549	0.02783	0.03021	0.03262
Cl^-	0.55511	0.67108	0.7888	0.90832	1.02963	1.15282	1.27792	1.40497	1.534	1.6651	1.79827
$SO_4{}^{-2d}$	0.02979	0.03601	0.04233	0.04874	0.05526	0.06187	0.06858	0.0754	0.08232	0.08936	0.0965
	0.00084	0.00101	0.00119	0.00137	0.00156	0.00174	0.00193	0.00213	0.00232	0.00252	0.00272
120 °C, Ca^{-2}	c	c	c	c	c	c	0.02214	0.0219	0.02153	0.02121	0.0208
$SO_4{}^{-2}$	c	c	c	c	c	c	0.06692	0.07113	0.07528	0.07955	0.0838
140 °C, Ca^{-2}	c	c	0.01448	0.01472	0.01478	0.01463	0.01444	0.01418	0.0139	0.01354	0.0094
$SO_4{}^{-2}$	c	c	0.04212	0.4654	0.05086	0.05503	0.05922	0.06341	0.06765	0.07188	0.0724
160 °C, Ca^{-2}	0.00871	0.00897	0.00914	0.00198	0.00928	0.00916	0.009	0.00881	0.00859	0.00836	0.00826
$SO_4{}^{-2}$	0.02816	0.03248	0.03678	0.041	0.04536	0.04956	0.05378	0.0584	0.06234	0.0667	0.07126
180 °C, Ca^{-2}	0.00501	0.00513	0.00522	0.00526	0.00528	0.00543	0.00507	0.00494	0.00481	0.00481	0.00452
$SO_4{}^{-2}$	0.02446	0.02864	0.03286	0.03708	0.04136	0.04583	0.04985	0.05417	0.05856	0.063	0.06752

[a] Chlorinity factor
[b] Salt concentration factor
[c] No precipitation occurs below this concentration and temperature limit
[d] At high temperatures, use values given in lower part of this table

Table 7.12 BPE data of Bromley (1973) for sea salt solutions

$T_0 = 60$ °C		$T_0 = 80$ °C		$T_0 = 100$ °C		$T_0 = 120$ °C	
wt. (%)	BPE (°C)	wt. (%)	BPE (°C)	wt. (%)	BPE (°C)	wt. (%)	BPE (°C)
0.6186	0.071	0.6239	0.081	0.7981	0.118	0.1901	0.0325
0.6369	0.072	1.253	0.163	1.001	0.147	0.6887	0.116
1.1263	0.128	1.8504	0.241	1.2566	0.185	1.2813	0.213
1.1735	0.134	2.5375	0.332	1.683	0.248	1.9457	0.324
1.5707	0.178	2.643	0.347	1.7117	0.251	2.5357	0.425
1.6867	0.192	3.467	0.461	2.151	0.318	2.6012	0.437
2.2542	0.258	3.469	0.461	3.03	0.454	3.0582	0.518
3.0542	0.356	4.0254	0.542	3.3541	0.505	3.3534	0.567
3.735	0.44	4.779	0.654	3.473	0.524	3.9086	0.67
4.5285	0.542	5.3929	0.749	3.618	0.547	4.4208	0.754
5.178	0.629	5.8352	0.816	4.281	0.657	4.911	0.856
5.7635	0.709	5.9126	0.83	4.876	0.756	5.4038	0.951
6.4212	0.803	6.3231	0.894	5.0555	0.789	5.9291	1.056
7.1354	0.907	6.4064	0.91	5.0648	0.79	6.4108	1.154
		6.571	0.937	5.45	0.857	6.5937	1.1925
		6.575	0.936	6.034	0.96	6.8656	1.265
		6.7474	0.964	6.521	1.049		
		6.9931	1.008	6.5884	1.061		
				6.6215	1.062		
				6.7165	1.088		
				6.7369	1.09		
				7.095	1.157		
				7.136	1.167		

Table 7.13 BPE of seawater experimental data by Hakuta et al. (1974)

$T_0 = 60$ °C		$T_0 = 110$ °C	
wt. (%)	BPE (°C)	wt. (%)	BPE (°C)
1.862	0.204	1.891	0.293
3.428	0.386	3.389	0.534
5.854	0.699	5.733	0.958
8.146	1.040	8.883	1.583
10.780	1.450	10.977	2.057
$T_0 = 80$ °C		$T_0 = 120.6$ °C	
wt. (%)	BPE (°C)	wt. (%)	BPE (°C)
1.864	0.238	1.863	0.299
3.574	0.468	3.381	0.559

(continued)

Table 7.13 (continued)

$T_0 = 80$ °C		$T_0 = 120.6$ °C	
wt. (%)	BPE (°C)	wt. (%)	BPE (°C)
6.086	0.831	5.288	0.912
8.841	1.297	7.99	1.467
11.196	1.739	10.863	2.13
$T_0 = 100$ °C		$T_0 = 130$ °C	
wt. (%)	BPE (°C)	wt. (%)	BPE (°C)
1.723	0.241	1.646	0.283
3.453	0.468	3.355	0.59
5.447	0.849	5.711	1.06
9.177	1.533	8.604	1.695
10.938	1.905	10.877	2.272

References

Al-Shammiri M (2002) Evaporation rate as a function of water salinity. Desalination 150(2):189–203. https://doi.org/10.1016/S0011-9164(02)00943-8

Aminian A (2010) Prediction of temperature elevation for seawater in multi-stage flash desalination plants using radial basis function neural network. Chem Eng J 162(2):552–556. https://doi.org/10.1016/j.cej.2010.05.060

Brogioli D, La Mantia F, Yip NY (2018) Thermodynamic analysis and energy efficiency of thermal desalination processes. Desalination 428:29–39. https://doi.org/10.1016/j.desal.2017.11.010

Bromley LA (1973) Thermodynamic properties of strong electrolytes in aqueous solutions. AIChE J 19(2):313–320. https://doi.org/10.1002/aic.690190216

El-Dessouky HT, Ettouney HM (2002) Fundamentals of salt water desalination. https://doi.org/10.1016/b978-0-444-50810-2.x5000-3

Fabuss BM, Korosi A (1966) Boiling point elevations of sea water and its concentrates. J Chem Eng Data 11(4):606–609. https://doi.org/10.1021/je60031a049

Generous MM, Qasem NAA, Qureshi BA, Zubair SM (2020) A comprehensive review of saline water correlations and data-part I: thermodynamic properties. Arab J Sci Eng 45(11):8817–8876. https://doi.org/10.1007/s13369-020-05019-y

Hakuta T, Goto T, Ishizaka S (1974) Boiling point elevation of sea water. Bull Soc Sea Water Sci Jpn 28(3):156–161. https://doi.org/10.11457/swsj1965.28.156

Obille A (2017) Investigating the relationship between variance of transition temperatures in an undergraduate laboratory setting, pp 48–51

Sharqawy MH, Lienhard V JH, Zubair SM (2011) Thermophysical properties of seawater: a review of existing correlations and data. Desalin Water Treat 29(1–3):355–355. https://doi.org/10.5004/dwt.2011.2947

Xinlei G, Wang X (2009) Calculations of freezing point depression, boiling point elevation, vapor pressure and enthalpies of vaporization of electrolyte solutions by a modified three-characteristic parameter correlation model. J Solution Chem 38(9):1097–1117. https://doi.org/10.1007/s10953-009-9433-0

Yue C, Wang B, Zhu B (2018) Thermal analysis for the evaporation concentrating process with high boiling point elevation based exhaust waste heat recovery. Desalination 436:39–47. https://doi.org/10.1016/j.desal.2018.02.010

Chapter 8
Vapor Pressure

Abstract The vapor pressure of saline water is an essential thermodynamic property for desalination processes that depend mainly on evaporation. This chapter introduces multicomponent models, correlation, and data used to estimate vapor pressure at various salinity and temperature values. Besides, the calculated vapor pressure values from the correlations and methods are compared and assessed against the experimental data of seawater. Based on their accuracy and operating conditions, some correlations and methods are recommended in this chapter. Furthermore, the codes used for the investigated methods are provided.

Keywords Vapor pressure · Correlations · Methods · Vaporization · Saline water

8.1 Introduction

The vapor pressure of saline water is an important thermodynamic property for desalination processes that depend mainly on the evaporation process. The boiling point is, in fact, a point where vapor pressure equals the external pressure. The most straightforward equation to approximate the vapor pressure of pure water or a simple single electrolyte solution was approximated by the Antoine equation given below (Rodgers and Hill 1978):

$$\log(\mathrm{P}) = \mathrm{A} + \frac{\mathrm{B}}{\mathrm{C} + \mathrm{T}} \tag{8.1}$$

The significance of this equation is its ability to be applied to a solution of a single volatile compound and pure water. Morillon et al. (1999) applied this equation to a saturated salt solution. Similarly, an extended developed the equation of state for aqueous electrolyte solutions was developed by Zuo and Fürst (2002) and Han (2012) to non-aqueous electrolyte solutions.

N. A. A. Qasem et al., *Thermodynamic and Thermophysical Properties of Saline Water*, Springer Water, https://doi.org/10.1007/978-3-031-35193-8_8

The vapor pressure is a strong function of the temperature and would be variable with the salinity. As the salinity increases, the vapor pressure may decrease. On the other hand, the temperature accelerates the rate of evaporation. The relevant correlations and methods for the vapor pressure of saline water are summarized below.

8.2 Multiple Components Models

Multicomponent models were proposed to be general for all types of saltwater under different salinity values. Major constituents of the aqueous ionic solutions are cations, anions, and water. Shiah and Tseng (1994) proposed a model for a multicomponent aqueous solution. They suggested the use of a mean spherical approximation (MSA) model to develop an expression for vapor pressure.

$$p = p_w \exp\left(-\Phi\left(\frac{\sum_{i=1}^{n} m_i M_w}{1000}\right)\right) \tag{8.2}$$

p is the vapor pressure of a solution in Pa, p_w is the vapor pressure of pure water in Pa, Φ is a molar osmotic coefficient, M_w is the molecular weight of water in g/mol, and m_i is the molality of ith ion.

The Shiah and Tseng (1994) model is valid in the temperature range of 25–70 °C and pressure of $P \le 1.01$ MPa with an accuracy of $\pm$ 0.42%. It can be used up to the moderate ionic strength range, but the source has not explicitly stated this limitation. A sample of the data upon which the model is based can be found in Appendix 1.

Cisternas and Lam (1989) proposed a model for multicomponent solutions. For temperature (T) in K and ionic strength (I) in mol/kg, the vapor pressure (P) in kPa is given by Eq. (8.3).

$$\mathrm{Log}(p) = KI\left(A - \frac{B}{T - E_s}\right) + \left(C - \frac{D}{T - E_s}\right) \tag{8.3}$$

where,

$$A = A_s + 3.60591 \times 10^{-4} I + \frac{M_s}{2303}$$
$$B = B_s + 1.382982\, I - 0.031185\, I^2$$
$$C = C_s - 3.99334 \times 10^{-3} I - 1.11614 \times 10^{-4} I^2 + \frac{M_s I(1 - x)}{2303}$$
$$D = D_s - 0.138481\, I + 0.027511\, I^2 - 1.79277 \times 10^{-3}\, I^3$$

Values A_s, B_s, C_s, and D_s are solvent-dependent parameters, and their values for different solvents are provided in Table 8.1. x can be determined by the following expression:

$$x = \frac{2(v_c + v_a)}{v_c Z_c^2 + v_a Z_a^2}$$

K is a salt-dependent parameter, and its values for different salts are also provided in Table 8.2. Also, total ionic strength can be estimated by using Eq. (8.4).

$$I_t = 1/2 \sum_{i=1}^{n} m_i Z_i^2 \tag{8.4}$$

where m_i and z_i are molality and ionic charge of the ith ion, respectively. In the case of a mixture of electrolytes, the following relations are used to calculate K and m of the mixture:

$$K_m = \sum_{j=1}^{n} y_j K_j^o$$

$$X_m = \sum_{j=1}^{n} y_j X_j^o$$

And vapor pressure of the solution will be equal to:

$$\text{Log}\,(p) = K_m = \sum_{j=1}^{n} y_j \log\left(P_j^o\right)$$

Table 8.1 Values of A_s, B_s, C_s, D_s and E_s (Cisternas and Lam 1989)

Solvent	A_s	B_s	C_s	D_s	E_s
Water	– 0.021302	– 5.390915	7.192959	1730.2857	39.53
Methanol	– 0.027430	– 1.524026	7.158563	1550.7179	36.50
Ethanol	– 0.038218	– 6.384607	7.338230	1652.0475	41.68

Table 8.2 K-values

Electrolyte	K	Electrolyte	K
$CaCl_2$	0.32138	NaBr	0.45116
KNO_3	– 0.18396	NaCl	0.39015
K_2CO_3	0.15324	Na_2SO_4	0.04490
LiBr	1.00000	$(NH_4)_2SO_4$	0.00000
LiCl	0.72567		

Here, y_j is a fraction of ionic strength given by:

$$y_j = \frac{I_j}{I_t}$$

where I_t, and I_j are total ionic strength and individual ionic strength of the jth electrolyte component. The following expression for the calculation of vapor pressure was also proposed:

$$p = \left(- \ln p_s + \frac{\Phi \nu m M_s}{1000} \right) \tag{8.5}$$

where m is the molality in mol/kg, M_s is the molar mass of solvent, ν is the total number of ions of salt, p_s is the pure solvent vapor pressure, and Φ is the osmotic pressure that can be determined as follows:

$$\Phi = - \frac{1000 \ln a_s}{\nu m M_s} \tag{8.6}$$

where a_s is the activity of the solvent. The Cisternas and Lam (1989) model is valid up to 44 mol/kg of ionic strength. The temperature must be in the range of 0–140 °C. The accuracy depends upon a combination of solutes. For example, it is reported ± 2.6% for aqueous solution constituting NaCl and $MgCl_2$.

8.3 Correlations and Data

Correlations are the easiest way to calculate the saltwater vapor pressure as they are fitted from the experimental data. Most of the available vapor pressure correlations in the literature are for seawater. Chou (1971) proposed a correlation (Eq. (8.7)) for the vapor pressure of seawater. For temperature (T) in K and chlorinity (Cl) in g/kg,

$$\ln(p) = \left(1 + a\,Cl + b\,Cl^{1.5}\right) \ln(P_w) + \left(c\,Cl + d\,Cl^{1.5} + e\,Cl^2\right) \tag{8.7}$$

Vapor pressure for pure water (p_w) can be calculated using Eqs. (8.17), (8.18), or Eq. (8.19), and the required constants are provided in Table 8.3. Cl is chlorinity that is linked with salinity by Eq. (8.8).

$$S(‰) = 1.80655 Cl(‰) \tag{8.8}$$

The correlation is valid in the salinity range of 0–289 g/kg, and temperature range of 0–175 °C with an accuracy of ± 2.3%.

Table 8.3 Constants for the correlation of Chou (1971) for vapor pressure calculations

Constants	Values
a	$0.54675794 \times 10^{-4}$
b	$-0.15410306 \times 10^{-5}$
c	$-0.99478110 \times 10^{-3}$
d	$0.11895489 \times 10^{-3}$
e	$-0.13716509 \times 10^{-4}$

Arons and Kientzler (1966) proposed a correlation for the vapor pressure of seawater (see Appendix 2 for data). For temperature (T) in K and chlorinity (Cl) in g/kg, the vapor pressure can be calculated using Eq. (8.9), and the required constants are provided in Table 8.4.

$$\ln\left(\frac{p}{p_w}\right) = \left(a\,Cl + b\,Cl^{1.5} + c\,Cl^2\right) + \frac{d\,Cl + e\,Cl^{1.5} + f\,Cl^2}{T} + \left(g\,Cl + h\,Cl^{1.5} + i\,Cl^2\right) \quad (8.9)$$

Vapor pressure for pure water (p_w) can be calculated using Eqs. (8.17), (8.18), or Eq. (8.19). Cl is the chlorinity that is linked with salinity by Eq. (8.8).

The following expressions in terms of salinity (in ppt) and chlorinity (in ppt) are also reported:

$$p = p_w(1 - 0.000537\,S) \quad (8.10)$$

$$p = p_w(1 - 0.000696\,Cl) \quad (8.11)$$

The most accurate data for p_w can be obtained from Eq. (8.17). Vapor pressure for pure water (p_w) can also be calculated using Eq. (8.18) or Eq. (8.19) at a given temperature. The correlation is valid in the salinity range of 18–290 g/kg, temperature range of 0–175 °C with an accuracy of ± 0.00423.

Emerson and Jamieson (1967) proposed a correlation for the vapor pressure of seawater based on data shown in Appendix 3. For the salinity (S) in g/kg,

Table 8.4 Constants for Arons and Kientzler (1966) correlation

Constants	Values	Constants	Values
a	-0.051807014	f	-0.037231985
b	$-0.66809897 \times 10^{-2}$	g	$0.75964905 \times 10^{-2}$
c	$0.68897827 \times 10^{-3}$	h	$0.97136709 \times 10^{-3}$
d	2.153011	i	$-0.10158384 \times 10^{-3}$
e	0.38508979		

Table 8.5 Constants for Emerson and Jamieson (1967) correlation for vapor pressure

Constants	Values
a_1	-2.1609×10^{-4}
a_2	-3.5015×10^{-7}

$$\log\left(\frac{p}{p_w}\right) = a_1 S - a_2 S^2 \tag{8.12}$$

Vapor pressure for pure water (p_w) can also be calculated using Eqs. (8.17), (8.18) or Eq. (8.19), and the required constants are provided in Table 8.5.

Vapor pressure for pure water (p_w) can be calculated using Eqs. (8.17), (8.18), or Eq. (8.19). These values can also be obtained from NEL steam tables 1964. Emerson and Jamieson (1967) have not explicitly stated its temperature range, but the temperature range of $100 \leq T \leq 180$ °C is based on the tables provided. The correlation is valid in the salinity range of 35–170 g/kg, temperature range of 100–180 °C with an accuracy of ± 0.07%.

Nayar et al. (2016) proposed a correlation between the vapor pressure of seawater. For a salinity (S) in g/kg and temperature (T) in K,

$$\log\left(\frac{p}{p_w}\right) = a_1 S - a_2 S^2 \tag{8.13}$$

P_w can be calculated using Eq. (8.14), and the required constants are provided in Table 8.6.

$$P_w = \frac{a_3}{T} + a_4 + a_5 T + a_6 T^2 + a_7 T^3 + a_8 \ln(T) \tag{8.14}$$

The correlation is valid in the salinity range of 0–160 g/kg, and temperature range of 0–180 °C with an accuracy of ± 0.026%.

Robinson's (1954) correlation, coupled with Chou (1971) correlation for pure water, is used for seawater vapor pressure calculation. The main correlation is also reported by Sharqawy et al. (2011). For salinity (S) in g/kg,

$$\frac{p}{p_w} = 1 + 0.57357\left(\frac{S}{1000 - S}\right) \tag{8.15}$$

Table 8.6 Constants for the correlation of Nayar et al. (2016) for the vapor pressure of seawater

Constants	Values	Constants	Values
a_1	-4.58180×10^{-4}	a_5	-4.8640×10^{-2}
a_2	-2.04430×10^{-6}	a_6	4.1765×10^{-5}
a_3	-5800	a_7	-1.4452×10^{-8}
a_4	1.3915	a_8	6.5460

Table 8.7 Constants of Millero (1983) correlation for the vapor pressure of seawater

Constants	Values	Constants	Values
a_1	-2.331×10^{-3}	a_5	-1.1320×10^{-5}
a_2	-1.4799×10^{-4}	a_6	-8.7086×10^{-6}
a_3	-7.520×10^{-6}	a_7	7.4936×10^{-7}
a_4	-5.5185×10^{-8}	a_8	-2.6327×10^{-8}

Vapor pressure for pure water (p_w) can be calculated using Eqs. (8.17), (8.18) or Eq. (8.19). The correlation is valid in the salinity range of 20–45 g/kg, temperature range of 0–100 °C with an accuracy of ± 0.2%.

Millero (1983) proposed a correlation (See Appendix 4 for data) for the vapor pressure of seawater (Eq. (8.16)). For temperature (T), in K, and salinity in g/kg,

$$\ln(p) = p_w + AS + BS^2 \tag{8.16}$$

A and B can be calculated from the following expressions, and the required constants are provided in Table 8.7.

$$A = a_1 + a_2(T - 273.15) + a_3(T - 273.15)^2 + a_4(T - 273.15)^3$$
$$B = a_5 + a_6(T - 273.15) + a_7(T - 273.15)^2 + a_8(T - 273.15)^3$$

Vapor pressure for pure water (p_w) can be calculated using Eqs. (8.17), (8.18), or Eq. (8.19) at a given temperature. The correlation is valid in the salinity range of 0 – 40 g/kg, temperature range of 0–40 °C with an accuracy of ± 0.015%.

A correlation for pure water is proposed by Dessouky and Ettouney (2002) in terms of temperature (T) in °C,

$$\ln\left(\frac{P_w}{P_c}\right) = \left(\frac{T_c}{T + 273.15} - 1\right)\sum_{i=1}^{8} f_i(0.01(T + 273.15 - 338.15))^{(i-1)} \tag{8.17}$$

$T_c = 647.286$ K and $P_c = 22089$ kPa, while the rest of the constants are provided in Table 8.8.

The correlation of Dessouky and Ettouney (2002) is valid in the temperature range of 5–200 °C with an accuracy of ± 0.05% (from steam table values).

Table 8.8 Constants to implement Dessouky and Ettouney (2002) correlation

Constants	Values	Constants	Values
f_1	– 7.419242	f_5	0.001094098
f_2	0.29721	f_6	– 0.00439993
f_3	– 0.1155286	f_7	0.002520658
f_4	0.008685635	f_8	– 0.000521868

Table 8.9 Constant of the Chou (1971) correlation for the vapor pressure of pure water

Constants	Values
a_1	71.023834
a_2	− 7380.4001
a_3	− 9.0977349
a_4	0.0070871862

Table 8.10 Constant of Eq. (8.19)

Constants	Values
a_1	11.809452
a_2	− 3902.2336
a_3	− 42.407613

Chou (1971) also provided a correlation for the vapor pressure of pure water. For temperature (T) in K, the correlation is summarized in Eq. (8.18), and the required constants are provided in Table 8.9.

$$\ln(p_w) = a_1 + \frac{a_2}{T} + a_3 \ln(T) + a_4 T \tag{8.18}$$

The correlation of Chou (1971) is valid in the temperature range of 0–170 °C with an accuracy of ± 0.0147 MPa.

Antoine equation (Morillon et al. 1999) can also be used to calculate the vapor pressure of pure water. For temperature (T) in K,

$$\ln(p_w) = a_1 + \frac{a_2}{a_3 + T} \tag{8.19}$$

Temperature (T) in K. The constants in Eq. (8.19) are summarized in Table 8.10. The Antoine equation (Morillon et al. 1999) is valid in the temperature range of 0–170 °C with an accuracy of ± 0.0525%.

8.4 Case Study

Figure 8.1 shows the effect of salinity on the accuracy of different correlations and models for the vapor pressure of saline water. These include the Shiah and Tseng model (1994), Cisternas and Lam (1989), Jamieson et al. (1969), Chou (1971), Kientzler and Arons (1966), Nayar et al. (2016), Millero (1983) and Robinson (1954) with the experimental data of Chou (1968) at 25 °C (see Appendix 5 for complete data set). It is obvious from Fig. 8.1a that the Cisternas and Lam model (1989), Jamieson et al. (1969), Chou (1971), Kientzler and Arons (1966), and Nayar et al. (2016) correlations have values close to the experimental data (1968) while the Shiah and Tseng model (1994), Robinson (1954) and Millero (1983) correlations have a different trend. The difference from the experimental data is shown in Fig. 8.1b. It can be noticed that Jamieson et al. (1969), Kientzler and Arons (1966), Chou (1971), and Nayar et al. (2016) correlations accurately predict the vapor pressure values within ± 1% difference for the investigated salinity range (up to 200 g/kg). Also, the Cisternas and Lam (1989) and Shiah and Tseng (1994) models have a difference of ± 5% up to a salinity of 150 and 75 g/kg, respectively.

The effect of temperature on the vapor pressure depends on the applied pressure. The vapor pressure increases slightly with the increase in temperature up to 100 °C and then increases sharply when T > 100 °C. When the vapor pressure is 1 bar (0.1 MPa), the vaporization starts at 100 °C. The results of different methods, including Robinson (1954), Millero (1983), Jamieson et al. (1969), Chou (1971), Nayar et al. (2016), and Kientzler and Arons (1966) at 58.5 g/kg are plotted in Fig. 8.2a against the experimental data of Kientzler and Arons (1966). It can be seen from the figure that the results of Robinson (1954), Millero (1983), Jamieson et al. (1969), Chou (1971), Nayar et al. (2016), and Kientzler and Arons (1966) overlap along the temperature range of 0–100 °C.

Figure 8.2b shows the difference in these correlations compared to the experimental data of Kientzler and Arons (1966). Except for the Robinson (1954) correlation, the other correlations can accurately predict the vapor pressure within a difference of ± 1% for the temperature range of 20–100 °C. Jamieson et al. (1969) and Nayar et al. (2016) correlations have the most accurate results.

Based on the above discussion, some methods and correlations could be recommended to be used, as summarized in Table 8.11. Vapor pressure is a very important aspect of thermal-driven desalination technologies. As saline water may have various compositions and types based on the source conditions, therefore, an adaptable model can accommodate multicomponent composition. Very few studies can fulfill these needs, although partially. It is therefore recommended to have more studies for the development of universal models for vapor pressure estimation that are supported by experimental investigations.

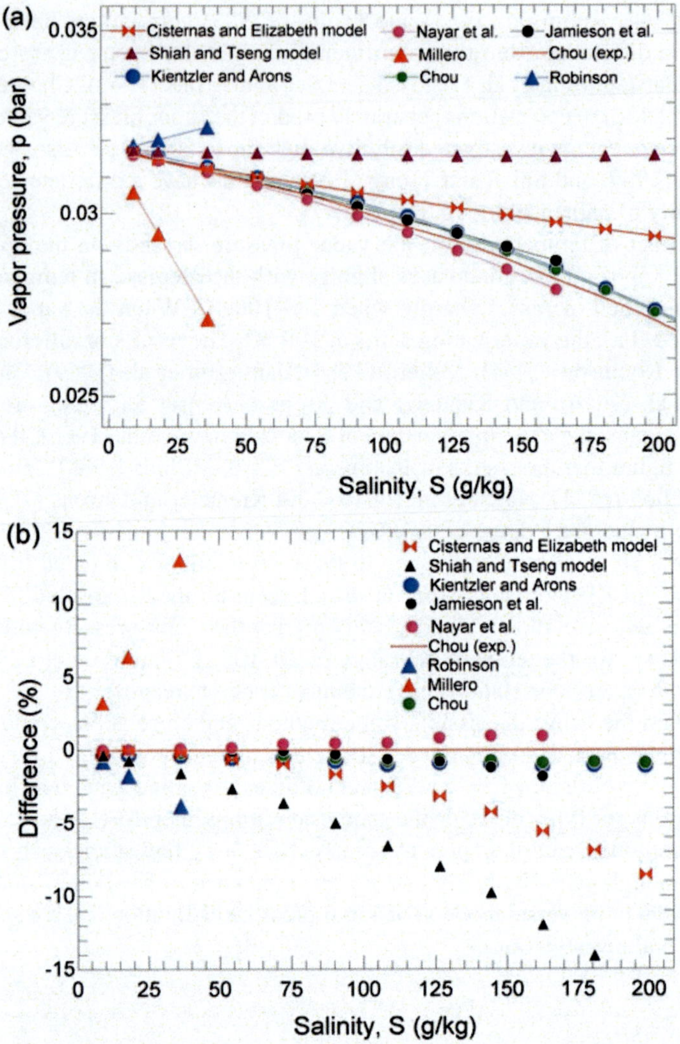

Fig. 8.1 **a** Salinity versus vapor pressure and **b** the difference against the experimental data of Chou (1968) for the methods of Chou (1968) for the methods of Shiah and Tseng model (1994), Cisternas and Elizabeth model (1989), Emerson and Jamieson (1967), Chou (1971), Kientzler and Arons (1966), Millero (1983), Robinson (1954) and Nayar et al. (2016) at 25 °C

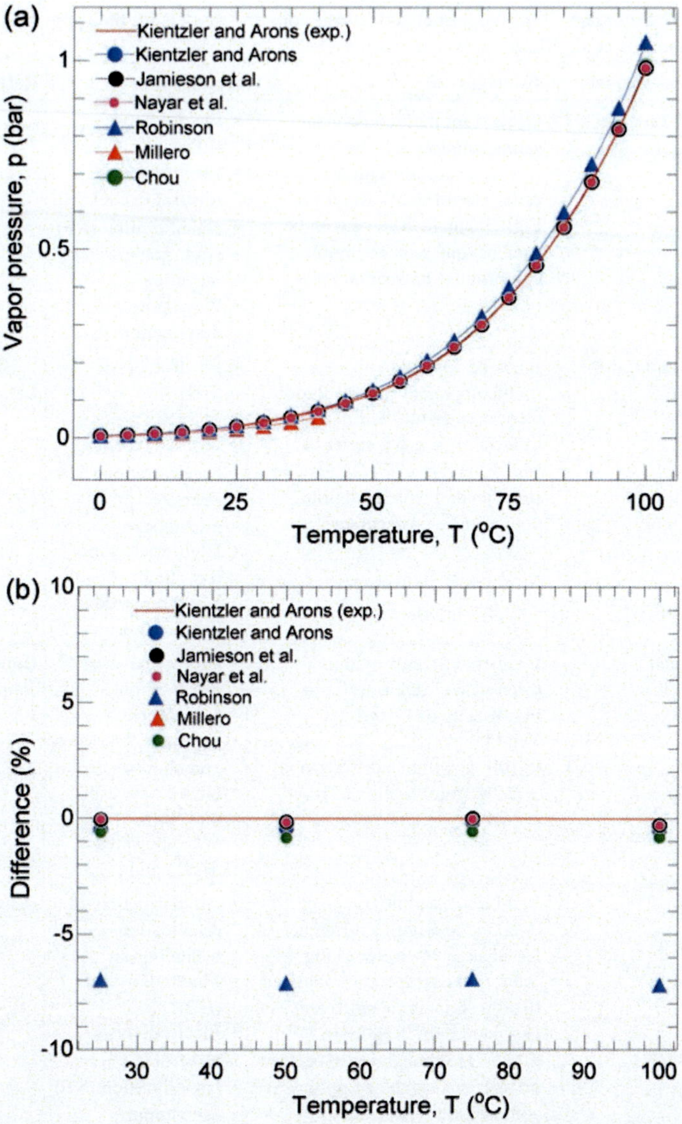

Fig. 8.2 **a** Temperature versus vapor pressure and **b** the difference against the experimental data of Kientzler and Arons (1966) for the methods of Robinson (1954), Millero (1983), Kientzler and Arons (1966), Emerson and Jamieson (1967), Chou (1971) and Nayar et al. (2016) at 58.5 g/kg

Table 8.11 Summary of the recommended correlations and models for the vapor pressure of saline water (Generous et al. 2020)

Correlation—model	Advantages	Applications	References
Shiah and Tseng model [Eq. (8.2)]	• General for multicomponent saline solution • Generates results within a difference of ± 5% (refer to Fig. 8.1) up to 90 g/kg (can be used beyond with negligible compromise on accuracy)	• Electrodialysis (ED) • Electrodialysis reversal (EDR) • Nanofiltration (NF) • High-grade water recycling • Wave-powered desalination	(Shiah and Tseng 1994)
Cisternas and Lam model [Eq. (8.3)]	• It can be applied to a multicomponent saline solution • Generates results within a difference of ± 5% (refer to Fig. 8.1) up to 170 g/kg (can be used beyond with negligible compromise on accuracy)	• Electrodialysis (ED) • Electrodialysis reversal (EDR) • Nanofiltration (NF) • Seawater greenhouse • High-grade water recycling • Wave-powered desalination	(Cisternas and Lam 1989)
Emerson and Jamieson [Eq. (8.12)]	• It generates results within a difference of ± 2% (refer to Fig. 8.1) up to 170 g/kg • Valid in the temperature range of 0–100 °C within a difference of ± 0.2% (refer to Fig. 8.2)	• All thermal-based desalination technologies • Seawater greenhouse • High-grade water recycling • Wave-powered desalination	(Emerson and Jamieson 1967)
Chou [Eq. (8.7)]	• Valid in the salinity range of 0–200 g/kg within a difference of less than ± 1% (refer to Fig. 8.1) • Valid in the temperature range of (0–100 °C) within a difference of less than ± 0.7% (refer to Fig. 8.2) (although it can be used up to 100 °C with negligible compromise on accuracy)	• All thermal-based desalination technologies • Electrodialysis (ED) • Electrodialysis reversal (EDR) • Nanofiltration (NF) • Membrane distillation (MD) • Seawater greenhouse • High-grade water recycling • Wave-powered desalination	(Chou 1971)

(continued)

Table 8.11 (continued)

Correlation—model	Advantages	Applications	References
Kientzler and Arons [Eq. (8.9)]	• Valid in a salinity range of 0–200 g/kg within a difference of ± 1% (refer to Fig. 8.1) • Valid in a temperature range of 0–100 °C within a difference of ± 0.5% (refer to Fig. 8.2) (although it can be used up to 100 °C with negligible compromise on accuracy)	• All thermal-based desalination technologies • Electrodialysis (ED) • Electrodialysis reversal (EDR) • Nanofiltration (NF) • Membrane distillation (MD) • Seawater greenhouse • High-grade water recycling • Wave-powered desalination	(Kientzler and Arons 1966)
Nayar et al. [Eq. (8.13)]	• Valid in a salinity range of 0 to 160 g/kg within a difference of less than ± 1% (refer to Fig. 8.1) • Valid in the temperature range of 0–100 °C within a difference of ± 0.2% (refer to Fig. 8.2) (although it can be used up to 100 °C with negligible compromise on accuracy)	• All thermal-based desalination technologies • Seawater greenhouse • High-grade water recycling • Wave-powered desalination	(Nayar et al. 2016)

8.5 EES Codes

The EES codes for the above correlations are:

Shiah and Tseng Model for Multicomponent Solution

"osmotic coefficient"

```
m_s = 0.6 "mol/kg"
m_w = 18.05 "g/mol"

phi = (1/10^0.1924) * (m_s^0.5/(1 + 2.798 * m_s^0.5)) − 0.01019 * m_s −
0.001201 * m_s^2 − 0.001324 * m_s^3

P_o = 3173.072 "Pascal"

P = P_o * exp(− phi * m_s * m_w/1000)

P_bars = P/10^5
```

Cisterns and Lam Model

T = 298.5 "K"

log10(P) = K * I * (A – B/(T – E_s)) + (C – D/(T – E_s))

A = a_s + 3.60591/10^4 * I + M_s/2303
B = b_s + 1.382982 – 0.031185 * I^2
C = c_s – 3.99334/10^3 * I – 1.11614/10^4 * I^2 + M_s * I * (1 – x)/2303
D = d_s – 0.138481 * I + 0.027511 * I^2 – 1.79277/10^3 * I^3

M_s = 18 "g/mol"

"For water as solvent"

a_s = – 0.021302
b_s = – 5.390915
c_s = 7.192959
d_s = 1730.2857
E_s = 39.53

"Let, for example, for NaCl."

K = 0.39015

x = 2 * (v_c + v_a)/(v_c * Z[1]^2 + v_a * Z[2]^2)

v_c = 1
v_a = 1
Z[1] = 1
z[2] = 1

I = 0.5 * sum(m[i] * z[i]^2, i = 1, 2)

m[1] = 0.469 "mol/kg"
m[2] = 0.546 "mol/kg"

"Let us convert the unit to bar."

P_bars = P/100

Chou Correlation

"Temperature in Kelvin"

T = 298.15 "K"

```
S = 35 "g/kg"

S = 1.80655 * X

"X is chlorinity."

P_w = exp(a1 + (a2/(a3 + T)))

a1 = 11.809452
a2 = – 3902.2336
a3 = – 42.407613

P = exp((1 + a * X + b * X^1.5) * ln(P_w) + (c * X + d * X^1.5 + e * X^2))

a = 0.54675794/10^4
b = – 0.15410306/10^5
c = – 0.99478110/10^3
d = 0.11895489/10^3
e = – 0.13716509/10^4
```

Arons and Kientzler Correlation

```
"Temperature in Kelvin"

T = 298.15 "K"
S = 35 "g/kg"

S = 1.80655 * X

"X is chlorinity"

P_w = exp(a1 + (a2/(a3 + T)))

a1 = 11.809452
a2 = – 3902.2336
a3 = – 42.407613

P = P_w * exp(a * X + b * X^1.5 + c * X^2 + ((d * X + e * X^1.5 + f * X^2)/
T) + (g * X + h * X^1.5 + i * X^2) * ln(T))

"Constants"

a = – 0.051807014
b = – 0.66809897/10^2
c = 0.68897827/10^3
d = 2.153011
```

e = 0.38508979
f = − 0.037231985
g = 0.75964905/10^2
h = 0.97136709/10^3
i = − 0.10158384/10^3

Emerson and Jamieson Correlation

S = 35 "g/kg"
T = 298.15 "K"
log10(P/P_w) = − 2.1609/10^4 * S − 3.5012/10^7 * S^2

"bars"

P_bars = P/100

"Pure water vapor pressure in kPa"

P_w = 10^2 * **exp**(71.023834 − (7380.4001/T) − (9.0977349 * **ln**(T)) + (0.0070871862 * T))

Nayar et al. Correlation

T = 298.15 "C"
S = 35 "g/kg"

P = P_w * **exp**(− 4.58180/10^4 * S − 2.04430/10^6 * S^2)

P_w = exp(a_3/T + a_4 + a_5 * T + a_6 * T^2 + a_7 * T^3 + a_8 * **ln**(T))

a_3 = − 5800
a_4 = 1.3915
a_5 = − 4.8640/10^2
a_6 = 4.1765/10^5
a_7 = − 1.4452/10^8
a_8 = 6.5460

"Pressure in bars"

P_bars = P/10^5

Robinson Correlation Coupled with Chou Correlation for Pure Water (Chou 1971)

S = 35 "g/kg"
T = 298.15 "K"

"Pure water vapor pressure in kPa"

```
P_w = 10^2 * exp(71.023834 – (7380.4001/T) – (9.0977349 * ln(T)) + (0.0070871862 * T))

P/P_w = 1 + 0.57357 * (S/(1000 – S))
```

"bars"

```
P_bars = P/100
```

Millero Correlation

"Temperature in Kelvin"

```
T = 298.15 "K"
S = 35 "g/kg"

P = P_w + A * S + B * S^1.5

A = – 2.331/10^3 – 1.4799/10^4 * (T – 273.15) – 7.520/10^6 * (T – 273.15)^2 – 5.5185/10^8 * (T – 273.15)^3
B = – 1.1320/10^5 – 8.7086/10^6 * (T – 273.15) + 7.4936/10^7 * (T – 273.15)^2 – 2.6327/10^8 * (T – 273.15)^3
```

"Pure water vapor pressure in kPa"

```
P_w = 10^2 * exp(71.023834 – (7380.4001/T) – (9.0977349 * ln(T)) + (0.0070871862 * T))
```

"vapor pressure in bars"

```
P_bars = P/100
```

Pure Water Correlation (Dessouky and Ettouney)

```
T = 25 "C"
```

"Constansts"

```
f[1] = – 7.419242
```

```
f[2] = 0.29721
f[3] = – 0.1155286
f[4] = 0.008685635
f[5] = 0.001094098
f[6] = – 0.00439993
f[7] = 0.002520658
f[8] = – 0.000521868

T_c = 647.286 "K"
P_c = 22,089 "kPa"

ln(P/P_c) = (T_c/(T + 273.15) – 1) * sum(f[i] * (0.01 * (T + 273.15 –
338.15))^(i – 1), i = 1, 8)
```

"bars"

```
P_bars = P/100
```

Pure Water Correlation (Chou)

"Temperature in Kelvin"

```
T = 298.15 "K"

P_w = exp(71.023834 – (7380.4001/T) – (9.0977349 * ln(T)) + (0.0070871862
* T))
```

Pure Water Correlation (Antoine Equation)

"Temperature in Kelvin"

```
T = 298.15 "K"

P_w = exp(a1 + (a2/(a3 + T)))

a1 = 11.809452
a2 = – 3902.2336
a3 = – 42.407613
```

Appendix 1

See Tables 8.12 and 8.13.

Table 8.12 Vapor pressure (kPa) data used for the validation of Shiah and Tseng (1994) model

m	303.15 K	313.15 K	323.15 K	333.15 K	343.15 K
LiCl					
3.010	3.68	6.42	10.78	17.47	27.41
4.999	3.19	5.56	9.34	15.13	23.75
5.998	2.89	5.06	8.51	13.82	21.75
7.856	2.34	4.12	6.98	11.40	18.05
10.168	1.72	3.07	5.27	8.70	13.93
12.693	1.20	2.18	3.80	6.37	10.36
15.661	0.80	1.48	2.63	4.48	7.40
18.456	0.58	1.09	1.96	3.37	5.66
LiBr					
2.000	3.82	6.67	11.21	18.27	28.84
3.494	3.49	6.09	10.22	16.58	26.14
5.092	2.98	5.23	8.82	14.40	22.72
7.026	2.43	4.28	7.20	11.75	18.55
9.192	1.74	3.09	5.27	8.66	13.75
11.318	1.18	2.12	3.64	6.02	9.66
13.777	0.70	1.29	2.26	3.82	6.25
15.971	0.39	0.74	1.33	2.31	3.88
LiI					
0.800	4.11	7.16	11.98	19.34	30.26
2.508	3.69	6.44	10.82	17.54	27.53
3.798	3.38	5.90	9.90	15.99	25.12
4.987	3.03	5.29	8.88	14.38	22.57
7.493	2.27	3.95	6.61	10.67	16.69
10.126	1.36	2.41	4.06	6.64	10.49

Table 8.13 Vapor pressure (kPa) data of aqueous solutions used for the validation of Shiah and Tseng (1994) mode at various temperatures

m	303.15 K	313.15 K	323.15 K	333.15 K	343.15 K	m	313.15 K	313.15 K	313.15 K	313.15 K	313.15 K
NaBr											
2.003	3.94	6.85	11.45	18.49	28.92	4.989	3.38	5.87	9.78	15.75	24.62
2.994	3.73	6.5	10.89	17.61	27.56	5.801	3.2	5.55	9.27	14.95	23.34
4.012	3.55	6.18	10.32	16.67	26.06	7.981	2.6	4.59	7.78	12.69	20.15
NaI											
1.004	4.1	7.13	11.92	19.25	30.1	3.958	3.5	6.08	10.16	16.43	25.72
1.673	3.99	6.93	11.58	18.71	29.25	5.073	3.24	5.63	9.4	15.19	23.79
2.158	3.89	6.77	11.31	18.25	28.65	6.693	2.84	4.95	8.26	13.35	20.91
3.303	3.65	6.34	10.59	17.12	26.83	8.398	2.46	4.28	7.16	11.57	18.06
KCl											
1.275	4.1	7.13	11.89	19.21	30.13	2.917	3.88	6.71	11.23	18.12	28.41
2.01	3.99	6.92	11.58	18.69	29.33	3.687	3.8	6.56	10.93	17.6	27.57
2.555	3.93	6.8	11.36	18.32	28.71	4.286	3.68	6.37	10.63	17.18	26.97
KBr											
1.255	4.04	7.05	11.81	19.08	29.92	3.021	3.82	6.62	11.08	17.9	28.1
1.608	4.01	6.98	11.67	18.86	29.56	3.81	3.71	6.42	10.75	17.36	27.29
2.22	3.93	6.82	11.43	18.45	28.94	4.349	3.53	6.13	10.32	16.78	26.47
KI											
1.006	4.11	7.17	12.01	19.36	30.22	3.027	3.89	6.71	11.12	17.89	27.94
1.523	4.07	7.06	11.8	19.02	29.67	4.059	3.73	6.43	10.65	17.14	26.74
2.003	4.01	6.96	11.59	18.67	29.18	4.747	3.57	6.21	10.25	16.63	25.9

(continued)

Table 8.13 (continued)

m	303.15 K	313.15 K	323.15 K	333.15 K	343.15 K	m	313.15 K	313.15 K	313.15 K	313.15 K	313.15 K
2.351	3.97	6.86	11.44	18.42	28.78	5.648	3.42	5.95	9.87	15.86	24.75
RbCl											
1.493	4.04	7.03	11.76	19	29.73	4.463	3.62	6.32	10.59	17.09	26.67
2.968	3.83	6.67	11.17	18.02	28.19	5.966	3.37	5.92	9.96	16.08	25.11
3.55	3.74	6.54	10.94	17.66	27.63	6.949	3.15	5.58	9.41	15.25	23.82
CaCl											
1.774	3.95	6.88	11.55	18.65	29.26	5.492	3.53	6.13	10.19	16.42	25.58
2.471	3.85	6.72	11.3	18.27	28.66	6.444	3.41	5.91	9.84	15.83	24.63
3.522	3.75	6.53	10.92	17.65	27.58	7.435	3.24	5.66	9.41	15.14	23.57
4.402	3.66	6.35	10.63	17.12	26.72	8.59	2.91	5.18	8.8	14.27	22.37
CaBr											
1.57	4.02	7	11.75	19.01	29.62	3.485	3.83	6.62	11.03	17.84	27.85
2.244	3.95	6.86	11.47	18.54	28.98	4.32	3.75	6.48	10.77	17.35	27.09
2.768	3.91	6.77	11.3	18.23	28.52	5.888	3.59	6.22	10.32	16.57	25.88
CaI											
1.345	4.08	7.07	11.82	19.12	29.86	2.014	3.99	6.91	11.56	18.67	29.25
1.546	4.05	7.01	11.73	18.98	29.68	2.595	3.87	6.8	11.35	18.33	28.62
1.799	4.02	6.95	11.62	18.79	29.39						
$MgCl_2$											
1.047	3.98	6.94	11.63	18.8	29.45	3.167	2.98	5.22	8.81	14.32	22.51
1.303	3.85	6.74	11.31	18.35	28.8	3.737	2.74	4.8	8.07	13.08	20.59

(continued)

Table 8.13 (continued)

m	303.15 K	313.15 K	323.15 K	333.15 K	343.15 K	m	313.15 K	313.15 K	313.15 K	313.15 K	313.15 K
2.006	3.6	6.29	10.57	17.17	26.91	4.343	2.32	4.09	6.93	11.33	17.91
2.449	3.33	5.86	9.87	16.04	25.33	4.801	1.99	3.52	6.04	9.93	15.77
$CaCl_2$											
1.002	3.99	6.94	11.61	18.77	29.39	4.951	2.15	3.82	6.52	10.75	17.18
2.249	3.58	6.22	10.43	16.87	26.46	6.019	1.7	3.06	5.27	8.77	14.15
3.017	3.18	5.55	9.34	15.17	23.89	6.786	1.45	2.63	4.56	7.66	12.41
3.887	2.7	4.74	8.03	13.12	20.77	7.885	1.2	2.2	3.85	6.5	10.61
$CaBr_2$											
1.006	3.95	6.92	11.59	18.74	29.36	3.022	3.03	5.31	8.96	14.59	23.04
1.238	3.87	6.76	11.36	18.37	28.84	3.376	2.83	4.97	8.4	13.69	21.64
1.709	3.7	6.46	10.83	17.54	27.6	3.949	2.51	4.43	7.51	12.26	19.4
2.254	3.44	6.02	10.11	16.48	25.93	4.596	2.17	3.84	6.57	10.75	17.05
CaI_2											
1.138	3.94	6.91	11.59	18.72	29.34	2.273	3.58	6.24	10.45	16.88	26.41
1.464	3.84	6.72	11.25	18.2	28.47	2.915	3.22	5.63	9.45	15.32	24.06
1.875	3.69	6.46	10.83	17.5	27.46						
$SrCl_2$											
0.662	4.1	7.13	11.92	19.25	30.09	1.797	3.72	6.49	10.88	17.59	27.61
0.871	4.08	7.08	11.84	19.12	29.82	2.213	3.59	6.26	10.49	16.98	26.64
1.299	3.92	6.87	11.4	18.36	28.87	2.693	3.39	5.91	9.93	16.11	25.28
1.503	3.81	6.65	11.19	18.13	28.44	3.203	3.07	5.41	9.12	14.89	23.48

(continued)

Table 8.13 (continued)

m	303.15 K	313.15 K	323.15 K	333.15 K	343.15 K	m	313.15 K	313.15 K	313.15 K	313.15 K	313.15 K
$SrBr_2$											
0.8	3.97	6.98	11.74	18.99	29.71	1.345	3.88	6.78	11.31	18.27	28.51
1.013	3.95	6.92	11.59	18.76	29.26	1.746	3.73	6.5	10.88	17.55	27.43
1.996	3.59	6.29	10.5	17.05	26.66	2.656	3.27	5.7	9.59	15.52	24.31
2.395	3.44	6	10.04	16.25	25.4	3.34	2.91	5.11	8.58	13.93	21.89
SrI_2											
0.97	3.97	6.91	11.57	18.72	29.31	2.37	3.37	5.89	9.89	16.02	25.09
1.264	3.86	6.73	11.29	18.27	28.63	2.933	3.11	5.42	9.08	14.69	22.95
1.534	3.77	6.57	11	17.79	27.84	3.517	2.76	4.83	8.1	13.1	20.5
1.96	3.58	6.25	10.49	16.98	26.58	4.156	2.37	4.19	7.08	11.54	18.13
$BaCl_2$											
0.504	4.07	7.15	12.03	19.44	30.44	1.214	3.9	6.85	11.53	18.67	29.27
0.738	4.03	7.07	11.88	19.23	30.1	1.388	3.85	6.77	11.42	18.54	29.05
0.975	3.98	6.97	11.72	18.97	29.68						
$BaBr_2$											
1.001	3.98	6.96	11.69	18.88	29.54	1.888	3.7	6.48	10.85	17.54	27.43
1.265	3.91	6.84	11.47	18.55	28.95	2.245	3.59	6.29	10.54	17.01	26.61
1.445	3.85	6.73	11.31	18.27	28.53	2.503	3.5	6.14	10.26	16.53	25.9
1.706	3.76	6.59	11.05	17.89	27.95	3.398	3.11	5.58	9.23	14.98	25.56

Table 8.14 Vapor pressure lowering (in mm Hg) tabulated data by Arons and Kientzler (1966)

Chlorinity (‰)	Temperature (°C)									
	35°	30°	25°	20°	15°	10°	5°	0°	– 5°	– 10°
5	0.18	0.13	0.10	0.08	0.06	0.04	0.03	0.02	–	–
10	0.35	0.27	0.20	0.15	0.11	0.08	0.06	0.04	–	–
20	0.74	0.56	0.43	0.32	0.23	0.16	0.12	0.08	–	–
30	1.20	0.90	0.68	0.49	0.36	0.26	0.18	0.12	–	–
40	1.67	1.25	0.92	0.67	0.49	0.35	0.24	0.17	–	–
50	2.24	1.67	1.24	0.90	0.65	0.45	0.31	0.22	0.15	–
60	2.87	2.14	1.58	1.14	0.81	0.57	0.39	0.27	0.18	–
70	3.45	2.57	1.88	1.36	0.96	0.68	0.47	0.32	0.22	–
80	4.08	3.04	2.25	1.62	1.16	0.81	0.56	0.38	0.26	–
90	4.87	3.65	2.67	1.92	1.37	0.95	0.66	0.44	0.30	0.2
100	5.64	4.17	3.07	2.20	1.57	1.10	0.76	0.52	0.35	0.2
110	6.50	4.81	3.52	2.55	1.82	1.27	0.87	0.59	0.40	0.3
120	7.29	5.44	4.03	2.91	2.07	1.44	1.00	0.68	0.46	0.3
130	8.17	6.09	4.50	3.28	2.36	1.67	1.16	0.79	0.54	0.4
140	8.89	6.71	4.99	3.63	2.62	1.85	1.28	0.88	0.60	0.4
150	9.93	7.42	5.52	4.01	2.93	2.08	1.47	1.02	0.69	0.5
160	10.73	8.20	6.14	4.49	3.27	2.33	1.65	1.14	0.79	0.5

*Values are based on data shown in Table 8.15

Appendix 2

See Tables 8.14, 8.15, 8.16 and 8.17.

Appendix 3

See Table 8.18.

Appendix 4

See Table 8.19.

Table 8.15 Original vapor pressure lowering (in mm Hg) data used to obtain Table 8.14

Corrected temperature (°C)	Observed water—Ice* V.P. p_o	Observed sample V.P. p	Observed differential $p_o - p$	Average corrected $p_o - p$
Sample #2	Cl (‰) = 35			
20	17.57	17.05	0.49	0.5
14.98	12.79	12.41	0.4	0.39
19.99	17.57	16.79	0.83	0.8
19.99	17.55	16.83	0.7	0.71
20.01	17.56	17.06	0.51	0.5
24.99	23.78	22.89	0.86	0.88
20.01	17.57	17.02	0.53	0.54
24.99	23.79	22.8	0.95	0.97
15	12.81	12.46	0.36	0.36
10.01	9.23	8.98	0.25	0.25
4.98	6.53	6.39	0.14	0.14
0.49	4.75	4.66	0.03	0.08
15.01	12.83	12.21	0.57	0.6
9.96	9.2	8.98	0.25	0.24
Sample #6	Cl (‰) = 93.6			
20.01	17.59	15.44	2.08	2.12
15.02	12.8	11.41	1.39	1.39
9.99	9.19	8.24	0.97	0.96
4.97	6.53	5.88	0.64	0.64
0.47	4.75	4.3	0.44	0.44
− 8.02	2.32*	2.26	0.05	0.24
− 5.04	3.01*	2.79	0.2	0.36
− 2.5	3.73*	3.41	0.31	0.4
− 5.01	3.01*	2.89	0.13	0.28
25	23.78	20.93	2.86	2.86
Sample #4	Cl (‰) = 68.5			
20	17.54	16.43	1.1	1.1
14.99	12.79	12.02	0.77	0.77
9.99	9.2	8.54	0.63	0.64
14.98	12.78	11.99	0.79	0.79
9.98	9.2	8.6	0.6	0.6
4.99	6.54	6.08	0.45	0.46
0.48	4.76	4.42	0.33	0.34
− 3.99	3.29*	3.17	0.1	0.24

(continued)

Table 8.15 (continued)

Corrected temperature (°C)	Observed water—Ice* V.P. p_o	Observed sample V.P. p	Observed differential $p_o - p$	Average corrected $p_o - p$
19.98	17.51	16.18	1.32	1.32
25.02	23.82	22.02	1.82	1.81
15.01	12.82	12.01	0.86	0.84
15	12.81	12.01	0.81	0.8
19.99	17.54	16.21	1.31	1.32
14.99	12.79	12	0.83	0.81
12.51	10.91	10.15	0.75	0.76
17.49	15	13.94	1.11	1.08
Sample #8	Cl (‰) = 123.1			
20.01	17.56	14.63	2.92	2.92
15	12.81	10.67	2.13	2.14
10.01	9.22	7.79	1.43	1.43
4.98	6.55	5.48	1.06	1.06
0.47	4.77	4.05	0.71	0.72
− 9.99	1.97*	1.96	0	0.2
− 7.49	2.44*	2.21	0.22	0.4
− 4.97	3.03*	2.67	0.34	0.5
20	17.56	14.35	3.18	3.2
24.99	23.73	19.51	4.19	4.2
20.01	17.56	14.56	2.97	2.98
20	17.56	14.53	3.04	3.04
15	12.79	10.59	2.2	2.2
Sample #10	Cl (‰) = 152.3			
24.98	23.75	18.09	5.64	5.65
19.99	17.52	13.39	4.17	4.15
14.99	12.79	9.81	2.99	2.98
10	9.22	7.12	2.11	2.11
4.98	6.54	5.07	1.47	1.47
0.49	4.76	3.68	1.08	1.08
− 5	3.03*	2.46	0.58	0.72
− 9.99	1.95*	1.69	0.26	0.46
− 15.02	1.25*	1.19	0.05	0.25
Sample #1	Cl (‰) = 17.4			
20	17.54	17.28	0.27	0.26

(continued)

Table 8.15 (continued)

Corrected temperature (°C)	Observed water—Ice* V.P. p_o	Observed sample V.P. p	Observed differential $p_o - p$	Average corrected $p_o - p$
20.01	17.55	17.28	0.27	0.27
14.98	12.78	12.58	0.2	0.2
0.48	4.76	4.69	0.07	0.07
4.98	6.55	6.44	0.11	0.11
9.97	9.19	9.03	0.15	0.16
25	23.75	23.38	0.37	0.37
Sample #10-A	Cl (‰) = 144.1			
20	17.56	13.67	3.87	3.88
15.02	12.87	10.02	2.82	2.84
15.02	12.8	10.02	2.82	2.8
19.99	17.5	13.66	3.9	3.82
10.01	9.24	7.32	1.94	1.96
− 14.98	1.27*	1.22	0.03	2.24
− 10.07	1.96*	1.76	0.18	0.39
− 5.01	3*	2.57	0.42	0.58
0.48	4.76	3.72	1.05	1.04
4.98	6.53	5.17	1.38	1.37
− 4.97	–	2.55	–	0.62
25	23.79	18.3	5.5	5.5
30	–	–	7.4	7.4
− 2.49	–	3.01	–	0.81
Sample #2-A	Cl (‰) = 30.5			
20.01	17.57	17.07	0.48	0.49
24.99	23.76	23.05	0.68	0.7
29.99	–	–	0.95	0.95
14.94	12.74	12.39	0.36	0.36
10.03	9.22	8.96	0.26	0.26
4.99	6.54	6.34	0.18	0.19
0.53	4.76	4.63	0.13	0.13
15	12.8	12.4	0.38	0.39
20.02	17.56	17.05	0.51	0.51
Sample #4-A	Cl (‰) = 60.2			
24.99	23.77	22.31	1.44	1.45

(continued)

Table 8.15 (continued)

Corrected temperature (°C)	Observed water—Ice* V.P. p_o	Observed sample V.P. p	Observed differential p_o − p	Average corrected p_o − p
20.03	17.57	16.42	1.12	1.14
15.01	12.79	11.95	0.81	0.82
10.04	9.25	8.73	0.53	0.52
5.05	6.55	6.12	0.42	0.42
− 4.96	3.06*	3.03	0.03	0.18
0.45	4.75	4.48	0.28	0.28
30	–	–	2.15	2.15
24.98	23.76	22.23	1.51	1.52
Sample #3	Cl (‰) = 48.1			
19.98	17.49	16.59	0.88	0.89
14.90	12.77	12.10	0.65	0.66
10.03	9.33	8.84	0.47	0.48
5.10	6.57	6.24	0.38	0.32
0.59	4.74	4.50	0.24	0.24
24.98	23.69	22.34	1.31	1.33
29.96	–	–	1.79	1.79
Sample #1/2	Cl (‰) = 10.3			
24.97	23.73	23.55	0.23	0.21
19.79	17.29	17.15	0.15	0.15
15.28	12.98	12.86	0.14	0.13
9.98	9.17	9.09	0.08	0.08

Table 8.16 Vapor pressure lowering (in mm Hg) more tabulated data (Kientzler and Arons 1966)

Chlorinity (‰)	Temperature (°C)							
	175°	150°	125°	100°	75°	50°	25°	0°
10	33	18.5	9.7	4.75	1.99	0.74	0.2	0.04
20	72.6	41	21.1	10.2	4.4	1.54	0.43	0.08
30	118	65.8	34	16.4	7.08	2.5	0.68	0.12
40	168	94.5	49.3	23.9	9.99	3.58	0.92	0.17
50	227	128	66.5	32.1	13.2	4.75	1.24	0.22
60	298	168	86.2	41	16.6	5.96	1.58	0.27
70	374	213	108	52	21.1	7.41	1.88	0.32
80	457	257	132	62.2	25.1	8.85	2.25	0.38
90	551	308	157	74.7	30.1	10.3	2.67	0.44
100	640	363	184	87.4	34.8	11.7	3.07	0.52

(continued)

Table 8.16 (continued)

Chlorinity (‰)	Temperature (°C)							
	175°	150°	125°	100°	75°	50°	25°	0°
110	749	425	217	103	41.1	13.6	3.52	0.59
120	868	486	248	117	46.3	15.2	4.03	0.68
130	1000	568	287	135	53.7	17.2	4.5	0.79
140	1115	630	317	151	58.4	18.8	4.99	0.88
150	1260	722	365	170	65.4	20.9	5.52	1.02
160	1440	822	408	188	72.8	22.8	6.14	1.14

*Values are based on data shown in Table 8.17

Table 8.17 Original vapor pressure lowering (in mm Hg) data used to obtain Table 8.16

Chlorinity (‰)	Temperature (°C)	Vapor pressure lowering (mm Hg)
19.4	24.97	0.6
	49.86	1.4
	75.65	2.8
	99.19	11.2
	126.14	12.9
	150.26	34.6
19.87	24.17	0.7
	50.18	1.5
	75.03	5.4
	99.33	14.7
	125.09	25.8
	150.34	40.3
	175.22	72
53.04	24.85	1.1
	74.09	15.5
	100.02	40
	124.16	76.3
	149.24	162.8
	176.41	315
104.24	24.24	2.1
	50.15	9.9
	75.23	35.1
	124.67	206
	150.3	407

(continued)

Table 8.17 (continued)

Chlorinity (‰)	Temperature (°C)	Vapor pressure lowering (mm Hg)
119.16	75	46.1
	100.16	122.4
	149.06	474
151.82	25.66	5.2
	50	21.3
	74.94	62.6
	99.7	160.7
	125.99	386
	148.27	699
	176.09	1360
154.32	24.94	3.9
	49.67	20.8
	74.8	62.4
	100.05	169.1
	125.97	396
	148.36	739
	172.45	1295
155.69	25.14	5.1
	49.72	21.7
	75.64	69.6
	99.95	175.9
	126.22	412
	147.5	753

Table 8.18 Vapor pressure data used for the model of Emerson and Jamieson (1967)

Sample #	Salinity (g/kg)	Temperature (°C)	Pressure (10^5 Pa)
1	33.13	100.392	1.0098
	33.21	119.686	1.9317
	33.32	136.705	3.2318
	33.56	159.998	6.0698
	33.88	180.315	9.9134
2	33.13	99.887	0.9922
	33.21	120.155	1.9608
	33.32	136.542	3.2162

(continued)

Table 8.18 (continued)

Sample #	Salinity (g/kg)	Temperature (°C)	Pressure (10^5 Pa)
	33.54	159.808	6.041
	33.84	177.948	9.385
3	66.26	101.97	1.0484
	66.41	119.422	1.8806
	66.66	137.635	3.257
	67.09	158.705	5.7608
	61.72	179.031	9.4414
4	115.97	101.776	1.0074
	116.24	119.748	1.8385
	116.66	138.059	3.1907
	117.49	159.583	5.7069
	118.69	180.531	9.4616
5	116.27	121.25	1.9289
	116.67	138.354	3.2181
	117.49	160.463	5.8357
	118.69	181.24	9.617
6	165.68	102.8	1.0059
	166.03	119.03	1.7322
	166.68	137.52	3.0285
	167.7	158.33	5.3304
	169.27	178.76	8.7678

Table 8.19 Vapor pressure data used for the model of Millero (1983)

Temperature (°C)	Pressure (mm Hg)
0	4.4959
5	6.4196
10	9.0365
15	12.5508
20	17.2118
25	23.3225
30	31.2453
35	41.412
40	54.3286

Table 8.20 Vapor pressure experimental data from Chou (1968)

Molality	Temperature (°C)	Pressure (mm Hg)
0.1	25	23.7
0.1	50	92.2
0.1	75	288.2
0.1	100	757.5
0.1	125	1734.8
0.1	150	3557.6
1.0	25	23
1.0	50	89.5
1.0	75	280
1.0	100	734.6
1.0	125	1682.4
1.0	150	3450.1
2.0	25	22.1
2.0	50	86.1
2.0	75	269.1
2.0	100	707.3
2.0	125	1619.9
2.0	150	3322
2.5	25	21.7
2.5	50	84.4
2.5	75	263.8
2.5	100	693.4
2.5	125	1588
2.5	150	3256.7

Appendix 5

See Table 8.20.

References

Chou JCS (1968) Thermodynamic properties of aqueous sodium chloride solutions from 32 to 350 °F. Georgia Institute of Technology

Chou JCS (1971) Thermal properties of sea water. Technical report 49. Hawaii University, Water Resources Research Center

Cisternas LA, Lam EJ (1989) A analytic correlation of vapour pressure of aqueous and non-aqueous solutions of single and mixed electrolytes. Fluid Phase Equilib 53(C):243–249. https://doi.org/10.1016/0378-3812(89)80092-5

El-Dessouky HT, Ettouney HM (2002) Fundamentals of salt water desalination. Fundam Salt Water Desalin. https://doi.org/10.1016/b978-0-444-50810-2.x5000-3

Emerson WH, Jamieson DT (1967) Some physical properties of sea water in various concentrations. Desalination 3(2):213–224. https://doi.org/10.1016/0011-9164(67)80012-2

Generous MM, Qasem NAA, Qureshi BA, Zubair SM (2020) A comprehensive review of saline water correlations and data-part I: thermodynamic properties. Arab J Sci Eng 45(11):8817–8876. https://doi.org/10.1007/s13369-020-05019-y

Han Z (2012) An equation of state for nonaqueous electrolyte solutions. J Chem Pharm Res 4(8):3861–3864

Jamieson DT, Tudhope JS, Morris R, Cartwright G (1969) Physical properties of sea water solutions: heat capacity. Desalination 7(1):23–30. https://doi.org/10.1016/S0011-9164(00)80271-4

Kientzler CF, Arons AB (1966) Vapor pressure of sea salt solutions. Technical report no. 2. https://doi.org/10.1001/jamadermatol.2014.3593

Millero FJ (1983) The thermodynamics of seawater part II: thermochemical properties. Ocean Sci Eng 8(1):1–40

Morillon V, Debeaufort F, Jose J, Tharrault JF, Capelle M, Blond G, Voilley A (1999) Water vapour pressure above saturated salt solutions at low temperatures. Fluid Phase Equilib 155(2):297–309. https://doi.org/10.1016/S0378-3812(99)00009-6

Nayar KG, Sharqawy MH, Banchik LD, Lienhard V, JH (2016) Thermophysical properties of seawater: a review and new correlations that include pressure dependence. Desalination 390:1–24. https://doi.org/10.1016/j.desal.2016.02.024

Robinson RA (1954) The vapour pressure and osmotic equivalence of sea water. J Mar Biol Assoc UK 33(2):449–455. https://doi.org/10.1017/S0025315400008468

Rodgers RC, Hill GE (1978) Equations for vapour pressure versus temperature: derivation and use of the antoine equation on a hand-held programmable calculator. Br J Anaesth 50(5):415–424. https://doi.org/10.1093/bja/50.5.415

Sharqawy MH, Lienhard V, JH, Zubair SM (2011) Thermophysical properties of seawater: a review of existing correlations and data. Desalin Water Treat 29(1–3):355–355. https://doi.org/10.5004/dwt.2011.2947

Shiah IM, Tseng HC (1994) A vapor pressure model for aqueous electrolyte solutions based on mean spherical approximation. Fluid Phase Equilib 99(C):75–85. https://doi.org/10.1016/0378-3812(94)80023-5

Zuo Y-X, Fürst W (2002) Prediction of vapor pressure for nonaqueous electrolyte solutions using an electrolyte equation of state. Fluid Phase Equilib 138(1–2):87–104. https://doi.org/10.1016/s0378-3812(97)00145-3

Chapter 9
Viscosity

Abstract The viscosity of saline water is a crucial parameter for fluid flow systems, including desalination. Heat transfer is also a boundary layer phenomenon strongly affected by viscosity. Viscosity is a thermophysical property having an important role in determining the transport properties of water. This chapter explains the thermodynamic approaches, multicomponent models, correlation, and data used to estimate viscosity at various salinity, temperature, and pressure values. Farther, the calculated viscosity values from the correlations and methods are assessed against the experimental data of seawater. Some correlations and methods are recommended based on their accuracy and operating conditions. Besides, the codes used for the investigated viscosity methods are presented.

Keywords Viscosity · Thermodynamics · Multicomponent · Correlations · Methods · Saline water

9.1 Introduction

The viscosity of liquid measures the deformation resistance at a given rate. It can be conceptualized as evaluating the inner frictional force that emerges between adjoining layers of liquid that are in relative movement. On occasion, when a thick liquid is constrained through a tube, it streams more rapidly close to the tube's center than close to its walls. The viscosity has a significant effect on the velocity, pressure, pumping power, heat transfer, and so on.

The viscosity of saline water is a key parameter for fluid flow systems, including those of desalination. Also, heat transfer is a boundary layer phenomenon and is strongly affected by viscosity. The viscosity is important in determining the transport properties of water (Swindellsr et al. 1952). Another motive is the use of salt solutions as micro-biological nutrients where viscosity and other properties appear to be prime factors (Pereira et al. 2001).

N. A. A. Qasem et al., *Thermodynamic and Thermophysical Properties of Saline Water*, Springer Water, https://doi.org/10.1007/978-3-031-35193-8_9

It was found that an ionic compound dissociates in water resulting in an increment in the density of saline water. This, in turn, results in an increment in its viscosity, which is called residual viscosity (Huber et al. 2009). The viscosity can be determined by experimental techniques such as the capillary flow method (Gupta 2014), viscometer (Chen et al. 1973), and torsional pendulum (Penn and Kearsley 1971).

The viscosity of saline water was widely studied, and a lot of correlations and models were reported in the literature. It is reported to be a function of temperature, pressure, and salinity (Marvin 1971). However, the effect of pressure is not significant, especially for temperatures more than 20 °C (Patil et al. 2014). Due to this fact, many reported correlations considered the viscosity as a function of temperature and salinity only. Nevertheless, Ghallab and Elnahas (2016) considered the effect of pressure on viscosity values.

Goldsack and Franchetto (1977) considered the salinity effect on viscosity variations. Korson et al. (1969) included the temperature effect in detail. Some use the correlation of NaCl solution to determine the viscosity of saline water, such as those of Numbere et al. (1977) and Simion et al. (2015). On the other hand, other correlations focused on multicomponent solutions, such as that of Lencka et al. (1998). They considered electrostatic interaction between charged bodies and interaction between neutral and charged bodies separately and used a structural making and breaking concept to come up with a model for the viscosity of multicomponent electrolyte solutions.

The effect of dissolved gases in saline water, especially CO_2 and O_2, was also investigated. Islam and Carlson (2012) have reported that CO_2 alone can contribute 2–3% variation in density and viscosity. Unfortunately, there is no available simulator that accommodates the effect of dissolved CO_2 on the viscosity variation of saline water. This is because the effect is often ignored. All the methods and correlations summarized in this chapter neglect the effect of dissolved carbon dioxide, oxygen, and other miscellaneous gases in the air.

9.2 Thermodynamic Approach

Viscosity is thermophysical property; so that can be measured directly under certain conditions. There are two types of viscosity: dynamic viscosity and kinematic viscosity. The latter can be calculated by dividing the dynamic viscosity by the solution density. The viscosity appears in the momentum equations since it causes shear stress that hinders the flow. The thermodynamic approach to calculating viscosity can be expressed by Spikers (1990):

$$\left(\frac{\partial \log(\mu)}{\partial T}\right)_v = \left(\frac{\partial \log(\mu)}{\partial T}\right)_P + \left(\frac{\partial \log(\mu)}{\partial P}\right)_T \cdot \left(\frac{\partial P}{\partial T}\right)_v \tag{9.1}$$

where μ is the viscosity, v is the specific volume, P is the pressure, and T is the temperature. The thermodynamic approaches for viscosity estimation are challenging and need a lot of data and details. Therefore, a lot of research was conducted to propose simpler models and correlations. The simplicity of these methods depends upon the applied conditions and the number of constituents in the solution. Anyway, the suggested models and correlations could accurately predict the saline water viscosity for the necessary applications of desalination and water treatment.

9.3 Multiple Components Models

Different saltwater types were studied to obtain their thermophysical properties. The obtained methods are the function of ionic strength, neglecting the other concentrations, such as organic and pollutant materials. Thus, the available methods in this section are valid for saltwater that consists of salt ions and water. Multicomponent models were proposed to be general for all types of saltwater under different salinity values. Major constituents of the aqueous ionic solutions are cations, anions, and water.

Fabuss et al. (1969) proposed a model for the viscosity of multicomponent aqueous solutions in terms of ionic strength, as in Eq. (9.2).

$$\log(\mu_R) = 0.0428\mathrm{I} + 0.00123\mathrm{I}^2 + 0.000131\mathrm{I}^3 + \left(-0.03724\mathrm{I} + 0.01859 - 0.00271\mathrm{I}^3\right)\mathrm{Log}(\mu_\mathrm{w}) \quad (9.2)$$

where μ_R is the relative viscosity in kg/m s, μ_w is the viscosity of pure water in kg/m s, and I is the ionic strength in mol/L. It can be determined by using the concept of molarity of ions as in Eq. (9.3).

$$\mathrm{I} = \sum_{\mathrm{i}=1}^{\mathrm{n}} \mathrm{M_i Z_i^2} \quad (9.3)$$

where $\mathrm{M_i}$ is the molarity of ions in the solution (in mol/L), and $\mathrm{Z_i}$ is the ionic charge (in C). The viscosity of pure water (μ_W) is linked with saline water viscosity by Eq. (9.4). For temperature (T) in $^\circ$C,

$$\mu_\mathrm{W} = 4.2844 \times 10^{-5} + \frac{1}{0.157(\mathrm{T} + 64.993)^2 - 91.296} \quad (9.4)$$

This model is valid for ionic strength of 0.7–3.5 mol/L, and temperature of 25–150 °C at an accuracy of ± 0.3%.

The viscosity of calcium-free saltwater at I = 0.70989 can be calculated from Eq. (9.5) (Fabuss et al. 1969).

$$\log(\mu_R) = 0.03103 - 0.01803 \log(\mu_W) \tag{9.5}$$

And the absolute viscosity of standard seawater can be approximated from Eq. (9.6).

$$\log(\mu_{SW}) = 0.03103 - 0.9817 \log(\mu_W) \tag{9.6}$$

The model can be used for ionic strength in the range of 0.7–2.1338 (tested range) at STP with a standard deviation of ± 0.28%.

9.4 Correlations and Data

Correlations are the straightforward method to estimate the viscosity by fitting the experimental data. Most of the available viscosity correlations in the literature are for seawater. Numbere et al. (1977) proposed a correlation for viscosity of seawater (μ) in centi-Poise (cP) in terms of temperature (T) in °F, salinity (S) in wt%, and pressure (P) in MPa as,

$$\frac{\mu}{\mu_W} - 1 = \left(-1.87 \times 10^{-3} S^{0.5}\right) + \left(2.18 \times 10^{-4} S^{2.5}\right) + \left(T^{0.5} - 1.35 \times 10^{-2} T\right)\left(2.76 \times 10^{-3} S - 3.4410^{-4} S^{1.5}\right) \tag{9.7}$$

where $\frac{\mu}{\mu_W}$ is the ratio of the viscosity of brine to the viscosity of water at the same temperature and pressure. μ_W for saturation pressure can be estimated as in Eq. (9.8).

$$\mu_w(T, P_s) = 241.4 \times 10^{\left(\frac{247.8}{T-140}\right)} \tag{9.8}$$

The weight (%) can be calculated from Eq. (9.9).

$$S\,(\text{wt\%}) = \frac{100\, MW_m}{1000 + MW_m} \tag{9.9}$$

Here, P_s is the saturation pressure that can be calculated in terms of temperature using Eq. (9.10).

$$P_s = 0.468\left(\frac{T}{100}\right)^{4.52} \tag{9.10}$$

For high pressure (in MPa):

$$\frac{\mu(T,P)}{\mu_w(T,P_s)} = 1 + \frac{P-P_s}{10^6}(1.046T - 319.24) \tag{9.11}$$

The accuracy of this method is ± 0.9% for the salinity up to 173.8 g/kg. Although the difference beyond this point is not reported, its usage is valid to 200 g/kg (within a difference of 20% (Qasem et al. 2021)). The correlation is valid up to the saturation pressure and temperature in the range of 50–350 °F (10–176.66 °C).

Sharqawy et al. (2011) reported a correlation in terms of temperature (T) in °C and practical salinity (S_p) in g/kg. The seawater viscosity (μ_{SW}) in mPa s is,

$$\mu_{SW} = \mu_w\left(1 + AS_p + BS_p^2\right) \tag{9.12}$$

The viscosity of pure water can be estimated by Eq. (9.13).

$$\ln(\mu_W) = -0.00379418 + \left(\frac{0.604129}{139.18+T}\right) \tag{9.13}$$

where constants A and B can be approximated as,

$$A = 1.474 \times 10^{-3} + 1.5 \times 10^{-5}T - 3.927 \times 10^{-8}T^2$$

$$B = 1.073 \times 10^{-5} - 8.5 \times 10^{-8}T + 2.230 \times 10^{-10}T^2$$

The correlation is valid for the salinity range of 0–150 g/kg and temperature range of 0–180 °C within an accuracy of ± 1%.

Isdale et al. (1972) proposed a correlation for the viscosity of seawater in terms of temperature (T) in °C and salinity (S) in g/kg. The viscosity (μ) in cP can be calculated from Eq. (9.14).

$$\log\left(\frac{\mu_{20}}{\mu}\right) = \frac{T-20}{T+109}\left[A\left(1 + a_1S + a_2S^2\right) + B\left(1 + b_1S + b_2S^2\right)(T-20)\right] \tag{9.14}$$

where μ_{20} is the viscosity at 20 °C in mNs/m^2. All the required constants are summarized in Table 9.1.

$$\mu_{20} = 1.002 + C_1S + C_2S^2$$

The correlation is valid for the salinity range of 0–150 g/kg, the pressure range of 0.1 MPa, and the temperature range of 20–180 °C within an accuracy of ± 1%.

Table 9.1 Constants for Isdale et al. (1972) correlation

Constant	Value	Constant	Value
a_1	– 0.000619	A	1.37060
a_2	0.000003	B	0.000832
b_1	0.000943	C_1	0.001652
b_1	– 0.000015	C_2	0.0000083

Islam and Carlson (2012) proposed a correlation in terms of temperature (T) in K and pressure (P) in MPa. The seawater viscosity (μ_{sw}) in MPa.s is calculated from Eq. (9.15). The required constants are summarized in Table 9.2.

$$\mu_{SW} = a_o + \sum_{n=1}^{03} b_i \exp(-C_i) + P \sum_{n=0}^{03} d_i(T - 293.15) \tag{9.15}$$

The correlation is valid for the salinity of 35 g/kg, the range pressure of 0–140 MPa, and the temperature range of 0–100 °C within an accuracy of ± 0.9%.

Ghallab and Elnahas (2016) fitted a correlation from experimental data and expressed it in terms of temperature (T) in °C and salinity (S) in kg/kg. The viscosity (μ) in cP is calculated from Eq. (9.16) and required constants values are listed in Table 9.3.

$$\mu = \sum_{i=1}^{28} X_i Y_i Z_i \tag{9.16}$$

The correlation is valid up to the salinity range of 0–150 g/kg and temperature range of 20–180 °C. The accuracy of this correlation is not specified by the source.

IAPWS-2008 correlation is adapted from Sharqawy et al. (2011). For temperature (T) in °C and salinity (S) in kg/kg, the seawater viscosity (μ_{sw}) in Pa s is given by Eq. (9.17).

$$\mu_{SW} = \mu_w\left(1 + AS + BS^2\right) \tag{9.17}$$

Table 9.2 Constants for Islam and Carlson (2012) method

Constants	Values	Constants	Values
a_o	2.27027348×10^1	C_3	$4.41842469 \times 10^{-2}$
b_1	1.095129×10^4	d_o	$-2.08076250 \times 10^{-1}$
b_2	9.59424756×10^8	d_1	$2.02116962 \times 10^{-2}$
b_3	-8.60130112×10^8	d_2	$-2.8017105 \times 10^{-4}$
C_1	$9.89379527 \times 10^{-3}$	d_3	$1.19302430 \times 10^{-6}$
C_2	4.3876740×10^{-2}		

Table 9.3 Constants for the correlation of Ghallab and Elnahas (2016)

X_i	Y_i	Z_i
1.6704	S^0	T^0
0.792	S^1	T^0
55	S^2	T^0
− 239.5	S^3	T^0
− 912.8	S^4	T^0
13,321	S^5	T^0
− 34,660	S^6	T^0
-4.51×10^{-2}	S^0	T^1
-2.23×10^{-2}	S^1	T^1
− 1.4265	S^2	T^1
8.49284	S^3	T^1
− 27.396	S^4	T^1
28.7108	S^5	T^1
7.28×10^{-4}	S^0	T^2
9.23×10^{-4}	S^1	T^2
1.07×10^{-2}	S^2	T^2
-3.82×10^{-2}	S^3	T^2
6.61×10^{-2}	S^4	T^2
7.43×10^{-6}	S^0	T^3
-1.18×10^{-5}	S^1	T^3
-3.57×10^{-5}	S^2	T^3
5.12×10^{-5}	S^3	T^3
4.62×10^{-8}	S^0	T^4
6.15×10^{-8}	S^1	T^4
4.70×10^{-8}	S^2	T^4
-1.59×10^{-10}	S^0	T^5
-1.16×10^{-10}	S^1	T^5
2.28×10^{-10}	S^0	T^6

μ_W can be calculated from Eq. (9.18).

$$\mu_W = 4.2844 \times 10^{-5} + \frac{1}{0.157(T + 64993)^2 - 91.296} \tag{9.18}$$

Constants A and B in Eq. (9.17) can be calculated as,

$$A = 1.541 + 1.998 \times 10^{-2}T - 9.52 \times 10^{-5}T^2$$

$$B = 7.974 - 7.561 \times 10^{-2}T + 4.724 \times 10^{-4}T^2$$

The viscosity unit is in kg/m s. The correlation is valid up to the salinity range of 0–150 g/kg and temperature range of 20–180 °C within an accuracy of ± 1.5%.

El-Dessouky and Ettouney (2002) proposed a correlation for the viscosity of seawater in terms of temperature (T) in °C and salinity (S) in g/kg as in Eq. (9.19).

$$\mu_{SW} = \mu_w(\mu_R) \times 10^{-3} \tag{9.19}$$

μ_w can be calculated as:

$$\ln(\mu_w) = \frac{-3.79418 + 604.129}{(139.18 + T)} \tag{9.20}$$

The relative viscosity (μ_R) can be calculated as,

$$\mu_R = 1 + AS + BS^2 \tag{9.21}$$

Constants A and B can be calculated as,

$$A = 1.474 \times 10^{-3} + 1.5 \times 10^{-5}T - 3.927 \times 10^{-8}T^2$$
$$B = 1.0734 \times 10^{-5} - 8.5 \times 10^{-8}T - 3.23 \times 10^{-10}T^2$$

The correlation is valid up to the salinity range of 0–130 g/kg, pressure of 0.1 MPa, and temperature range of 10–130 °C. The accuracy is not specified by the source.

Millero (1974) reported a correlation for seawater viscosity in terms of temperature in °C and chlorinity (Cl) in g/kg. The seawater viscosity (μ_{sw}) in Pa s is given by Eq. (9.22).

$$\mu_{SW} = \mu_W\left(1 + A\,Cl^2 + B\,Cl\right) \tag{9.22}$$

and μ_W is the viscosity of pure water,

$$\mu_W = 4.2844 \times 10^{-5} + \frac{1}{0.157(T + 64.993)^2 - 91.296} \tag{9.23}$$

Constants A and B can be calculated as,

$$A = 5.185 \times 10^{-5}T + 1.0675 \times 10^{-4}$$
$$B = 3.300 \times 10^{-5}T + 2.591 \times 10^{-3}$$

The correlation is valid up to the salinity range of 0–40 g/kg, pressure of 0.1 MPa, and temperature range of 5–25 °C with an accuracy of ± 0.5%.

Laliberte (2007) proposed a correlation in terms of temperature (T) in °C, and the water viscosity (μ_w) in mPa s is expressed in Eq. (9.24).

$$\mu_W = \frac{T + 246}{(0.05594T + 5.2842) + 137.37} \tag{9.24}$$

The correlation is valid in the temperature range of 0–150 °C and pressure of 0.1 MPa with an accuracy of ± 0.01%.

9.5 Case Study

This section critically analyzes and compares the correlations that are summarized above. The viscosity of saline water is a strong function of salinity and temperature and a weak function of pressure. The calculated values of viscosity using El-Dessouky and Ettouney (2002), Fabuss et al. (1969), Isdale et al. (1972), IAPWS (2008), Ghallab and Elnahas (2016), and Cooper and Dooley (2008) methods and experimental data of Leyendekkers (1980) are plotted at 20 °C and 0.1 MPa, as shown in Fig. 9.1.

Figure 9.1a shows that viscosity values generated by all these correlations are close to the experimental data except for Cooper and Dooley (2008), Fabuss et al. (1969), and Ghallab and Elnahas (2016). The difference from the experimental work (Leyendekkers 1980) is shown in Fig. 9.1b. The correlation of Isdale et al. has the most accurate prediction for the viscosity values followed by IAPWS (2008), Isdale et al. (1972), and El-Dessouky and Ettouney (2002) within a difference of ± 4%, while other methods have errors exceeding ± 10%.

The effect of temperature on viscosity values is shown in Fig. 9.2 as a comparison between the methods of Fabuss et al. (1969), El-Dessouky and Ettouney (2002), Islam and Carlson (2012), IAPWS (2008), Cholorinity (Cooper and Dooley 2008) and Isdale et al. (1972) with the experimental data of Leyendekker (1980) at 32.33 g/kg and 0.1 MPa. It can be seen from Fig. 9.2a that viscosity values obtained by all these correlations largely overlap except for Cholorinity [by Cooper and Dooley (2008)] and Fabuss et al. (1969) methods. Figure 9.2b shows a difference in the estimated viscosity values against the experimental data of Leyendekker (1980). El-Dessouky and Ettouney (2002), Islam and Carlson (2012), and IAPWS (2008) have the most accurate results within ± 2% of the experimental data, while Cooper and Dooley (2008) have a difference of ± 7%.

The pressure-dependence viscosity values are plotted using Islam and Carlson correlation (2012) in a range of 0–180 MPa at 20 °C and 35 g/kg, as shown in Fig. 9.3. For the data obtained by this correlation, the viscosity values decrease in a parabolic trend with an increase in pressure. For pressure greater than 100 MPa, the influence of pressure on the viscosity is minor.

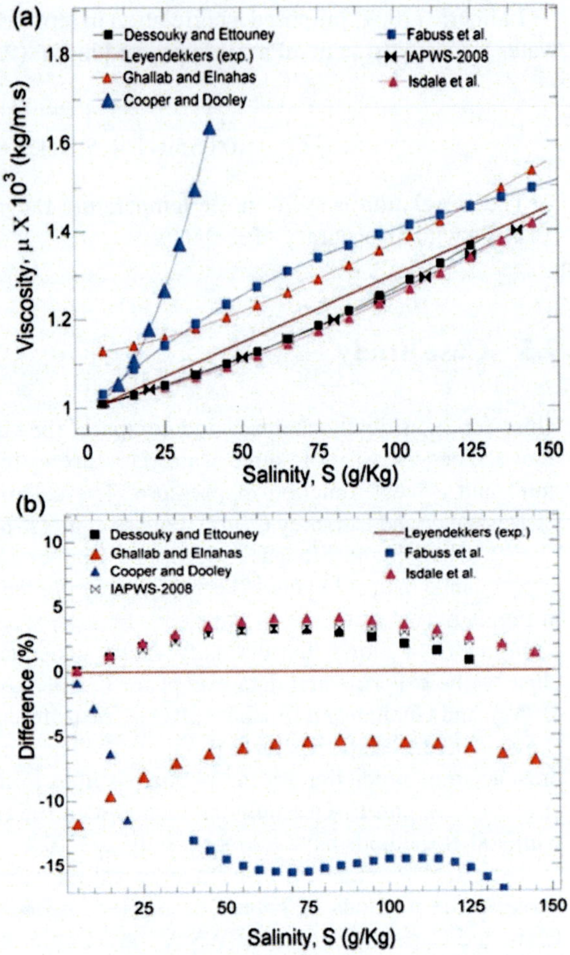

Fig. 9.1 **a** Salinity versus viscosity and **b** the difference against the experimental data of Leyendekkers (1980) for the methods of El-Dessouky and Ettouney (2002), Fabuss et al. (1969), Ghallab and Elnahas (2016), Isdale et al. (1972), IAPWS (2008), Cholorinity (by Cooper and Dooley (2008)) at 20 °C and 0.1 MPa

The viscosity of saline water is a property that depends on the type of saline water, temperature, and pressure. The selection of correlation for estimation is largely dependent upon the application and saline water type used. As for as seawater is concerned, sufficient correlation and data were reported. However, fewer accurate and flexible models are available for the estimation of the viscosity of saline water with changing composition. Therefore, it is recommended to formulate more studies on modeling viscosity that can accommodate constituents, either those ionic or covalent compounds and molecules. Based on the above discussion, some methods and correlations could be recommended to be used, as summarized in Table 9.4.

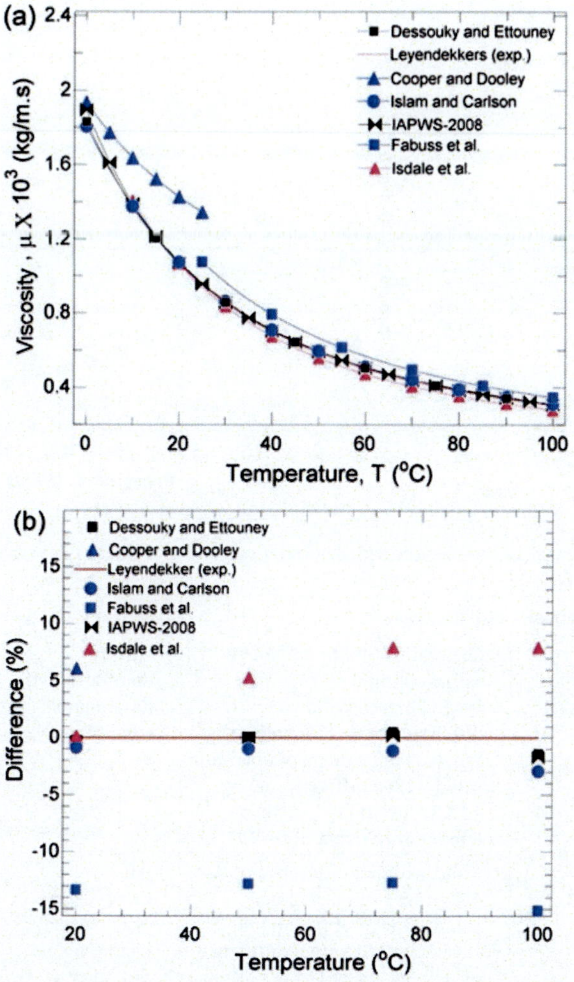

Fig. 9.2 **a** Temperature versus viscosity and **b** the difference against the experimental data of Leyendekkers (1980) for the methods of Fabuss et al. (1969), El-Dessouky and Ettouney (2002), Islam and Carlson (2012), IAPWS (2008), Cooper and Dooley (2008) and Isdale et al. (1972) at 32.33 g/kg and 0.1 MPa

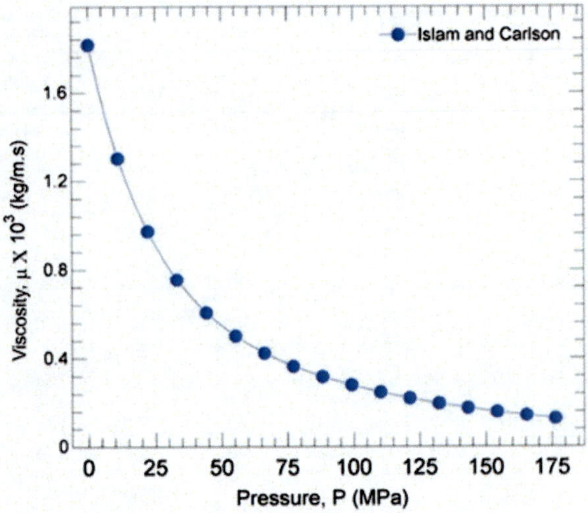

Fig. 9.3 Pressure versus viscosity at 20 °C and 35 g/kg for Islam and Carlson correlation (2012)

Table 9.4 Summary of the recommended correlations and models for the viscosity of saline water (Qasem et al. 2021)

Correlation—model	Advantages	Applications	References
Fabuss et al. (multi-species model) [Eq. (9.2)]	• General for multicomponent saline solution • It can be used within a difference of ± 15% (refer to Fig. 9.1) in its range of operation • Valid with a difference of ± 15% (refer to Fig. 9.2) in a temperature range of 0–100 °C	• All thermal-based desalination technologies	(Fabuss et al. 1969)
Isdale et al. [Eq. (9.14)]	• Valid within a difference of ± 4% (refer to Fig. 9.1) in a salinity range of 0–150 g/kg • It can be used with a difference of ± 0.5% (refer to Fig. 9.2) in a temperature range of 0–40 °C	• All thermal-based desalination technologies • Seawater greenhouse • High-grade water recycling • Wave-powered desalination	(Isdale et al. 1972)
El-Dessouky and Ettouney [Eq. (9.19)]	• Valid in the salinity range of 0–150 g/kg within a difference of ± 3% (refer to Fig. 9.1) • Covers a wide range of temperatures (0–100 °C) within a difference of ± 2% (refer to Fig. 9.2)	• All thermal-based desalination technologies • Seawater greenhouse • High-grade water recycling • Wave-powered desalination	(El-Dessouky and Ettouney 2002)

(continued)

Table 9.4 (continued)

Correlation—model	Advantages	Applications	References
Islam and Carlson [Eq. (9.15)]	• It can be used for high pressure of 0–180 MPa (refer to Fig. 9.3)	• All thermal-based desalination technologies • Seawater greenhouse • High-grade water recycling	(Islam and Carlson 2012)

9.6 EES Codes

The EES codes for the correlations above are:

Fabuss et al. Multicomponent Model

"Just plug in molar concentration of ions. The model will determine ionic strength for viscosity calculation."

```
T = 20 "C"

log10(mu_r) = (0.042808 * I_s + 0.00123 * I_s^2 + 0.000131 * I_s^3) + ((−
0.03724 * I_s) + (0.01859 * I_s^2) − (0.00271 * I_s^3)) * log10(mu_w)
mu_w = (4.2844/10^5) + (1/(0.157 * (T + 64.993)^2 − 91.296))
mu_r = mu_sw/mu_w
```

"Ionic Strength"

"Concentrations in mol/kg"

"Please note that 1: H_2O, 2: Cl^-, 3: Na^+, 4: Mg^{++}, 5: SO_4^{--}, 6: Ca^{++}, 7: K^+"

```
n[1] = 53.6
n[2] = 0.546
n[3] = 0.469
n[4] = 0.0528
n[5] = 0.0282
n[6] = 0.0103
n[7] = 0.0102
n_t = sum(n[i], i = 1, 7) "total no. of moles"
```

"Mole Fractions"

```
Duplicate i = 1, 7
X[i] = n[i]/n_t
End
```

"Molar Masses in kg/mol"

```
m[1] = 18/1000
m[2] = 35.453/1000
m[3] = 23/1000
m[4] = 24.305/1000
m[5] = 96/1000
m[6] = 40/1000
m[7] = 39.1/1000
```

"Summation of mole fraction of all ions"

```
X_ions = sum(X[i], i = 2, 7)
```

"Mole Fraction of Water"

```
X_water = 1 – X_ions
```

"Volume of Species in m^3/mol"

```
v[1] = 18/10^6
v[2] = 23.5130/10^6
v[3] = 14.6883/10^6
v[4] = 19.4707/10^6
v[5] = 69.4141/10^6
v[6] = 25.1865/10^6
v[7] = 27.0279/10^6
```

"Compensation Factor for Ions-water molecules interaction in m^3/mol"

```
alpha[2] = – 14.6221/10^6
alpha[3] = – 5.9081/10^6
alpha[4] = – 18.8749/10^6
alpha[5] = – 65.8818/10^6
alpha[6] = – 22.3772/10^6
alpha[7] = – 8.0763/10^6
```

"Average Molar Mass of Solution"

```
M_av = (sum(x[i] * m[i], i = 2, 7)) + (X_water * m[1])
```

"Molar Volume of Solution"

```
V = (sum(v[i] * x[i], i = 2, 7)) + (v[1] * X_water) + (sum(alpha[i] * x[i] * X_
water, i = 2, 7))
```

"Density"

```
rho = M_av/V
```

"Charge of each species"

```
z[2] = – 1
```

```
z[3] = + 1
z[4] = + 2
z[5] = − 2
z[6] = + 2
z[7] = + 1
```

"Molarity of Individual Ions (mol/L)"

```
Duplicate i = 2, 7
Molarity[i] = n[i] * rho/10^3
End
```

"Ionic Strength"

```
I_s = 0.5 * sum(Molarity[i] * (Z[i])^2, i = 2, 7)
```

Numbere et al. correlation:

"Numbere et al. Correlation in term salinity and temperature for the salinity of 17.38%"

```
T = 68 "F"
M = 35 "g/kg"
W = 58.5

C = (M * W)/((1000 + (M * W)))

(mu_p/mu_w) = ((− 1.87/10^3 * C^0.5) + (2.18/10^4 * C^2.5) + ((T^0.5 −
((1.35/10^2 * T))) * ((2.76/10^3 * C) − (3.44/10^4 * C^1.5)))) + 1

mu_w = 241.4 * 10^(247.8/(T − 140))
```

"Where C is salt concentration in weight percent."

"Note that viscosity will be in Centi-poise. let us convert it in kg/ms."

```
mu = mu_p/1000
```

Sharqawy et al. correlation:

"Salinity"

```
S_p = 35 "g/kg"
```

"Temperature"

```
T = 20 "C"

mu_sw = mu_w * (1 + (A * S_p) + (B * S_p^2)) "in mPa S"

ln(mu_w) = − 0.00379418 + (0.604129/(139.18 + T))
```

“Constants”

```
A = (1.474/10^3) + (1.5/10^5 * T) − (3.927/10^8 * T^2)
B = (1.073/10^5) − (8.5/10^8 * T) + (2.230/10^10 * T^2)
```

“Unit Conversion to SI”

```
mu_SI = mu_sw/1000
```

Isdale et al. Correlation

“Temperature Range 0–180 °C.”

“Salinity Range 0–150 g/kg”

“Temperature”

```
T = 20 “C”

S = 35 “g/kg”

log10(mu_20/mu) = (((T − 20)/(t + 109)) * (A * (1 + (a_1 * S) + (a_2 * S^2))))
+ (B * (1 + (b_1 * S) + (b_2 * S^2)) * (T − 20))
```

“correlation for viscosity at 20 C in mNs/m^−2”

```
mu_20 = 1.002 + (C_1 * S) + (C_2 * S^2)
```

“constants”

```
A = 1.37060
a_1 = − 0.000619
a_2 = 0.000003
B = 0.000832
b_1 = 0.000943
b_2 = − 0.000015
C_1 = 0.001652
C_2 = 0.0000083
```

“Unit Conversion”

```
mu_SI = mu/1000
```

Islam and Carlson (Islam and Carlson 2012) Correlation

“Temperature Range 0–120 °C”

“Salinity Range 35 g/kg”

“Temperature”

```
T = 283.15 “K”
```

P = 0.1 “MPa”

“Constants”

a_o = 2.27027348 * 10^1

b[1] = 1.0951298 * 10^4
b[2] = 9.59424756 * 10^8
b[3] = – 8.60130112 * 10^8

C[1] = 9.89379527/10^3
C[2] = 4.38767404/10^2
C[3] = 4.41842469/10^2

d[0] = – 2.08076250/10^1
d[1] = 2.02116962/10^2
d[2] = – 2.8017105/10^4
d[3] = 1.19302430/10^6

mu = (a_o) + (**sum**(b[i] * **exp**(-C[i] * T), i = 1, 3)) + ((P * (**sum**(d[i] * ((T – 293.15))^i, i = 0, 3))))

“Unit conversion from mega-pascal scale to standard SI unit”

mu_SI = mu/10^6 “kg/m s”

Ghallab and Elnahas Correlation

“Temperature”

T = 20 “C”

“Salinity”

S = 0.035 “kg/kg”

X[1] = 1.6704 * S^0 * T^0
X[2] = 0.792 * S^1 * T^0
X[3] = 55 * S^2 * T^0
X[4] = – 239.5 * S^3 * T^0
X[5] = – 912.8 * S^4 * T^0
X[6] = 13321 * S^5 * T^0
X[7] = – 34660 * S^6 * T^0
X[8] = – 4.51/10^2 * S^0 * T^1
X[9] = – 2.23/10^2 * S^1 * T^1
X[10] = – 1.4265 * S^2 * T^1
X[11] = 8.49284 * S^3 * T^1

```
X[12] = – 27.396 * S^4 * T^1
X[13] = 28.7108 * S^5 * T^1
X[14] = 7.28/10^4 * S^0 * T^2
X[15] = 9.23/10^4 * S^1 * T^2
X[16] = 1.07/10^2 * S^2 * T^2
X[17] = – 3.82/10^2 * S^3 * T^2
X[18] = 6.61/10^2 * S^4 * T^2
X[19] = 7.43/10^6 * S^0 * T^3
X[20] = – 1.18/10^5 * S^1 * T^3
X[21] = – 3.57/10^5 * S^2 * T^3
X[22] = 5.12/10^5 * S^3 * T^3
X[23] = 4.62/10^8 * S^0 * T^4
X[24] = 6.15/10^8 * S^1 * T^4
X[25] = 4.70/10^8 * S^2 * T^4
X[26] = – 1.59/10^10 * S^0 * T^5
X[27] = – 1.16/10^10 * S^1 * T^5
X[28] = 2.28/10^13 * S^0 * T^6

mu = sum(X[i], i = 1, 28) "Centipoise"
```

"let us know convert centipoise to SI unit."

```
mu_SI = mu/1000
```

IAPWS-2008 Correlation (Adapted from (Sharqawy et al. 2011))

"Please note that this correlation is valid only for relative viscosity. it is not correct of viscosity."

```
S = "kg/kg"

T = 20 "C"

mu_sw = mu_w * (1 + (A * S) + (B * S^2)) "in kg/ms"

mu_w = (4.2844/10^5) + (1/(0.157 * (T + 64.993)^2 – 91.296))
```

"Constants"

```
A = 1.541 + (1.998/10^2 * T) – (9.52/10^5 * T^2)
B = 7.974 – (7.561/10^2 * T) + (4.724/10^4 * T^2
```

El-Dessouky and Ettouney Correlation

"El-Dessouky's correlation for temperature 0–180 °C and salinity of 0–130 g/kg and pressure of 1 atm"

```
S = 35 "g/kg"
```

```
T = 20
mu = (mu_w * mu_r)/10^3
Ln(mu_w) = − 3.79418 + (604.129/(139.18 + T))
mu_r = 1 + (A * S) + (B * S^2)
A = (1.474/10^3) + (1.5/10^5 * T) − (3.927/10^8 * T^2)
B = (1.0734/10^5) − (8.5/10^8 * T) + (2.23/10^10 * T^2)
```

"It can be noted that if I have dynamic viscosity and density are known, then kinematic viscosity can be calculated by the ratio of these two quantities."

Millero Method

"Please note that this correlation is valid only for relative viscosity. It is not correct of viscosity."

```
S = 35 "g/kg"
T = 20 "C"

mu_sw = mu_w * (1 + (A * Cl^2) + (B * Cl))

S = 1.80655 * Cl

mu_w = (4.2844/10^5) + (1/(0.157 * (T + 64.993)^2 − 91.296))
```

"Constants"

```
A = (5.185/10^5 * T) + (1.0675/10^4)
B = (3.300/10^5 * T) + (2.591/10^3)
```

Fabuss et al. Correlation

"Please note that this correlation is valid only for relative viscosity. It is not correct of viscosity."

```
log10(mu_r) = 0.03103 − (0.01803 * log10(mu_w))
log10(mu_sw) = 0.03103 + (0.98197 * log10(mu_w))
mu_r = mu_sw/mu_w
```

Pure Water Correlation (Laliberte)

"Laliberte Correlation for Pure water as a function of temperature only in the range of 0–150 °C."

"Please note that temperature must be in degree C."

```
T = 20 "C"

mu = (T + 246)/((0.05594 + 5.2842) + 137.37)
```

"viscosity is mPa s. Let us convert in the kg/ms."

mu_SI = mu/1000

References

Chen SF, Chan RC, Read SM, Bromley LA (1973) Viscosity of sea water solutions. Desalination 13(1):37–51. https://doi.org/10.1016/S0011-9164(00)80090-9

Cooper JR, Dooley RB (2008) The international association for the properties of water and steam for the viscosity of ordinary water substance. International Organization

El-Dessouky HT, Ettouney HM (2002) Fundamentals of salt water desalination. https://doi.org/10.1016/b978-0-444-50810-2.x5000-3

Fabuss BM, Korosi A, Othmer DF (1969) Viscosities of aqueous solutions of several electrolytes present in sea water. J Chem Eng Data 14(2):192–197. https://doi.org/10.1021/je60041a025

Ghallab AO, Elnahas MH (2016) Determination of physical properties of saline water and different pressure and concentrations. Sci Proc Int Sci Conf 1(2):109–112

Goldsack DE, Franchetto R (1977) The viscosity of concentrated electrolyte solutions. I. Concentration dependence at fixed temperature. Can J Chem 55(6):1062–1072. https://doi.org/10.1139/v77-148

Gupta SV (2014) Viscometry for liquids, p 194. https://doi.org/10.1007/978-3-319-04858-1

Huber ML, Perkins RA, Laesecke A, Friend DG, Sengers JV, Assael MJ, Metaxa IN, Vogel E, Mareš R, Miyagawa K (2009) New international formulation for the viscosity of H_2O. J Phys Chem Ref Data 38(2):101–125. https://doi.org/10.1063/1.3088050

IAPWS (2008) Release on the IAPWS formulation 2008 for the viscosity of ordinary water substance. In: International association for the properties of water and steam, R12-08, pp 1–9. http://www.iapws.org/

Isdale JD, Spence CM, Tudhope JS (1972) Physical properties of sea water solutions: viscosity. Desalination 10(4):319–328. https://doi.org/10.1016/S0011-9164(00)80002-8

Islam AW, Carlson ES (2012) Viscosity models and effects of dissolved CO_2. Energy Fuels 26(8):5330–5336. https://doi.org/10.1021/ef3006228

Korson L, Drost-Hansen W, Millero FJ (1969) Viscosity of water at various temperatures. J Phys Chem 73(1):34–39. https://doi.org/10.1021/j100721a006

Laliberté M (2007) Model for calculating the viscosity of aqueous solutions. J Chem Eng Data 52(2):321–335. https://doi.org/10.1021/je0604075

Lencka MM, Anderko A, Sanders SJ, Young RD (1998) Modeling viscosity of multicomponent electrolyte solutions. Int J Thermophys 19(2):367–378. https://doi.org/10.1023/A:1022501108317

Leyendekkers JV (1980) Prediction of the heat capacities of seawater and other multicomponent solutions from the Tammann-Tait-Gibson model. Mar Chem 9(1):25–35. https://doi.org/10.1016/0304-4203(80)90004-3

Marvin RS (1971) The accuracy of measurements of viscosity of liquids. J Res A Phys Chem 75A(6). https://doi.org/10.6028/jres.075a.041

Millero FJ (1974) in The Sea, Vol. 5. Marine Chemistry, A. E. Maxwell et al., John Wiley and Sons eds., New York

Numbere D, Brigham WE, Standing MB (1977) Correlation for physical properties of petroleum reservoir brines. In: United States: N. P. https://doi.org/10.2172/6733264

Patil RS, Shaikh VR, Patil PD, Borse AU, Patil KJ (2014) The viscosity and coefficient (Jones–Dole equation) studies in aqueous solutions of alkyltrimethylammonium bromides at 298.15 K. J Mol Liq 200:416–424. https://doi.org/10.1016/j.molliq.2014.11.003

Penn RW, Kearsley EA (1971) An absolute determination of viscosity using channel flow 75(6):541–551

Pereira G, Moreira R, Vázquez J, Chenlo F (2001) Kinematic viscosity prediction for aqueous solutions with various solutes 81:35–40. https://doi.org/10.1016/S1385-8947(00)00203-5

Qasem NAA, Generous MM, Qureshi BA, Zubair SM (2021) A comprehensive review of saline water correlations and data: part II—thermophysical properties. Arab J Sci Eng 46(3):1941–1979. https://doi.org/10.1007/s13369-020-05020-5

Sharqawy MH, Lienhard V JH, Zubair SM (2011) Thermophysical properties of seawater: a review of existing correlations and data. Desalin Water Treat 29(1–3):355–355. https://doi.org/10.5004/dwt.2011.2947

Simion AI, Grigoraş C-G, Roșu A-M, Gavrilă L (2015) Mathematical modelling of density and viscosity of NaCl aqueous solutions. J Agroaliment Process Technol 21(1):41–52

Spikes HA (1990) A thermodynamic approach to viscosity. Tribol Trans 33(1):140–148. https://doi.org/10.1080/10402009008981940

Swindellsr F, Coer R, Godfrey TB (1952) Absolute viscosity of water at 20 °C. J Res Natl Bur Stan 48(1). https://doi.org/10.6028/jres.048.001

Chapter 10
Thermal Conductivity

Abstract The thermal conductivity of saline water is a vital thermophysical property for heat transfer applications, including thermal-based desalination and water treatment technologies. This chapter introduces multicomponent models, correlation, and data used to estimate thermal conductivity at different salinity, temperature, and pressure conditions. Also, the estimated thermal conductivity values from the correlations and methods are compared and assessed against the experimental data of seawater. A new multicomponent model for thermal conductivity is suggested with good accuracy. The chapter suggests some correlations and methods based on their accuracy and operating conditions. Furthermore, the codes used for the investigated methods to predict the thermal conductivity of saline water are provided.

Keywords Thermal conductivity · Correlations · Methods · Multicomponent · Heat transfer · Saline water

10.1 Introduction

The thermal conductivity of saline water is important for heat transfer applications. Sometimes, it becomes inevitable to find the thermal conductivity of saline water in laboratory exercises. The thermal conductivity of water improves when salt crystals are inserted into it. Upon adding salt crystals, water breaks salt units into individual ions. These ions could become an additional carrier of heat from one point to another. However, with the addition of salt, the thermal conductivity of the saline solution decreases. The second most important parameter that affects transport properties in general and thermal conductivity, in particular, is the temperature of saline water. Correlations reported also cover the effect of temperature on thermal conductivity. The thermal conductivity increases, in general, with an increase in temperature below

N. A. A. Qasem et al., *Thermodynamic and Thermophysical Properties of Saline Water*, Springer Water, https://doi.org/10.1007/978-3-031-35193-8_10

the boiling point. See and White (1997) studied the temperature dependence of concentrated potassium hydroxide solution and reported a linear increase of thermal conductivity for − 20 to 100 °C. Jamieson et al. (1975) experimentally investigated the effect of temperature on the thermal conductivity of seawater. They also reported a linear increase in thermal conductivity but up to around 100 °C.

Pressure is also a prime factor that influences the thermal conductivity of saline water. It is an important factor for high-pressure applications involving salt solutions as working fluid. The pressure effect is also important due to its linear relationship with temperature. Abdulagatov and Magomedov (2003) studied the effect of pressure on the aqueous solution of $SrBr_2$ at high pressure and found a proportional increase in thermal conductivity with an increase of pressure values.

10.2 Multiple Components Models

Because the methods reported in the literature for multicomponent thermal conductivity are complicated, an effort is made by the current work [represented by Qasem et al. (2021)] to obtain a very simple model with considerable accuracy. Linear expression in terms of molar concentration is obtained by assimilating the contribution of solvent (water), ionic species under the influence of the solvent, and ion-ion interaction. Therefore, thermal conductivity can be written as:

$$\mathrm{k} = k_{\mathrm{H_2O}} + k_{\mathrm{ions-solvent}} + k_{\mathrm{ion-ion}} \tag{10.1}$$

Contribution of thermal conductivity by solvent (water) as a function of temperature (T) in K and pressure (P) in MPa is that of Abdulagatov and Magomedov (2003).

$$k_{\mathrm{H2O}} = 7 \times 10^{-9} T^3 - 1.5113 \times 10^{-5} T^2 + 8.801 \times 10^{-3} T - 0.8624 + 1.57 \times 10^{-6} PT \tag{10.2}$$

Thermal conductivity by ions under the influence of a particular solvent can be obtained from the following relations (Wang and Anderko 2012, 2015).

$$k_{\mathrm{ions-solvent}} = \sum_{i=1}^{n} X_i \alpha_\mathrm{i} \tag{10.3}$$

where X_i is the molar fraction, and α_i is a function of temperature that can be expressed as (Wang and Anderko, 2012, 2015):

$$\alpha_\mathrm{i} = \alpha_1 + \alpha_2 \exp(-0.023(\mathrm{T} - \mathrm{T_o})) \tag{10.4}$$

where T is the temperature in K, and $\mathrm{T_o}$ is 273.15 K. The numerical values of α_1 and α_2 for different species are given in Tables 10.1 and 10.2.

Table 10.1 Individual aqueous species considered in this study and parameters for calculating their contributions to thermal conductivity (Wang and Anderko 2015)

Species	α_1, H_2O	α_2, H_2O	Molality in reference seawater
Na^+	0.0	0.0	0.4860597
Mg^{2+}	– 0.496250	0.052652	0.0547421
Ca^{2+}	– 0.052799	0.126519	0.0106568
K^+	– 0.382485	0.44932	0.0105797
Cl^-	– 0.360439	0.006076	0.5657647
SO_4^{2-}	– 0.029475	0.044903	0.0292643
HC	– 0.215690	– 0.167037	0.0017803
Br^-	– 1.147163	0.249998	0.0008728
CO_3^{2-}	0.341711	0.0	0.0002477

Table 10.2 Coefficient (α) for selected aqueous ions (Wang and Anderko 2015)

Cations	α_1, H_2O	α_2, H_2O	Anions	α_1, H_2O	α_2, H_2O
H_3O^+	– 0.716 880	0.289719	Cl^-	– 0.360439	0.006076
Li^+	– 0.207 041	0.057691	NO_3^-	– 0.422374	0.033717
Na^+	0.0	0.0	SO_4^{2-}	– 0.029457	0.044903
K^+	– 0.38285	0.044932	HSO_4^-	– 1.23553	0.12045
Mg^{2+}	– 0.496250	0.052652	OH^-	0.4923076	– 0.018014
Ca^{2+}	– 0.052799	0.126519	F^-	0.059371	– 0.10886
Ba^{2+}	– 0.651055	0.065461	HCO_3^-	– 0.215690	– 0.167037
NH_4^+	– 0.04580	– 0.35817	CO_3^{2-}	0.341711	0.0
Fe^{3+}	– 1.14149	– 0.53622	PO_4^{3-}	– 0.16858	– 1.14083
Fe^{2+}	– 1.52435	0.77168	HPO_4^{2-}	0.21776	– 0.27154
Ni^{2+}	– 0.281946	0.007328	$H_2PO_4^-$	– 0.54115	0.31969
Cu^{2+}	– 0.975205	– 0.09887			
Zn^{2+}	– 0.971744	0.055408			
Cr^{3+}	– 1.139483	– 0.536249			

$k_{ion-ion}$ could be considered, but Wang et al. (2004) recommended neglecting it within a difference of less than 0.094% (at T < 300 °C). When it is necessary to include it, the method of Wang and Anderko (2015) could be used. The model of Qasem et al. (2021) can be used in salinity range of 0–250 g/kg, pressure of 0.1–140 MPa, and temperature of 0–200 °C within an accuracy of ± 0.007 W/m K.

10.3 Correlations and Data

Jamieson and Tudhope (1970) proposed a correlation in terms of temperature (T) in °C and salinity (S) in g/kg, the thermal conductivity (k) in mW/m K is expressed in Eq. (10.5). The mandatory constants are provided in Table 10.3.

$$\log10(k) = \log10(a_1 + a_2 S_p) - a_3 \left(a_4 - \frac{a_5 + a_6 S_p}{T + 273.15} \right) \left(1 - \frac{T + 273.15}{a_7 + a_8 S_p} \right)^{a_9} \quad (10.5)$$

S_p is the practical salinity in g/kg. The Jamieson and Tudhope (1970) correlation can be used in salinity range of 0–160 g/kg, pressure of 0.1 MPa, and temperature of 0–180 °C with an accuracy of ± 3%.

Ghallab and Elnahas (2016) polynomial for the thermal conductivity (k) in mW/m K is expressed in terms of temperature (T) in °C and salinity (S) in kg/kg in Eq. (10.6). The required constants are provided in Table 10.4.

$$k(T, S) = \sum_{i=1}^{28} X_i Y_i Z_i \quad (10.6)$$

The correlation can be used in salinity range of 0–160 g/kg, pressure of 0.1 MPa, and temperature of 0–180 °C. However, the accuracy is not specified by the source.

Nayar et al. (2016) proposed a correlation (Eq. (10.7)) for seawater in terms of temperature (T) in °C, pressure (p) in MPa, and salinity (S) in g/kg, the thermal conductivity (k) in W/m K can be expressed as:

$$k = \frac{K_1}{1 + 0.00022S} \quad (10.7)$$

$$K_1 = K_2(1 + P_g(a_1 + a_2 T_g + a_3 T_g^2 + a_4 T_g^3 + a_5 T_g^4$$
$$K_2 = b_1 T_g^{-0.194} + b_2 T_g^{-4.717} + b_3 T_g^{-6.385} + b_4 T_g^{2.134}$$

where, T_g and P_g can be determined as follows, and the required constants are provided in Table 10.5.

Table 10.3 The constants of Jamieson and Tudhope (1970) correlation for thermal conductivity

Constant	Value	Constant	Value
a_1	240	a_6	3.7×10^{-2}
a_2	2×10^{-4}	a_7	647
a_3	0.434	a_8	3×10^{-2}
a_4	2.3	a_9	1/3
a_5	343.5		

Table 10.4 Values to implement Ghallab and Elnahas (2016) polynomial

X_i	Y_i	Z_i
570.893	S^0	T^0
-134.11	S^1	T^0
-941.08	S^2	T^0
16,714.9	S^3	T^0
-255.6×10^3	S^4	T^0
1838.07×10^3	S^5	T^0
-4651.5×10^3	S^6	T^0
1.77376	S^0	T^1
1.95142	S^1	T^1
-5.23×10^3	S^2	T^1
226.994	S^3	T^1
-1971	S^4	T^1
5112.74	S^5	T^1
-7.57×10^{-3}	S^0	T^2
7.092×10^{-3}	S^1	T^2
-15.36×10^{-3}	S^2	T^2
76.176×10^{-3}	S^3	T^2
-60.98×10^{-3}	S^4	T^2
-1.2×10^{-5}	S^0	T^3
-9.8×10^{-5}	S^1	T^3
-2.5×10^{-4}	S^2	T^3
7.94×10^{-4}	S^3	T^3
3.0797×10^{-7}	S^0	T^4
7.1038×10^{-7}	S^1	T^4
2.3667×10^{-7}	S^2	T^4
-1.90×10^{-9}	S^0	T^5
-1.74×10^{-9}	S^1	T^5
4.015×10^{-12}	S^0	T^6

$$T_g = \frac{T + 273.15}{300}$$

$$P_g = \frac{P - 0.1}{139.9}$$

The correlation can be used in salinity range of 0–120 g/kg, pressure of 0.1 MPa, and temperature of 10–90 °C with an accuracy of ± 2.57%. It is also valid in salinity range of 0–35 g/kg, pressure of 0.1–20 MPa, and temperature of 10–60 °C with the same accuracy.

Table 10.5 The constants for the correlation of Nayar et al. (2016)

Constant	Value	Constant	Value
a_1	21.942	b_1	0.797015
a_2	– 77.387	b_2	– 0.251242
a_3	102.81	b_3	0.096437
a_4	– 60.727	b_4	– 0.032696
a_5	13.464		

Caldwell (1974) correlation (also reported by Sharqawy et al. (2011)) in terms of temperature (T) in °C, pressure (P) in MPa, and practical salinity (S_p) in g/kg for the thermal conductivity of seawater (k) in W/m K is provided in Eq. (10.8). The mandatory constants are enlisted in Table 10.6.

$$k = 0.5715(1 + aT + bT + cP + dS_p) \tag{10.8}$$

Caldwell (1974) also reported another correlation for ocean water with an accuracy of around 0.5%. Thermal conductivity will be in cal/cm °C s. It is as in Eq. (10.9).

$$k\,(P(\text{kbar}), T(°C), S(\text{ppt})) = 0.0001365(1 + 0.003T - 1.025T^2 + 0.0653P - 0.00029) \tag{10.9}$$

The correlation can be used in salinity range of 0–60 g/kg, pressure of 0–40 MPa, and temperature of 0–60 °C with an accuracy of ± 0.5%.

Castelli et al. (1974) proposed a correlation for the calculation of thermal conductivity (k) in W/m. °C. It is in terms of temperature (T) in °C and pressure (P) in MPa as shown in Eq. (10.10). The mandatory constants are provided in Table 10.7.

$$k = a + bP + cT + dT^3 \tag{10.10}$$

The correlation can be used in salinity range of 35 g/kg, pressure of 0–140 MPa, and temperature of 0–30 °C with an accuracy of ± 0.4%.

Abdulagatov and Magomedov (2003) proposed a correlation (Eq. (10.11)) for the thermal conductivity of pure water (k_w) in W/m K in terms of temperature (T) in K and pressure (P) in MPa. The required constants are provided in Table 10.8.

$$k_w = b_1 + b_2T^2 + b_3T + b_4 + b_5PT \tag{10.11}$$

Table 10.6 The constants for Eq. (10.8)

Constant	Value
a	0.003
b	-1.025×10^{-5}
c	6.53×10^{-3}
d	0.00029

Table 10.7 The constants for the correlation of Castelli et al. (1974)

Constant	Value
a	0.55286
b	3.4025×10^{-4}
c	1.8346×10^{-3}
d	-3.3058×10^{-7}

Table 10.8 The constant for Eq. (10.11)

Constant	Value	Constant	Value
b_1	7×10^{-9}	b_4	-0.8624
b_2	-1.5113×10^{-5}	b_5	1.57×10^{-6}
b_3	8.801×10^{-3}		

The correlation can be used in pressure range of 0.1–140 MPa, and temperature of 0–200 °C with an accuracy of ± 0.1.6%.

10.4 Case Study

This section discusses the dependence of thermal conductivity on salinity, temperature, and pressure. Figure 10.1 shows the effect of various salinity values on the thermal conductivity values calculated by Wang and Anderko (2015), Caldwell (1974), Nayar et al. (2016), Jamieson et al. (1969) and Ghallab and Elnahas (2016) methods and the experimental data of Castelli et al. (1974) at 20 °C and 0.1 MPa. It can be seen from Fig. 10.1a that Caldwell (1974), Nayar et al. (2016), Jamieson et al. (1969), and Ghallab and Elnahas (2016) predict thermal conductivity values close to those of the experimental data (1974). Ghallab and Elnahas (2016), Wang and Anderko (2015), and Caldwell (1974) methods generate the most accurate results within a difference of 1%. The Riedel equation by Wang and Anderko (2015) is found to have an offset of about 0.08 W/m K (about ± 14% difference, as shown in Fig. 10.1b).

The thermal conductivity of saline water increases with an increase in temperature. The values of thermal conductivity calculated using the Riedel equation [by Wang and Anderko (2015)], Caldwell (1974), Nayar et al. (2016), Castelli et al. (1974), Ghallab and Elnahas (2016), and Jamieson et al. (1975) methods for various temperatures at 35 g/kg and 0.1 MPa are compared with the experimental data of Ghallab and Elnahas (2016), as shown in Fig. 10.2. Figure 10.2a shows that the methods of Caldwell (1974), Jamieson et al. (1975), and Nayar et al. (2016) have the same trend as that of the experimental data. The difference with respect to the experimental

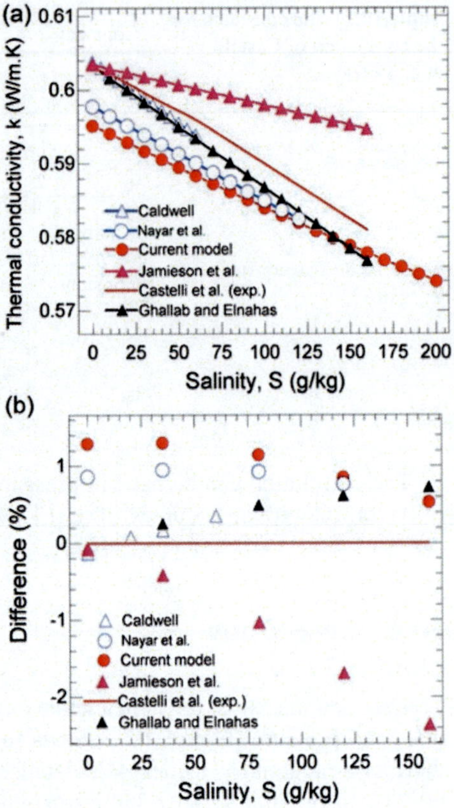

Fig. 10.1 **a** Salinity versus thermal conductivity and **b** the difference against the experimental data of Castelli et al. (1974) for the methods of Riedel equation [by Wang and Anderko (2015)], Caldwell (1974), Nayar et al. (2016), Jamieson et al. (1969) and Ghallab and Elnahas (2016) at 20 °C and 0.1 MPa

data is shown in Fig. 10.2b. Caldwell (1974), Jamieson et al. (1975), and Nayar et al. (2016) correlations are within a difference of ± 1%. The Ghallab and Elnahas (2016) correlation has a difference of ± 9%, while others have more than a ± 10% difference.

The thermal conductivity of saline water has a slight increase with the increase in pressure. The results obtained for correlations of Caldwell (1974), Nayar et al. (2016) and Castelli et al. (1974) for various pressure ranges, limited by their ranges of operation, are plotted with the experimental data of Castelli et al. (1974) at 20.25 °C and 35 g/kg, as shown in Fig. 10.3a. It can be seen from the figure that Nayar et al. (2016) and Castelli et al. (1974) methods generate overlapping values with the experimental data, while Caldwell (1974) diverges from the expected trend. The correlation of Castelli et al. (1974) has a difference of ± 2%, while Sharqawy (2013) correlation generates values within ± 4% difference, as shown in Fig. 10.3b.

The thermal conductivity of saline water is of great importance for many thermal desalination and cooling technologies where saline water is used as a coolant fluid. As the thermal conductivity is very much dependent upon saline water type (NaCl solution, seawater, multicomponent water) therefore, the selection of an appropriate

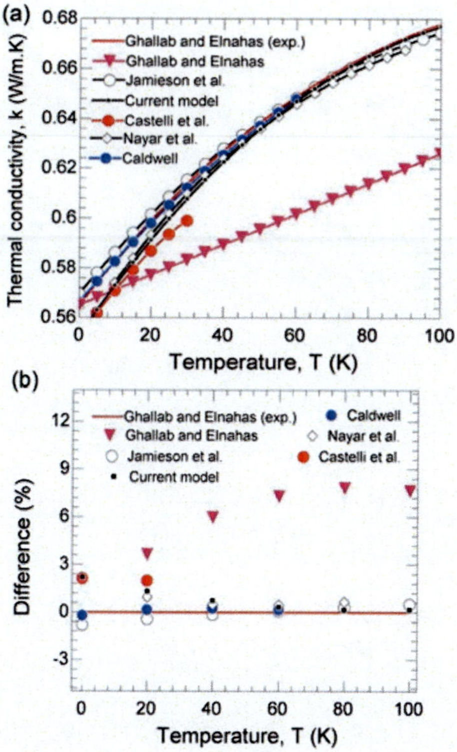

Fig. 10.2 **a** Temperature versus thermal conductivity and **b** the difference against the experimental data of Ghallab and Elnahas (2016) for the methods of Riedel equation [by Wang and Anderko (2015)], Caldwell (1974), Nayar et al. (2016), Castelli et al. (1974), Jamieson et al. (1969) and Ghallab and Elnahas (2016) at 0.1 MPa and 35 g/kg

correlation or model is of great significance. In accordance with these requirements, Table 10.9 proposes the recommended correlations and methods to be used to calculate the thermal conductivity of saline water.

It has been observed that few studies are reported on the thermal conductivity of saline water with multicomponent composition. one such model is proposed by Qasem et al. (2021). However, more studies for the development of a universal model are required and recommended.

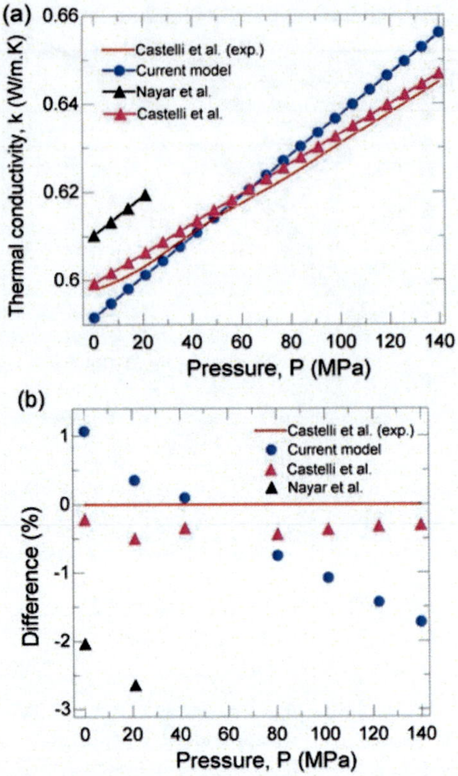

Fig. 10.3 **a** Pressure versus thermal conductivity and **b** the difference against experimental data of Castelli et al. (1974) for the correlations of Caldwell (1974), Castelli et al. (1974) and Nayar et al. (2016) at 30.25 °C and 35 g/kg

Table 10.9 Summary of the recommended correlations and models for the thermal conductivity of saline water (Qasem et al., 2021)

Correlation—model	Advantages	Applications	References
Qasem et al. [Eq. (10.1)])	• General for multicomponent saline solution • Valid within ± 15% (refer to Fig. 10.1) in a salinity range of 0–160 g/kg • Valid in the temperature range of (0–100 °C) with a difference of less than ± 5% (refer to Fig. 10.2)	• All thermal-based desalination technologies • Seawater greenhouse	(Qasem et al. 2021)

(continued)

Table 10.9 (continued)

Correlation—model	Advantages	Applications	References
Jamieson et al. [Eq. (10.5)]	• One of the most accurate correlations in salinity up to 160 g/kg within a difference of ± 2.5% (refer to Fig. 10.1) • Covers a wider range of temperatures (0–100 °C) within a difference of ± 0.8% (refer to Fig. 10.2)	• All thermal-based desalination technologies	(Jamieson et al. 1969)
Nayar et al. [Eq. (10.7)]	• One of the most accurate correlations in salinity up to 120 g/kg within a difference of ± 1% (refer to Fig. 10.1) • Covers a broader temperature range of 10–90 °C within a difference of ± 1% (refer to Fig. 10.2) • Accommodates the pressure effect with ± 4% difference (refer to Fig. 10.3) for a pressure range of 0–20 MPa	• All membrane-based technologies • All thermal-based desalination technologies • Seawater greenhouse • High-grade water recycling • Wave-powered desalination	(Nayar et al. 2016)
Caldwell [Eq. (10.8)]	• It is a function of temperature, pressure, and salinity • It can be used up to 60 g/kg within a difference of ± 0.5% (refer to Fig. 10.1) • Equally accurate for temperature effects. It can be used in a range of 0–60 °C within a difference of ± 0.8% (refer to Fig. 10.2)	• All thermal-based desalination technologies	(Caldwell 1974)
Castelli et al. [Eq. (10.10)]	• The best correlation to study the effect of pressure (0–140 MPa) within a difference of ± 2% (refer to Fig. 10.3)	• All membrane-based desalination technologies • All thermal-based desalination technologies • Seawater greenhouse • High-grade water recycling • Wave-powered desalination	(Castelli et al. 1974)

(continued)

Table 10.9 (continued)

Correlation—model	Advantages	Applications	References
Ghallab and Elnahas [Eq. (10.6)]	• One of the most accurate correlations in salinity up to 160 g/kg (data range tested) with a difference of ± 0.3% (refer to Fig. 10.1) • Covers a wider range of temperatures (0–100 °C) within a difference of ± 8% (refer to Fig. 10.2)	• All membrane-based desalination technologies • All thermal-based desalination technologies • Seawater greenhouse • High-grade water recycling • Wave-powered desalination	(Ghallab and Elnahas 2016)

10.5 EES Codes

The EES codes used for the above correlations are:

New Multicomponent Model

```
T = 293.40 "K"
P = 0.1 "MPa"
f = 1
```

"Concentrations in mol/kg"

"Please note that 1: H_2O, 2: Cl^-, 3: Na^+, 4: Mg^{++}, 5: SO_4^{--}, 6: Ca^{++}, 7: K^+"

```
n[1] = 53.6
n[2] = f * 0.546
n[3] = f * 0.469
n[4] = f * 0.0528
n[5] = f * 0.0282
n[6] = f * 0.0103
n[7] = f * 0.0102
n_t = sum(n[i], i = 1, 7) "total no. of moles"
```

"Mole Fractions"

```
Duplicate i = 1, 7
X[i] = n[i]/n_t
End

alpha_i[2] = – 0.360439
alpha_i[3] = 0
alpha_i[4] = – 0.496250
alpha_i[5] = – 0.029457
```

```
alpha_i[6] = – 0.052799
alpha_i[7] = – 0.382485

alpha_j[2] = 0.006076
alpha_j[3] = 0
alpha_j[4] = 0.052652
alpha_j[5] = 0.044903
alpha_j[6] = 0.126519
alpha_j[7] = 0.044932

K = 0.023
T_o = 273.15 "K"

Duplicate i = 2, 7
alpha[i] = alpha_i[i] + (alpha_j[i] * exp(– K * (T – T_o)))
End
```

"Thermal Conductivity"

```
k_thermal = k_t + sum(alpha[i] * x[i], i = 2, 7)
k_t = (7/10^9) * T^3 – (1.5113/10^5) * T^2 + (8.801/10^3) * T – 0.8624 +
(1.57/10^6) * P * T
```

Jamieson Correlation

```
S = 35 "g/kg"

T = 25 "C"

log10(K) = log10(240 + (0.0002 * S)) + 0.434 * (2.3 – (343.5 + (0.037 * S))/
(T + 273.15)) * (1 – (T + 273.15)/(647 + (0.03 * S)))^(1/3)

K_SI = K/1000
```

Ghallab and Elnahas Polynomial

"Temperature"

```
T = 25 "C"
```

"Salinity"

```
S = 0.035 "kg/kg"

X[1] = 570.893 * S^0 * T^0
X[2] = – 134.11 * S^1 * T^0
```

```
X[3] = – 941.08 * S^2 * T^0
X[4] = 16,714.9 * S^3 * T^0
X[5] = – 255.6 * 10^3 * S^4 * T^0
X[6] = 1838.07 * 10^3 * S^5 * T^0
X[7] = – 4651.5 * 10^3 * S^6 * T^0
X[8] = 1.77376 * S^0 * T^1
X[9] = 1.95142 * S^1 * T^1
X[10] = – 5.23 * 10^3 * S^2 * T^1
X[11] = 226.994 * S^3 * T^1
X[12] = – 1971 * S^4 * T^1
X[13] = 5112.74 * S^5 * T^1
X[14] = – 7.57/10^3 * S^0 * T^2
X[15] = 7.092/10^3 * S^1 * T^2
X[16] = – 15.36/10^3 * S^2 * T^2
X[17] = 76.176/10^3 * S^3 * T^2
X[18] = – 60.98/10^3 * S^4 * T^2
X[19] = – 1.2/10^5 * S^0 * T^3
X[20] = – 9.8/10^5 * S^1 * T^3
X[21] = – 2.5/10^4 * S^2 * T^3
X[22] = 7.94/10^4 * S^3 * T^3
X[23] = 3.0797/10^7 * S^0 * T^4
x[24] = 7.1038/10^7 * S^1 * T^4
X[25] = 2.3667/10^7 * S^2 * T^4
X[26] = – 1.90/10^9 * S^0 * T^5
X[27] = – 1.74/10^9 * S^1 * T^5
X[28] = 4.015/10^12 * S^0 * T^6

K = sum(X[i], i = 1, 28) "mW/mk"
```

"let us know convert centi-poise to SI unit"

```
K_SI = K/1000
```

Nayar et al. (2016) ***Correlation***

```
S = 35 "g/kg"

T = 20 "C"

P = 0.1 "MPa"
```

"Thermal conductivity at given temperature, pressure and salinity"

```
K_tsp = K_tp/(1 + (0.00022 * S))

K_tp = K_t * (1 + Pg * (a1 + a2 * Tg + a3 * Tg^2 + a4 * Tg^3 + a5 * Tg^4))

k_t = b1/Tg^0.194 + b2/Tg^4.717 + b3/Tg^6.385 + b4/Tg^2.134
```

```
"Relations for Tg and Pg"

  Tg = (T + 273.15)/300
  Pg = (P-0.1)/139.9

"Constants"

  a1 = 21.942
  a2 = – 77.387
  a3 = 102.81
  a4 = – 60.727
  a5 = 13.464

  b1 = 0.797015
  b2 = – 0.251242
  b3 = 0.096437
  b4 = – 0.032696
```

***Caldwell Correlation (Also Reported by Sharqawy et al.* (2011))**

```
  S = 35 "g/kg"

  T = 28.6 "C"

  P = 0.1 "MPa"

  K = 0.5715 * (1 + (0.003 * T) – (1.025/10^5 * T^2) + (6.53/10^3 * P) –
  (0.00029 * S))
```

Castelli et al. Correlation

```
  S = 35 "g/kg"

  T = 25 "C"

  P = 0.1 "MPa"

  K = 0.55286 + (3.4025/10^4 * P) + (1.8364/10^3 * T) – (3.3058/10^7 * T^3)
```

References

Abdulagatov IM, Magomedov UB (2003) Thermal conductivity of pure water and aqueous $SrBr_2$ solutions at high temperatures and high pressures. High Temp High Press 35–36(2):149–168. https://doi.org/10.1068/htjr097

Caldwell DR (1974) Thermal conductivity of sea water. Deep Sea Res Oceanogrc Abstr 21(2):131–137. https://doi.org/10.1016/0011-7471(74)90070-9

Castelli VJ, Stanley EM, Fischer EC (1974) The thermal conductivity of seawater as a function of pressure and temperature. Deep Sea Res Oceanogr Abstr 21(4):311–319. https://doi.org/10.1016/0011-7471(74)90102-8

Ghallab AO, Elnahas MH (2016) Determination of physical properties of saline water and different pressure and concentrations. Sci Proc Int Sci Conf 1(2):109–112

Jamieson DT, Tudhope JS (1970) Physical properties of seawater solutions: thermal conductivity. Desalination 8:393–401

Jamieson DT, Tudhope JS, Morris R, Cartwright G (1969) Physical properties of sea water solutions: heat capacity. Desalination 7(1):23–30. https://doi.org/10.1016/S0011-9164(00)80271-4

Jamieson DT, Irving JB, Tudhope JS (1975) Prediction of the thermal conductivity of petroleum products. Wear 33(1):75–83. https://doi.org/10.1016/0043-1648(75)90225-2

Nayar KG, Sharqawy MH, Banchik LD, Lienhard V JH (2016) Thermophysical properties of seawater: a review and new correlations that include pressure dependence. Desalination 390:1–24. https://doi.org/10.1016/j.desal.2016.02.024

Qasem NAA, Generous MM, Qureshi BA, Zubair SM (2021) A comprehensive review of saline water correlations and data: part II—thermophysical properties. Arab J Sci Eng 46(3):1941–1979. https://doi.org/10.1007/s13369-020-05020-5

See DM, White RE (1997) Temperature and concentration dependence of the specific conductivity of concentrated solutions of potassium hydroxide. J Chem Eng Data 42(6):1266–1268. https://doi.org/10.1021/je970140x

Sharqawy MH (2013) New correlations for seawater and pure water thermal conductivity at different temperatures and salinities. Desalination 313:97–104. https://doi.org/10.1016/j.desal.2012.12.010

Sharqawy MH, Lienhard V JH, Zubair SM (2011) Thermophysical properties of seawater: a review of existing correlations and data. Desalin Water Treat 29(1–3):355–355. https://doi.org/10.5004/dwt.2011.2947

Wang P, Anderko A (2012) Modeling thermal conductivity of electrolyte mixtures in wide temperature and pressure ranges: Seawater and its main components. Int J Thermophys 33(2):235–258. https://doi.org/10.1007/s10765-012-1154-8

Wang P, Anderko A (2015) Revised model for the thermal conductivity of multicomponent electrolyte solutions and seawater. Int J Thermophys 36(1):5–24. https://doi.org/10.1007/s10765-014-1756-4

Wang P, Anderko A, Young RD (2004) Modeling electrical conductivity in concentrated and mixed-solvent electrolyte solutions. Ind Eng Chem Res 43(25):8083–8092. https://doi.org/10.1021/ie040144c

Chapter 11
Activity Coefficient

Abstract The activity coefficient is an important thermophysical property to consider the non-ideality of solutions. It is a corrective parameter for the solution concentration. This chapter presents multicomponent models, correlation, and data used to estimate activity coefficients at different salinity values. Also, the estimated activity coefficient values from the correlations and methods are compared and assessed against the experimental data of seawater. This chapter recommends some correlations and methods based on their accuracy and operating conditions. In addition, the codes used to predict the activity coefficient of saline water are provided.

Keywords Activity coefficient · Correlations · Methods · Multicomponent · Ideality · Saline water

11.1 Introduction

The activity coefficient is an important parameter to evaluate the non-ideality of solutions. In an ideal solution, the microscopic interactions between each pair of species are the same; thus, the properties of mixtures can directly be evaluated as a function of concentrations or species partial pressures. A concentration is modified by considering an activity coefficient. The most accurate method to evaluate the activity coefficient of a multi-species solution was provided by the Pitzer model (1987), which was extended by Harvie and Weare (1980) and Harvie et al. (1984). The model is based on a virial expansion of the excess Gibbs free energy and takes into account binary and ternary ion interactions. Due to the complexity of this model, some authors suggested an activity coefficient for a binary solution such as NaCl (Mistry and Lienhard V 2013). This chapter summarizes the most useful models and correlations for the activity coefficient.

N. A. A. Qasem et al., *Thermodynamic and Thermophysical Properties of Saline Water*, Springer Water, https://doi.org/10.1007/978-3-031-35193-8_11

11.2 Multiple Components Models

Multiple components models were proposed to be general for all types of saltwater under different salinity values. Major constituents of the aqueous ionic solutions are cations, anions, and water. Pitzer model (extended) (Harvie et al. 1984; Harvie and Weare 1980; Pitzer 1987; Pitzer and Kim 1974; Thiel and Lienhard V 2014) is a multicomponent model that can be used for all salinity values. For an individual cation, *M*,

$$\begin{aligned}\ln\gamma_M = {} & z_M^2 F + \sum_a m_a(2B_{Ma} + ZC_{Ma}) \\ & + \sum_c m_c\left(2\Phi_{Mc} + \sum_a m_a \Psi_{Mca}\right) \\ & + \sum\sum_{a<a'} m_a m_{a'} \Psi_{aa'M} + |z_M| \sum_c \sum_a m_c m_a C_{ca} \\ & + \sum_n m_n(2\lambda_{nM})\end{aligned} \tag{11.1}$$

For an individual anion, X,

$$\begin{aligned}\ln\gamma_X = {} & z_X^2 F + \sum_c m_c(2B_{cX} + ZC_{cX}) \\ & + \sum_a m_a\left(2\Phi_{Xa} + \sum_c m_c \Psi_{Xac}\right) \\ & + \sum\sum_{c<c'} m_c m_{c'} \Psi_{cc'X} + |z_X| \sum_c \sum_a m_c m_a C_{ca} \\ & + \sum_n m_n(2\lambda_{nX})\end{aligned} \tag{11.2}$$

Here, $Z = \sum_i |z_i| m_i$, m is molality, and z_i is the charge number of ion. The subscript "c" represents cations other than M, "a" represents anions other than X, and "n" indicates neutral (uncharged) solutes. The summation over all "i" denotes a sum over all solutes; similarly, summation over all c, a, and n represents a sum over all cations, anions, and neutral solutes, respectively. The summation notations $c < c'$ and $a < a'$ denote that the sum should be implemented over all distinct cation pairs and anion pairs, respectively.

The values of Ψ_{ijk} have been tabulated in many sources (Harvie et al. 1984; Kim and Frederick 1988; Zemaitis et al. 1986b) with a slight difference in values. For electrolyte cation and anion:

$$\ln \gamma_{MX} = \frac{1}{v}(v_M \ln \gamma_M + v_X \ln \gamma_X) \quad (11.3)$$

where ν is the stoichiometric coefficient. For uncharged species, N,

$$\ln \gamma_N = \sum_c m_c(2\lambda_{Nc}) + \sum_a m_a(2\lambda_{Na}) \quad (11.4)$$

For water, the activity can be taken as:

$$\ln a_W = -\frac{M_W \sum_i m_i}{1000}\phi \quad (11.5)$$

The osmotic coefficient, ϕ, is explained in the next table (see Chap. 12). Pitzer parameters can be determined as:

$$\begin{aligned} F = {} & -A^{\phi}\left(\frac{\sqrt{I}}{1+1.2\sqrt{I}} + \frac{2}{1.2}\ln(1+1.2\sqrt{I})\right) \\ & + \sum_c \sum_a m_c m_a B'_{ca} + \sum\sum_{c<c'} m_c m_{c'} \Phi'_{cc} \\ & + \sum\sum_{a<a'} m_{a'} m_{a'} \Phi'_{aa'} \end{aligned} \quad (11.6)$$

$$I = \frac{1}{2}\sum_i m_i z_i^2 \quad (11.7)$$

where I is the ionic strength in mol/kg, and A^{ϕ} is related to the Debye–Hückel limiting law (Thiel and Lienhard V 2014). The interactions between cations and anions are evaluated by the functions B_{ij},

$$B_{MX} = \beta_{MX}^{(0)} + \beta_{MX}^{(1)} g\left(\alpha_{MX}\sqrt{I}\right) + \beta_{MX}^{(2)} g\left(12\sqrt{I}\right)$$

$$C_{MX} = \frac{C_{MX}^{\phi}}{2|z_M z_X|^{1/2}}$$

$$B'_{MX} = \beta_{MX}^{(1)} g'\left(\alpha_{MX}\sqrt{I}\right)/I + \beta_{MX}^{(2)} g'(12\sqrt{I})/I$$

Here, $\alpha_{MX} = 2.0$ for $j - 1$ electrolyte and $\alpha_{MX} = 1.4$ for 2–2 and more electrolytes. The values of ${\beta_{MX}}^{(i)}$ are tabulated for a given ion pair, and ${\beta_{MX}}^{(2)}$ is associated with the complex formation and generally only non-zero for 2–2 electrolytes. The functions $g(x)$ and $g'(x)$ are given by:

$$g(x) = \frac{2(1 - (1 + x)e^{-x})}{x^2}$$

$$g'(x) = -\frac{2}{x^2}\left[1 - \left(1 + x + \frac{x^2}{2}\right)e^{-x}\right]$$

The interactions between like-charged are given by:

$$\Phi_{ij} = \theta_{ij} + E_{\theta_{ij}}(I)$$

$$\Phi'_{ij} = E_{\theta'_{ij}}(I)$$

Here, θ_{ij} is the only adjustable parameter for a given ion pair. The terms $E_{\theta ij}(I)$ and $E_{\theta ij'}(I)$ are functions of I only:

$$E_{\theta_{ij}} = \frac{z_i z_j}{4I}\left(J_0(x_{ij}) - \frac{1}{2}J_0(x_{ii}) - \frac{1}{2}J_0(x_{jj})\right)$$

$$E_{\theta'_{ij}} = \frac{z_i z_j}{8I^2}\left(J_1(x_{ij}) - \frac{1}{2}J_1(x_{ii}) - \frac{1}{2}J_1(x_{jj})\right) - \frac{E_{\theta_{ij}}}{I}$$

where,

$$J_0(x) = \frac{1}{4}x - 1 + \frac{1}{x}\int_0^\infty \left[1 - \exp\left(-\frac{x}{y}e^{-y}\right)\right]y^2\,dy$$

$$J_1(x) = \frac{1}{4}x - 1 + \frac{1}{x}\int_0^\infty \left[1 - \left(1 + \frac{x}{y}e^{-y}\right) \times \exp\left(-\frac{x}{y}e^{-y}\right)\right]y^2\,dy$$

$$x_{ij} = 6 z_i z_j A^{\phi}\sqrt{I}$$

More details about Pitzer parameters and their dependence on temperature are available in (Zemaitis et al. 1986a). Other detailed methods, including Bromley, Meissner, and Chen were also explained in (Zemaitis et al. 1986a). The temperature for the model is 25 °C. However, the Pitzer parameters could be taken as a function of temperature (Thiel and Lienhard V 2014).

Bromley (1973) proposed a model for single electrolyte solutions like NaCl. The model is summarized in Eq. (11.8).

$$\log\gamma_{\pm} = \frac{-A|z_+ z_-|\sqrt{I}}{1 + \sqrt{I}} + \frac{(0.06 + 0.6\,B)|z_+ z_-|I}{\left(1 + \frac{1.5}{|z_+ z_-|}I\right)^2} + BI \tag{11.8}$$

where A is the Debye–Huckel constant equal to 0.5, and B is 0.1433 for HCl and 0.0574 for NaCl (Kraaijeveld et al. 1995). For temperature dependence, the parameters as a function of temperature were explained in (Zemaitis et al. 1986b). The ionic strength can be found using Eq. (11.7). The model can be used for all values of salinity with $R^2 = 0.9986$ (Qasem et al. 2018).

Güntelberg (1926) proposed a simple correlation for the calculation of activity coefficients of single electrolyte solutions in the form of Eq. (11.9). The model is valid for the salinity of less than 3.55 g/kg.

$$\log \gamma_{\pm} = \frac{-A|z_+ z_-|\sqrt{I}}{1+\sqrt{I}} \tag{11.9}$$

Davies (1938) proposed a similar model in terms of activity coefficient as in Eq. (11.10).

$$\log \gamma_{\pm} = -A|z_+ z_-|\left(\frac{\sqrt{I}}{1+\sqrt{I}} - 0.2\,I\right) \tag{11.10}$$

The model of Davies (1938) is valid for the salinity of less than 29 g/kg.

11.3 Correlations and Data

Correlations are the easiest way to calculate the saltwater activity coefficient as they are fitted from the experimental data. Most of the available activity correlations in the literature are for seawater or NaCl aqueous solutions. Mistry and Lienhard V (2013) provided the best-fitting from experimental data in the form of (Eq. (11.11)

$$\gamma_{\pm} = 10^{-0.5131\left(\frac{\sqrt{m_s}}{1+1.408\sqrt{m_s}} - 0.17\,m_s\right)} - 0.09262\,m_s + 0.002\,m_s^2 - 0.001259\,m_s^3 \tag{11.11}$$

The solution molality, m_s, is in mol/kg. The fitting is valid for all salinity values but only for the NaCl case with an accuracy of $R^2 = 1$.

Similarly, Fidaleo and Moresi (2005) fitting is a good approximation of the activity coefficient. The fitting is in terms of the molality of salt, as expressed in Eq. (11.12)

$$\ln \gamma_{\pm} = \frac{-0.979\sqrt{m_s}}{1+0.732\sqrt{m_s}} + 0.139\,m_s \tag{11.12}$$

The fitting is valid for all salinity values but limited to the solutions of NaCl only with an accuracy of R-square = 0.9986 (Qasem et al. 2018).

Nayar et al. (2016) proposed a correlation for the calculation of the activity of seawater. It is provided in Eq. (11.13).

$$a_W = \frac{P_{v,sw}}{P_{v,w}} \tag{11.13}$$

The activity of water and seawater are related, as given in Eq. (11.14)

$$\ln\left(\frac{P_{v,sw}}{P_{v,w}}\right) = -4.58180 \times 10^{-4}\,S - 2.04430 \times 10^{-6}\,S^2 \tag{11.14}$$

where superscripts v, sw, and v, w are vapor of seawater and vapor of water, respectively. S is salinity in g/kg. The correlation is valid for seawater (or NaCl) for salinity range of 0–120 g/kg, temperature range of 0–180 °C within an accuracy of ± 1.56%.

11.4 Case Study

A comparison between the activity coefficient calculated using some listed methods above and the experimental data of Robinson and Stokes (1970) is shown in Fig. 11.1a exhibits the real values of the activity coefficient against the salinity up to ~ 260 g/kg (6 mol/kg (molal)). All estimated values are close to those of the experimental data. The difference between them is shown in Fig. 11.1b as the difference with respect to the experimental work. In general, all listed models (see Fig. 11.1) and fitting correlations demonstrate a small difference of less than 1.5%. To sum up, the Pitzer model is an excellent approach to evaluating the activity coefficient for any solution, whereas Mistry and Lienhard's fitting overlaps (for NaCl) with the experimental data. Notwithstanding, the Pitzer model predicts an accurate value of the activity coefficient while the other models and fittings provide reasonably accurate and straightforward methods.

Based on the above discussion, some methods and correlations could be recommended to be used, as summarized in Table 11.1.

To select a proper correlation to estimate the activity coefficient, some criteria should be considered, such as the type of saline water (seawater, multicomponent, high saline, and so on), operating limits (salinity, temperature, and pressure), and fitting accuracy. Besides these, a comparison with the existing experimental data (which were mostly reported for seawater) can be used to test and select the recommended correlation. Some correlations are valid for a particular type of desalination process, whereas others are not. For example, humidification-dehumidification desalination systems operate at 0.1 MPa so that the dependent-pressure correlations may not be needed. However, reverse osmosis (RO) desalination needs variable pressure properties. In accordance with these criteria, Table 11.1 proposes the recommended correlations and methods to be used to calculate the activity coefficient of saline water.

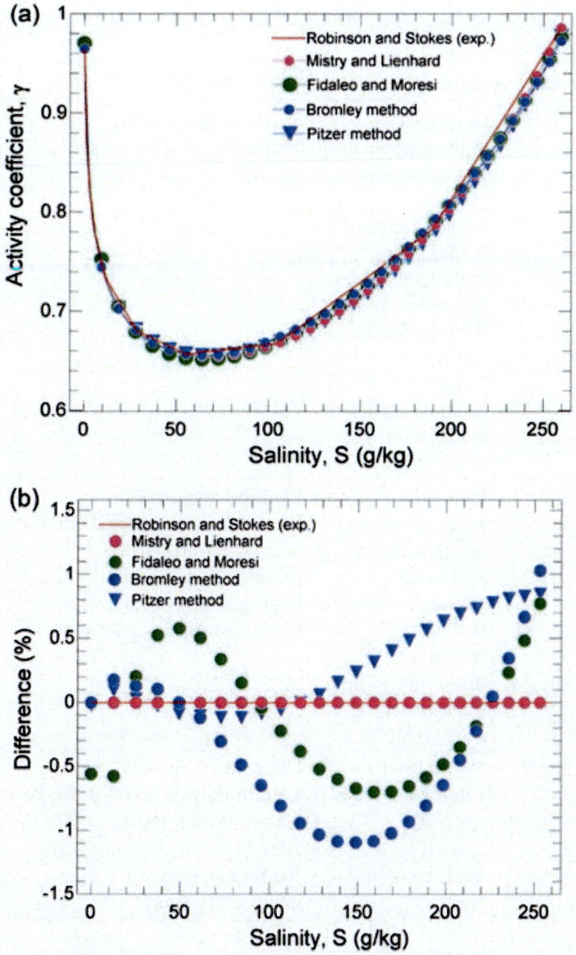

Fig. 11.1 Comparison of activity coefficient values of different models [Pitzer (Harvie et al. 1984; Harvie and Weare 1980; Pitzer 1987; Pitzer and Kim 1974), Bromley (1973), Mistry and Lienhard V (2013), and Fidaleo and Moresi (2005)] against experimental NaCl data of Robinson and Stokes (1970): **a** estimated values and **b** the difference from the experimental data

It is also noticed that most correlations are for NaCl properties, while the multicomponent correlations are limited. It is strongly recommended to conduct large-scale experimental investigations to evaluate multicomponent saline water properties at different values of salinity, temperature, and pressure. The corresponding correlations of the tested properties could then be obtained.

Table 11.1 Summary of the recommended correlations and models for the activity coefficient of saline water (Qasem et al. 2021)

Correlation—model	Advantages	Applications	References
Pitzer model [Eq. (11.3)]	• General for multicomponent saline solution • Valid in a salinity range of 0–260 g/kg within a difference of ± 0.9% (refer to Fig. 11.1) • The most accurate model, but it is not straightforward	• All membrane-based desalination technologies • All thermal-based desalination technologies	(Harvie et al. 1984; Harvie and Weare 1980; Pitzer 1987; Pitzer and Kim 1974)
Bromley method [Eq. (11.8)]	• General for multicomponent saline solution • Valid in a salinity range of 0–260 g/kg within a difference of ± 1.3% (refer to Fig. 11.1)	• Electrodialysis (ED) • Electrodialysis reversal (EDR) • Nanofiltration (NF) • Membrane distillation (MD) • High-grade water recycling • Wave-powered desalination	(Bromley 1973; Kraaijeveld et al. 1995)
Mistry and Lienhard V [Eq. (11.11)	• Facilitates the calculation of the NaCl activity coefficient in the concentration range of 0–260 g/kg with $R^2 = 1$ (refer to Fig. 11.1) • Simple to be used under ambient conditions	• Electrodialysis (ED) • Electrodialysis reversal (EDR) • Nanofiltration (NF) • Membrane distillation (MD) • Humidification–dehumidification (HDH) • High-grade water recycling • Wave-powered desalination	(Mistry and Lienhard V 2013)

11.5 EES Codes

The EES codes used for the above correlations are:

Pitzer Model

"Pitzer model is simplified in this code to be suitable for NaCl at 25 °C; fittings are generated to avoid using data from tables and to be the only function of concentration."

"for m_c in molal"

```
gamma_Pitzer_2 = (exp(−(623817971511577 * m_c^(1/2) * exp(− 2 * m_
c^(1/2)))/2251799813685248) * exp((2433947485683789 * m_c^2)/
1152921504606846976) * exp(−(7052476696057303 * m_c^(1/2))/
(18014398509481984 * ((6 * m_c^(1/2))/5 + 1))) * exp((5431261216814931 *
m_c)/36028797018963968) * exp(−(623817971511577 * exp(− 2 * m_c^(1/
2)))/4503599627370496) * exp(623817971511577/4503599627370496) *
exp((623817971511577 * m_c * exp(− 2 * m_c^(1/2)))/2251799813685248))/
((6 * m_c^(1/2))/5 + 1)^(3526238348028651 5/54043195528445952)
```

"for concentration C_c in mol/m^3"

```
C_c = 1000 "mol/m³"
```

"To change between C_c and by m_c (molal), use the following relations"

```
"dens = 997.202 + 0.0403301 * C_c
m_c = (C_c)/(dens-58.4/1000 * (C_c)"
gamma_Pitzer = exp((6012155292713906 25 * C_c^2)/(281474976710656 *
((29042553869648316 * C_c)/1770887431076117 −
17708874310761169551 36/1770887431076117)^2) − (35253379179008765 *
ln(12 * 10^(1/2) * (− C_c/((29042553869648316 * C_c)/1770887431076117 −
17708874310761169551 36/1770887431076117))^(1/2) + 1))/
54043195528445952 − (1000 * C_c * ((2494053059009051 *
((29042553869648316 * C_c)/1770887431076117 −
17708874310761169551 36/1770887431076117) * (exp(− 20 * 10^(1/2) * (−
C_c/((29042553869648316 * C_c)/1770887431076117 −
17708874310761169551 36/1770887431076117))^(1/2)) * ((2000 * C_c)/
((29042553869648316 * C_c)/1770887431076117 −
17708874310761169551 36/1770887431076117) + 20 * 10^(1/2) * (− C_c/
((29042553869648316 * C_c)/1770887431076117 −
17708874310761169551 36/1770887431076117))^(1/2) + 1) − 1))/
(18014398509481984000 * C_c) + 84738370821429/562949953421312))/
((29042553869648316 * C_c)/1770887431076117 −
17708874310761169551 36/1770887431076117) − (35253379179008765 *
10^(1/2) * (− C_c/((29042553869648316 * C_c)/1770887431076117 −
17708874310761169551 36/1770887431076117))^(1/2))/(9007199254740992 *
```

(12 * 10^(1/2) * (− C_c/((29042553869648316 * C_c)/1770887431076117 − 1770887431076116955136/1770887431076117))^(1/2) + 1))).

Bromley Method for NaCl

m_c = 1 "molal".
gamma_ Bromley = 10^(− 0.51 * **sqrt**(m_c)/(1 + **sqrt**(m_c)) + (0.06 + 0.6 * 0.0574) * m_c/(1 + 1.5 * m_c)^2 + 0.0574 * m_c)

Mistry and Lienhard V Fitting from Experimental Data for NaCl

gamma _mistry_fitting = 10^(− 0.5131 * (**sqrt**(m_c)/(1 + 1.408 * **sqrt**(m_c))− 0.17 * m_c)) − 9.262e−2 * m_c + 2e−3 * m_c^2 − 12.59e−4 * m_c^3

Fidaleo and Moresi Fittings

gamma_Fidaleo_fitting = exp(− 0.979 * sqrt(m_c)/(1 + 0.732 * sqrt(m_c)) + 0.139 * m_c)

References

Bromley LA (1973) Thermodynamic properties of strong electrolytes in aqueous solutions. AIChE J 19(2):313–320. https://doi.org/10.1002/aic.690190216

Davies CW (1938) The extent of dissociation of salts in water. Part VIII. An equation for the mean ionic activity coefficient of an electrolyte in water, and a revision of the dissociation constants of some sulphates. J Chem Soc (Resumed) 2093–2098

Fidaleo M, Moresi M (2005) Optimal strategy to model the electrodialytic recovery of a strong electrolyte. J Membr Sci 260(1–2):90–111. https://doi.org/10.1016/j.memsci.2005.01.048

Güntelberg E (1926) Untersuchungen über Ioneninteraktion. Z Phys Chem 123(1):199–247

Harvie CE, Weare JH (1980) The prediction of mineral solubilities in natural waters: the Na–K–Mg–Ca–Cl–SO_4–H_2O system from zero to high concentration at 25 °C. Geochim Cosmochim Acta 44(7):981–997. https://doi.org/10.1016/0016-7037(80)90287-2

Harvie CE, Møller N, Weare JH (1984) The prediction of mineral solubilities in natural waters: the Na–K–Mg–Ca–H–Cl–SO_4–OH–HCO_3–CO_3–CO_2–H_2O system to high ionic strengths at 25 °C. Geochim Cosmochim Acta 48(4):723–751. https://doi.org/10.1016/0016-7037(84)90098-X

Kim H, Frederick WJ (1988) Evaluation of Pitzer ion interaction parameters of aqueous electrolytes at 25 °C. J Chem Eng Data 33(5):174–188

Kraaijeveld G, Sumberova V, Kuindersma S, Wesselingh H (1995) Modelling electrodialysis using the Maxwell-Stefan description. Chem Eng J Biochem Eng J 57(2):163–176. https://doi.org/10.1016/0923-0467(94)02940-7

Mistry KH, Lienhard V JH (2013) Effect of nonideal solution behavior on desalination of a sodium chloride solution and comparison to seawater. J Energy Res Technol 135(4):042003. https://doi.org/10.1115/1.4024544

Nayar KG, Sharqawy MH, Banchik LD, Lienhard V JH (2016) Thermophysical properties of seawater: a review and new correlations that include pressure dependence. Desalination 390:1–24. https://doi.org/10.1016/j.desal.2016.02.024

Pitzer KS (1987) A thermodynamic model for aqueous solutions of liquid-like density. Rev Mineral 17:97–142

Pitzer KS, Kim JJ (1974) Thermodynamics of electrolytes. IV. Activity and osmotic coefficients for mixed electrolytes. J Am Chem Soc 96(18):5701–5707. https://doi.org/10.1021/ja00825a004

Qasem NAA, Qureshi BA, Zubair SM (2018) Improvement in design of electrodialysis desalination plants by considering the Donnan potential. Desalination 441. https://doi.org/10.1016/j.desal.2018.04.023

Qasem NAA, Generous MM, Qureshi BA, Zubair SM (2021) A comprehensive review of saline water correlations and data: part II—thermophysical properties. Arab J Sci Eng 46(3):1941–1979. https://doi.org/10.1007/s13369-020-05020-5

Robinson RA, Stokes RH (1970) Electrolyte solutions, 2nd rev edn. Butterworth and Co

Thiel GP, Lienhard V JH (2014) Treating produced water from hydraulic fracturing: composition effects on scale formation and desalination system selection. Desalination 346:54–69. https://doi.org/10.1016/j.desal.2014.05.001

Zemaitis JF, Clark DM, Rafal M, Scrivner NC (1986a) Activity coefficients of multicomponent strong electrolytes. In: Handbook of aqueous electrolyte thermodynamics. Wiley, pp 294–298. https://doi.org/10.1002/9780470938416.ch4

Zemaitis JF, Clark DM, Rafal M, Scrivner NC (1986b) Activity coefficients of single strong electrolytes. In: Handbook of aqueous electrolyte thermodynamics. Wiley, pp 46–203. https://doi.org/10.1002/9780470938416.ch4

Chapter 12
Osmotic Coefficient

Abstract The osmotic coefficient is an important thermophysical property used to consider the non-ideality of a solvent, such as water. It is usually used to predict the actual osmotic pressure. This chapter presents multicomponent models, correlation, and data used to estimate the osmotic coefficient at different salinity values. Besides, the calculated osmotic coefficient values from the correlations and methods are assessed and compared to the experimental data of seawater. Some correlations and methods are suggested based on their accuracy and operating conditions. In addition, the codes used in this chapter to predict the osmotic coefficient of saline water are supplied.

Keywords Osmotic coefficient · Correlations · Methods · Multicomponent · Osmotic pressure · Saline water

12.1 Introduction

The osmotic coefficient is used to consider the non-ideality of a solvent, such as water. It is usually used to predict the actual osmotic pressure, as expressed in the following equation in Eq. (12.1)

$$\pi = R\,T\phi\,C \tag{12.1}$$

The osmotic coefficient, ϕ, can be calculated using water activity or some models and fittings such as Pitzer and Kim (1974), Bromley (1973), and Mistry and Lienhard (2013) as listed in the following sections.

N. A. A. Qasem et al., *Thermodynamic and Thermophysical Properties of Saline Water*, Springer Water, https://doi.org/10.1007/978-3-031-35193-8_12

12.2 Multiple Components Models

Multiple components models were proposed to be general for all types of saltwater under different salinity values. Major constituents of the aqueous ionic solutions are cations, anions, and water. Pitzer and Kim, model (1974) can be used to calculate the osmotic coefficient multicomponent solution. The model for the osmotic coefficient is summarized in in Eq. (12.2).

$$\begin{aligned}\phi = 1 + \frac{2}{\sum_i m_i}\Bigg[&\frac{-A^{\phi} I^{3/2}}{1+1.2\sqrt{I}} \\ &+\sum_c\sum_a m_c m_a\left(B^{\phi}_{ca} + ZC_{ca}\right) \\ &+\sum\sum_{c<c'} m_c m_{c'}\left(\Phi^{\phi}_{cc'} + \sum_a m_a \Psi_{cc'a}\right) \\ &+\sum\sum_{a<a'} m_a m_{a'}\left(\Phi^{\phi}_{aa'} + \sum_c m_c \Psi_{aa'c}\right) \\ &+\sum_n\sum_a m_n m_a \lambda_{na} + \sum_n\sum_c m_n m_c \lambda_{nc}\Bigg]\end{aligned} \tag{12.2}$$

The parameters used here are those used to determine the activity coefficient as described in Chap. 11 (Pitzer model). Additionally,

$$B^{\phi}_{MX} = \beta^{(0)}_{MX} + \beta^{(1)}_{MX}\exp\left(-\alpha_{MX}\sqrt{I}\right) + \beta^{(2)}_{MX}\exp\left(12\sqrt{I}\right)$$

$$\Phi^{\phi}_{ij} = \theta_{ij} +^{E} \theta_{ij}(I) + I\, E_{\theta_{ij}'}(I)$$

The interactions between cations and anions are evaluated by the functions B_{ij},

$$B_{MX} = \beta^{(0)}_{MX} + \beta^{(1)}_{MX} g\left(\alpha_{MX}\sqrt{I}\right) + \beta^{(2)}_{MX} g\left(12\sqrt{I}\right)$$

$$C_{MX} = \frac{C^{\phi}_{MX}}{2|z_M z_X|^{1/2}}$$

$$B'_{MX} = \beta^{(1)}_{MX} g'\left(\alpha_{MX}\sqrt{I}\right)/I + \beta^{(2)}_{MX} g'\left(12\sqrt{I}\right)/I$$

where $\alpha_{MX} = 2.0$ for $j-1$ electrolyte and $\alpha_{MX} = 1.4$ for 2–2 and more electrolytes. $\beta_{MX}{}^{(i)}$ values are tabulated for a given ion pair, and $\beta^{(2)}_{MX}$ is associated with the complex formation and generally only non-zero for 2–2 electrolytes. $g(x)$ and $g'(x)$ are described in Chap. 11 (Pitzer model). The model can be used for all salinity values at 25 °C. However, the Pitzer parameters could be taken as a function of temperature (Thiel and Lienhard 2014).

Bromley (1973) proposed a model for the osmotic coefficient of multicomponent solutions. The model in terms of ionic strength is given in Eq. (12.3).

$$1-\phi = 2.303\,A\frac{|z_+z_-|}{I}\left[1+\sqrt{I}-\frac{1}{1+\sqrt{I}}-2\ln(1+\sqrt{I}\right] - 2.303(0.06+0.6\,B)\frac{(z_+z_-)^2}{1.5}\left[\frac{1+2\frac{1.5}{|z_+z_-|}I}{\left(1+\frac{1.5}{|z_+z_-|}I\right)^2} - \frac{\ln\left(1+\frac{1.5}{|z_+z_-|}I\right)}{\frac{1.5}{|z_+z_-|}I}\right] - 2.303\,B\frac{I}{2} \tag{12.3}$$

Here, A is in $kg^{1/2}/mol^{1/2}$ when T = 25 °C. B values for solutions and ions and the effect of temperature were listed in (Bromley 1973). The model is valid for all salinity values.

The osmotic coefficient can be determined from the following relation if the activity coefficient is known (Mazloumi 2015; Thiel and Lienhard 2014). The formula can be used for single electrolyte solutions.Eq. (12.4).

$$\phi = -1000\frac{\ln a_W}{M_W\sum_i m_i} \tag{12.4}$$

where a_w is the activity of water, Mw is the molar mass of water (g/mol), and m is molality (mol/kg). It is also valid for all salinity values.

12.3 Correlations and Data

Correlations are the easiest way to calculate the saltwater osmotic coefficient as they are fitted from the experimental data. Most of the available osmotic coefficient correlations in the literature are for seawater. Mistry and Lienhard (2013) provided the best-fitting obtained from the experimental data of (Robinson and Stokes 1970),

$$\phi = 10^{-0.1924\left(\frac{\sqrt{m_s}}{1+2.798\sqrt{m_s}}-0.3506m_s\right)} - 0.01019\,m_s - 0.001201\,m_s^2 - 0.001324\,m_s^3 \tag{12.5}$$

The solution molality, m_s, is in mol/kg. The fitting can be used for all salinity values with an accuracy of $R^2 \sim 1$.

12.4 Case Study

To compare some of the mentioned methods, NaCl experimental data (Robinson and Stokes 1970) of the osmotic coefficient is selected over a wide range of salinity (up to 6 molal, ~ 260 g/kg). Figure 12.1a shows the estimated values of the osmotic coefficient versus the molal concentration up to 260 g/kg. All compared methods demonstrate an excellent agreement with those of the experimental data. The difference from the experimental data is shown in Fig. 12.1b. Pitzer model proves to be an excellent approach with a difference of less than 0.3%, Mistry and Lienhard's (2013) fitting overlaps with the experimental data. However, this fitting correlation is only valid for NaCl. The Bromley model can also be used with a difference of less than 1%.

Based on the above discussion, some methods and correlations could be recommended to be used as summarized in Table 12.1.

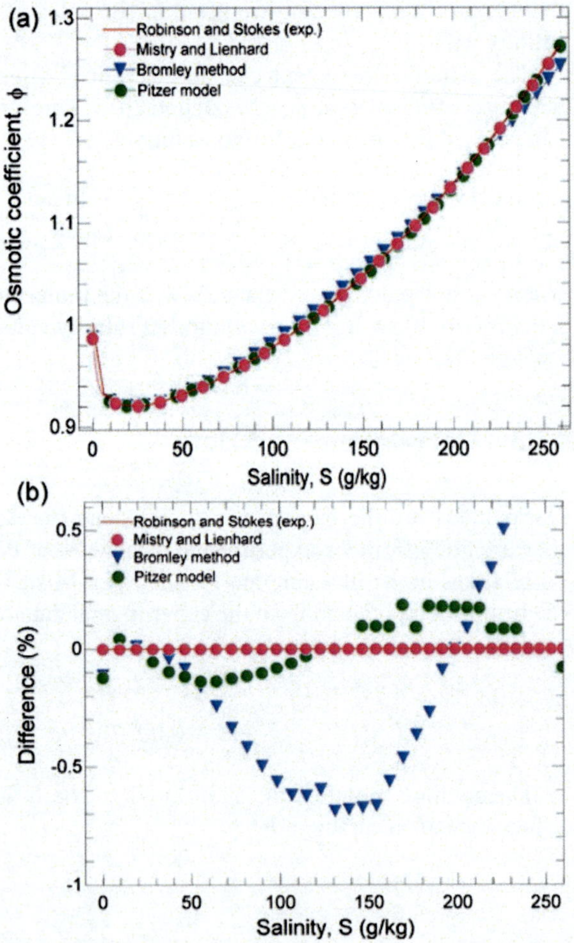

Fig. 12.1 Comparison of osmotic coefficient values of different models (Pitzer and Kim (1974), Bromley (1973), Mistry and Lienhard (2013), and Fidaleo and Moresi (2005)) against experimental NaCl data of Robinson and Stokes (1970): **a** estimated values and **b** the difference from the experimental data

Table 12.1 Summary of the recommended correlations and models for the osmotic coefficient of saline water (Qasem et al. 2021)

Correlation-model	Advantages	Applications	Refs.
Pitzer model [No. # 1]	• General for multicomponent saline solution • It is one of the most accurate models, with a salinity range of 0–260 g/kg of saline water with a difference of approximately ± 0.2% (refer to Fig. 12.1)	• All membrane-based desalination technologies • All thermal-based desalination technologies	Pitzer and Kim (1974), Thiel and Lienhard (2014)
Bromley model [Eq. (12.3)]	• General for multicomponent saline solution • Valid in the salinity range of 0–6 mol/kg within an approximated difference of ± 0.9% (refer to Fig. 12.1)	• All membrane-based desalination technologies • All thermal-based desalination technologies	Bromley (1973)
Mistry and Lienhard [Eq. (12.5)]	• Facilitates the calculation of the NaCl osmotic coefficient in the concentration range of 0–260 g/kg within $R^2 = 1$ (refer to Fig. 12.1) • Simple to be used under ambient conditions	• Electrodialysis (ED) • Electrodialysis reversal (EDR) • Nanofiltration (NF) • Membrane distillation (MD) • Humidification–dehumidification (HDH) • High-grade water recycling • Wave-powered desalination	Mistry and Lienhard (2013)

To select a proper correlation to estimate the osmotic coefficient, some criteria should be considered, such as the type of saline water (seawater, multicomponent, high saline, and so on), operating limits (salinity, temperature, and pressure), and fitting accuracy. Besides these, a comparison with the existing experimental data (which were mostly reported for seawater) can be used to test and select the recommended correlation. Some correlations are valid for a particular type of desalination process, whereas others are not. For example, humidification-dehumidification desalination systems operate at 0.1 MPa so that the dependent-pressure correlations may not be needed. However, reverse osmosis (RO) desalination needs variable pressure properties. In accordance with these criteria, Table 12.1 proposes the recommended correlations and methods to be used to calculate the osmotic coefficient of saline water.

It is also noticed that most correlations are for NaCl properties, while the multicomponent correlations are limited. It is strongly recommended to conduct large-scale experimental investigations to evaluate multicomponent saline water properties at different values of salinity, temperature, and pressure. The corresponding correlations of the tested properties could then be obtained.

12.5 EES Codes

The EES codes used for calculating the above correlations are:

Pitzer Model

"for NaCl at T = 298 K; the model here is obtained by fitting some values from tables for simplicity"

m_c = 1 "molal"

phi_Pitzer = (5431261216814931 * m_c)/72057594037927936 + (623817971511577 * m_c * exp(−2 * m_c^(1/2)))/2251799813685248 + 176311917401432575/(648518346341351424 * (m_c^(1/2) + 5/6)) + (811315828561263 * m_c^2)/576460752303423488 + 72824007576605389/ 108086391056891904

Bromley Method

"for NaCl"

m_c = 1 "molal"

phi_Bromley = 1 − 2.303 * 0.511/m_c * (1 + **sqrt**(m_c) − 1/(1 + **sqrt**(m_c)) − 2 * **ln**(1 + **sqrt**(m_c))) − 2.303 * (0.06 + 0.6 * 0.0574)/1.5 * ((1 + 2 * 1.5 * m_c)/(1 + 1.5 * m_c)^2 − **ln**(1 + 1.5 * m_c)/(1.5 * m_c)) − 2.303 * 0.0574 * m_c/2)

Mistry and Linhard V Fitting from Experimental Data for NaCl

phi_Mistry = 10^(−0.1924 * (m_c^0.5/(1 + (2.798 * m_c^0.5)) − 0.3506 * m_c)) − (0.1019 * m_c) − (0.001201 * m_c^2) − (0.001324 * m_c^3)

References

Bromley LA (1973) Thermodynamic properties of strong electrolytes in aqueous solutions. AIChE J 19(2):313–320. https://doi.org/10.1002/aic.690190216

Fidaleo M, Moresi M (2005) Optimal strategy to model the electrodialytic recovery of a strong electrolyte. J Membr Sci 260(1–2):90–111. https://doi.org/10.1016/j.memsci.2005.01.048

Mazloumi SH (2015) Representation of activity and osmotic coefficients of electrolyte solutions using non-electrolyte Wilson-NRF model with ion-specific parameters. Fluid Phase Equilib 388:31–36. https://doi.org/10.1016/j.fluid.2014.12.035

Mistry KH, Lienhard VJH (2013) Effect of nonideal solution behavior on desalination of a sodium chloride solution and comparison to seawater. J Energy Resour Technol 135(4):042003. https://doi.org/10.1115/1.4024544

Pitzer KS, Kim JJ (1974) Thermodynamics of electrolytes. IV. Activity and osmotic coefficients for mixed electrolytes. J American Chem Soc 96(18):5701–5707. https://doi.org/10.1021/ja00825a004

Qasem NAA, Generous MM, Qureshi BA, Zubair SM (2021) A comprehensive review of saline water correlations and data: part II—thermophysical properties. Arab J Sci Eng 46(3):1941–1979. https://doi.org/10.1007/s13369-020-05020-5

Robinson RA, Stokes RH (1970) Electrolyte solutions, 2nd. Butterworth and Co.

Thiel GP, Lienhard VJH (2014) Treating produced water from hydraulic fracturing: composition effects on scale formation and desalination system selection. Desalination 346:54–69. https://doi.org/10.1016/j.desal.2014.05.001

Chapter 13
Surface Tension

Abstract Surface tension is a physical property to be considered for designing thermal desalination technologies, as evaporation is the core process at the water–air interface. It is sensitive to liquid temperature and salinity. Thus, this chapter explains multicomponent models, correlation, and data used to estimate surface tension at different salinity and temperature values. Besides, the calculated surface tension values from the correlations and methods are compared and discussed against the experimental data of seawater. This chapter recommends some correlations and methods based on their accuracy and operating conditions. Also, the codes used for the investigated surface tension methods are provided.

Keywords Surface tension · Multicomponent · Correlations · Methods · Water–air interface · Saline water

13.1 Introduction

Surface tension is a phenomenon that arises at the interface of different fluids as the result of an imbalance of forces between the molecules of the two fluids. This delicate phenomenon can be observed at the interface of the water and air in ordinary life as the forces between the water molecules are stronger than the forces between water and air molecules at the interface. The phenomenon is at the back of many interesting real-world phenomena, such as bubble formation and drop formation (Peter and Robert 1996; Sato et al. 2010). It is one of the important physical properties to be considered while designing thermal desalination technologies, as evaporation is the core process that happens at the water–air interface (Nayar 2014).

The surface tension of pure water is 0.0728 N/m (at ambient conditions), and it is sensitive to many parameters, including temperature, pressure, and salinity (for saline water). It increases with an increase in salinity. Non-volatile solutes, primarily ionic compounds, dissociate into the water and produce positive and negative ions. The dipole nature of water molecules (due to the electronegativity difference of hydrogen and oxygen in the water molecules) makes this phenomenon more interesting. Hydrogen with a partial positive charge surrounds negative ions, and partial

N. A. A. Qasem et al., *Thermodynamic and Thermophysical Properties of Saline Water*, Springer Water, https://doi.org/10.1007/978-3-031-35193-8_13

negative oxygen surrounds positive ions generated by the ionic compounds. This dipole–dipole interaction results in an even stronger imbalance of forces, which is the reason for the proportional relation of salinity and surface tension of saline water (Boniewicz-Szmyt and Pogorzelski 2016; Lima et al. 2013; Nayar 2014; Nayar et al. 2014; Okur et al. 2017; Salameh et al. 2016; Wang et al. 2018). Studies reported by Marcus (2015) and Bromley et al. (Leroy et al. 2010) reiterated and confirmed the situation described above.

Temperature is another important thermodynamic property that affects this phenomenon. Average kinetic energy and random motion of molecules accelerate with the increase in temperature of saline water. This results in a reduction of forces between water molecules themselves, and the imbalance of forces declines to a lower value, which results in lower tension on the water–air interface. That is why surface tension decreases with an increase in temperature. Součková et al. (2008) and Leonard et al. (2018) studied surface tension relation with temperature and pressure at the gas–liquid interface and reported the reduction of this force with increasing temperature.

The objective of this chapter is to summarize various correlations and models to approximate this surface force and their relative accuracy.

13.2 Multiple Components Models

Multicomponent models were proposed to be general for all types of saltwater under different salinity values. Major constituents of the aqueous ionic solutions are cations, anions, and water. The Pitzer equation can be used for the calculated surface tension of a multicomponent aqueous solution (Kamali et al. 2015; Stairs 2006). The main relation is provided in Eq. (13.1).

$$\sigma = \sigma_w + \frac{\nu RT}{55.51\, A_w}\left(\Phi^b \sum_{i=1}^{n} V_i m_i^b - \Phi^S \sum_{i=1}^{n} V_i m_i^s\right) \tag{13.1}$$

where A_w is the molal area of water that can be calculated as:

$$A_w = V_w^{\frac{2}{3}}\, N_A^{\frac{1}{3}}$$

Here, V_w and N_A are the molar volume of water and Avogadro number, respectively. σ is the surface tension of saline water in N/m, σ_w is the surface tension of pure water in N/m, R is the gas constant in J/mol.K, T is the temperature in K, Φ^b is the osmotic coefficient of the bulk liquid phase, Φ^s is the osmotic efficient of surface phase, V_i is the summation of stoichiometric coefficients, g_i is the interface parameter (see Table 13.1), m_i^b is the molality of electrolyte in bulk phase in mol/kg, $m_i{}^s$ is the molality of electrolyte in surface phase in mol/kg, and ν is the stoichiometric coefficient of ions.

Table 13.1 Calculated results of surface tension and parameters for single–electrolyte aqueous solutions (Li et al. 1999)

System	T(K)	g	% AAD	Max. concentration (m)
HCl	293.15	1.06935	0.35	14.98
NaCl	293.15	0.10898	0.52	6.00
HNO_3	293.15	1.23725	0.96	19.08
KOH	29I.15	0.01199	0.73	3.80
NaOH	291.15	0.53682	2.51	14.00
$HClO_4$	298.15	1.00647	2.69	25.92
KCl	293.15	0.05712	0.09	4.39
LiCl	298.15	0.31607	0.56	3.82
NaBr	293.15	0.20303	0.22	2.90
$NaNO_3$	293.15	0.06280	0.16	7.00
CsCl	298.15	0.16498	0.57	8.89
LiBr	303.15	0.84835	4.08	17.27
KI	298.15	0.39899	0.17	7.29
KNO_2	293.15	0.21082	1.41	30.06
NH_4Cl	298.15	0.19310	0.21	5.63
NH_4NO_3	293.15	0.07258	0.60	19.36
HCOONa	303.15	0.46171	0.59	9.80
NaAc	303.15	1.07064	1.14	10.31
C_2H_5COONa	303.15	1.83275	0.31	4.40
NaI	298.15	0.70873	1.17	8.82
RbCl	298.15	0.10332	0.23	6.93
KAc	303.15	1.07634	4.47	23.79
KSCN	298.15	0.95574	0.04	6.83
$AgNO_3$	293.15	0.17869	1.61	6.02
KNO_3	298.15	0.00014	0.41	2.18
KBr	293.15	0	0.13	1.05
$ZnCl_2$	298.15	0.24109	0.19	2.15
H_2SO_4	298.15	0.84411	0.31	5.16
$BaCl_2$	303.15	0	1.15	1.60
$CaCl_2$	303.15	0.60296	2.43	6.00
$CuSO_4$	303.15	0	2.51	1.10
K_2CrO_4	298.15	0	0.89	1.37
K_2SO_4	298.15	0	0.56	1.05
$LaCl_3$	298.15	0	0.30	1.03
$K_3Fe(CN)_6$	298.15	0	0.49	1.45

(continued)

Table 13.1 (continued)

System	T(K)	g	% AAD	Max. concentration (m)
$K_4Fe(CN)_6$	298.15	0	0.26	0.54
$MgCl_2$	293.15	0.60967	1.84	3.50
$MgSO_4$	283.15	0	1.64	2.32
$MnCl_2$	291.15	0.40118	1.48	5.62
$(NH_4)_2SO_4$	303.15	0	1.27	5.06
Na_2CrO_4	303.15	0.35885	3.49	6.59
Na_2SO_4	303.15	0	1.46	1.24
$SrCl_2$	293.15	0	0.65	2.81
UO_2SO_4	303.15	0	2.04	2.34
$Zn(NO_3)_2$	313.15	0.76658	2.51	7.01
$ZnSO_4$	298.15	0	2.11	2.86

And, m_i^s is related m_i^b by the interface parameter (g_i) as:

$$m_i^s = g_i m_i^b$$

The expression for surface tension can be modified by substituting for molality of electrolyte in terms of molality of electrolyte in the bulk phase as:

$$\sigma_s = \sigma_w + \frac{\nu RT}{55.51}\left(\Phi^b \sum_{i=1}^{n} V_i m_i^b - \Phi^S \sum_{i=1}^{n} g_i V_i m_i^b\right) \tag{13.2}$$

The osmotic coefficients (Φ^S and Φ^b) can be calculated from Pitzer equations (see Chap. 12). The model is valid for all salinity values at 20 °C and 0.1 MPa with an accuracy of $\pm$ 1.22%. However, temperature limitation depends on the data for interface parameters.

Using the activity of water, the following expression can be used for the calculation of the surface tension of saline water (Kamali et al. 2015).

$$\sigma = \sigma_w + \frac{RT}{A_w} \ln\left(\frac{a_w^s}{a_w^b}\right) \tag{13.3}$$

where σ is the surface tension of saline water in N/m, σ_w is the surface tension of pure water in N/m, R is the gas constant in J/mol.K, T is the temperature in K, a_w^s is the activity of water in the surface phase, a_w^b is the activity of water in the liquid phase, and A_w is the molal area of water in m^2 and can be expressed as,

$$A_w = V_w^{\frac{2}{3}} N_A^{\frac{1}{3}}$$

Here, V_w and N_A are the molar volume of water and Avogadro number, respectively. The model is valid for all salinity and temperature values at 0.1 MPa with an average absolute deviation of 1.7%.

13.3 Correlations and Data

Correlations are the easiest way to calculate the saltwater surface tension as they are fitted from the experimental data. Most of the available surface tension correlations in the literature are for seawater. Nayar et al. (2014) proposed a correlation for the surface tension of seawater in terms of temperature in °C and salinity in g/kg. The surface tension of saline water (σ) in mN/m can be obtained from Eq. (13.4), and the required constants are provided in Table 13.2.

$$\sigma = \sigma_w(1 + a_1S + a_2ST) \tag{13.4}$$

The surface tension of pure water (σ_w) can be determined in Eq. (13.11). The correlation is valid in the salinity range of 0–130 g/kg, pressure range of 0.1–1 MPa, and temperature range of 0–100 °C with an accuracy of ± 0.60%.

Sharqawy et al. (2011) correlation (reported in IAPWS 1992 also) in terms of temperature in °C and salinity in g/kg, the surface tension of seawater (σ) in mN/m is expressed in Eq. (13.5), and the required constants are provided in Table 13.3.

$$\sigma = \sigma_w(1 + (a_1T + a_2)\ln(1 + bS)) \tag{13.5}$$

The surface tension of pure water (σ_w) can be determined using Eq. (13.11). The correlation of Sharqawy et al. (2011) is valid in a salinity range of 0–40 g/kg, pressure range of 0.1–1 MPa, and temperature range of 0–40 °C within an accuracy of ± 0.18%.

Table 13.2 The constants for Eq. (13.4)

Constants	Values
a_1	3.766×10^{-4}
a_2	2.347×10^{-6}

Table 13.3 The constants for Eq. (13.5)

Constants	Values
a_1	0.000226
a_2	0.00946
b	0.0331

Table 13.4 The constants for Eq. (13.7)

Constants	Values
a_1	75.59
a_2	− 0.13476
a_3	0.021352
a_4	0.00029529

Guohua et al. (1994) fitted a correlation for the surface tension of seawater. It is a function of temperature (T) in °C and salinity (S) in g/kg. The surface tension of seawater (σ) in mN/m is expressed in Eq. (13.6), and mandatory constants are provided in Table 13.4.

$$\sigma = a_1 + a_2 T + a_3 S_p + a_4 S_p T \tag{13.6}$$

Salinity is related to the practical salinity by Eq. (13.7).

$$S = 1.0047 S_p \tag{13.7}$$

The correlation is valid in the salinity range of 0–35 g/kg, pressure range of 0.1 MPa, and temperature range of 15–35°C within an accuracy of ± 0.1%.

Krummel (1907) proposed a correlation for the surface tension of seawater (σ) in mN/m. It is in terms of temperature (T) in °C and salinity (S) in g/kg as expressed in Eq. (13.8). The required constants are provided in Table 13.5.

$$\sigma = a_1 + a_2 T + a_3 S \tag{13.8}$$

The Krummel (1907) correlation is valid in the salinity range of 10–35 g/kg, pressure range of 0.1 MPa, and temperature range of 0–40 °C within an accuracy of ± 0.1%.

Fleming and Revelle (1939) proposed a correlation for the surface tension of seawater (σ) in mN/m. It is in terms of temperature (T) in °C and salinity (S) in g/kg as expressed in Eq. (13.9). The required constants are provided in Table 13.6.

$$\sigma = a_1 + a_2 T + a_3 S \tag{13.9}$$

Table 13.5 The constants of Eq. (13.8)

Constants	Values
a_1	77.09
a_2	− 0.17788
a_3	0.0221

Table 13.6 The constants of Eq. (13.9)

Constants	Values
a_1	75.64
a_2	− 0.144
a_3	− 0.0221

Table 13.7 Constants for Eq. (13.11)

Constants	Values
a_1	0.2358
a_2	647.096
a_3	− 0.625

The Fleming and Revelle (1939) correlation is valid in the salinity range of 10–35 g/kg, pressure range of 0.1 MPa, and temperature range of 0–40 °C within an accuracy of ± 0.5%.

Dessouky and Ettouney (2002) proposed a correlation for the surface tension of pure water (σ_w) in N/m in the form of Eq. (13.10). It is a function of temperature in °C. The required constants are provided in Table 13.7.

$$\sigma_w = a_1 + a_2 T + a_3 T^2 \tag{13.10}$$

The correlation is valid in the temperature range of 0–136 °C at 0.1 MPa within an accuracy of ± 0.3%.

Nayar et al. (2014) provided a correlation for the surface tension of pure water (σ_w) in mN/m. It is in terms of temperature in °C as can be seen in Eq. (13.11). The required constants are provided in Table 13.7.

$$\sigma_w = a_1 \left(1 - \frac{T + 273.15}{a_2}\right)^{1.256} \left(1 + a_3 \left(1 - \frac{T + 273.15}{a_2}\right)^{1.256}\right) \tag{13.11}$$

The correlation is valid in the temperature range of 0.01–370 °C at 0.1 MPa within an accuracy of ± 0.08%.

13.4 Case Study

The surface tension of saline water is affected by salinity and temperature. A comparison between different correlations and methods in terms of salinity is shown in Fig. 13.1. These correlations are Fleming and Revelle (1939), Guohua et al. (1994), Krummel (1907), and Nayar et al. (2016), while the experimental data of Nayar et al. (2016) is taken as a reference. Figure 13.1a shows the surface tension values against various salinity values at 30 °C and 0.1 MPa. It can be seen from Fig. 13.1a that the correlations are showing an increasing trend with increasing salinity, as expected. The Nayar et al. (2014) correlation shows closer values to those of Nayar et al. (2016), while others have slightly lower values. Figure 13.1b shows the difference between these correlations against the experimental data of Nayar et al. (2016). The correlation of Nayar et al. (2014) has a difference of less than 0.2%, and the correlation of Krummel (1907), Fleming and Revelle (1939), and Guohua et al. (1994) demonstrate an excellent accuracy (± 0.5% error) within their operating range.

The effect of temperature is highlighted, as shown in Fig. 13.2. A comparison between values of the surface tension calculated using the correlations and methods of Fleming and Revelle (1939), IAPWS (1992), Krummel (1907), Nayar et al. (2014), and Guohua et al. (1994) for various temperature ranges at 35 g/kg and 0.1 MPa is plotted in Fig. 13.2a. Nayar et al. (2016) method is taken as a reference in this comparison. All correlations expected the same decreasing trend of the surface

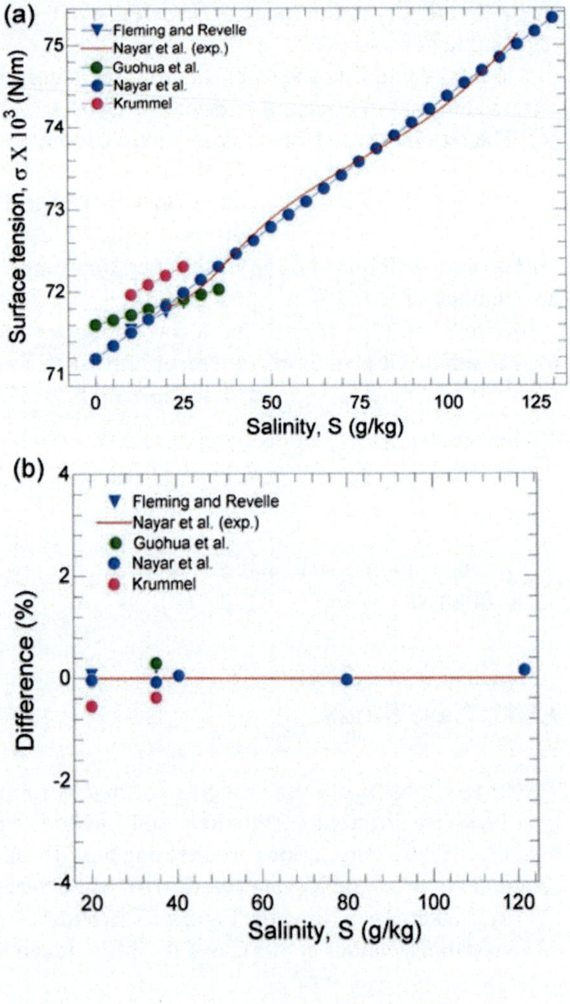

Fig. 13.1 **a** Salinity versus surface tension and **b** the difference against the experimental data of Nayar et al. (2016) for the methods of Fleming and Revelle (1939), Guohua et al. (1994), Krummel (1907) and Nayar et al. (2014) at 30 °C and 0.1 MPa

tension against the increase in temperature values. Interestingly, those of Guohua et al. (1994) and Fleming and Revelle (1939) overlap with the data of Nayar et al. (2016). All the correlations estimate surface tension values within ± 2% error, as shown in Fig. 13.2b. The most accurate methods are Guohua et al. (1994) and Fleming and Revelle (1939), with a difference of 0.05% and 0.8%, respectively. Guohua et al. (1994) correlation is strongly recommended under the investigated conditions.

Properties like surface tension are equally important for thermal-based desalination technologies and electrically driven technologies such as electrodialysis. As surface tension arises due to the polar nature of water, that can also hinder the motion of ionic species. Therefore, highly viscous fluid may have more surface tension and require energy to separate ions from water. Table 13.8 proposes the recommended correlations and methods to be used to calculate the surface tension of saline water.

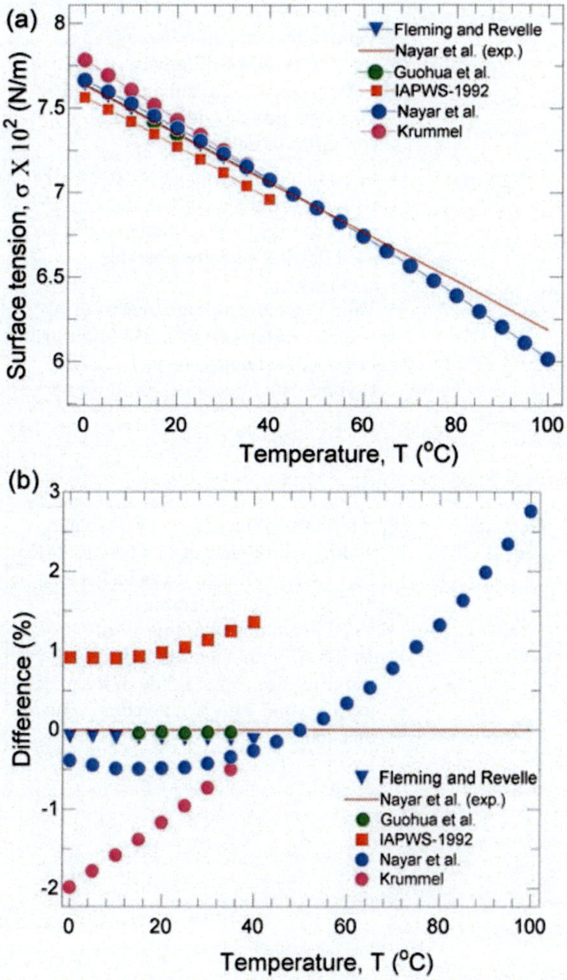

Fig. 13.2 **a** Temperature versus surface tension and **b** the difference against the experimental data of Nayar et al. (2016) for the methods of Fleming and Revelle (1939), IAPWS (1992), Krummel (1907), Nayar et al. (2014) and Guohua et al. (1994) at 35 g/kg and 0.1 MPa

Table 13.8 Summary of the recommended correlations and models for the surface tension of saline water (Qasem et al., 2021)

Correlation -model	Advantages	Applications	Refs.
Pitzer model [Eq. (13.1)]	• It is valid for a multicomponent solution • It was reported to have a small difference of less than 0.6%	• All membrane-based desalination technologies • All thermal-based desalination technologies	Kamali et al. (2015), Li et al. (1999), Stairs (2006)
Nayar et al [Eq. (13.4)]	• Most accurate in a salinity range of 0–120 g/kg within a difference of 0.5% (refer to Fig. 13.1); although it can be used beyond with considerable accuracy • Valid in the temperature range of 0–90 °C within a difference of ± 2.5% (refer to Fig. 13.2) (although it can be used beyond this with a minute compromise on accuracy)	• Electrodialysis (ED) • Electrodialysis reversal (EDR) • Nanofiltration (NF) • High-grade water recycling • Wave-powered desalination	Nayar et al. (2014)
Krummel [Eq. (13.8)]	• Valid in a salinity range of 10–35 g/kg within a difference of less than ± 0.5% (refer to Fig. 13.1) (although it can be used beyond with considerable accuracy) • Valid in a temperature range of 0–40 °C within a difference of ± 2% (refer to Fig. 13.2) (although it can be used beyond with minute compromise on accuracy)	• Electrodialysis (ED) • Electrodialysis reversal (EDR) • Nanofiltration (NF) • All thermal desalination technologies • High-grade water recycling • Wave-powered desalination	Krummel (1907)
Guohua [Eq. (13.6)]	• Valid in a salinity range of 0–35 g/kg within a difference of ± 0.5% (refer to Fig. 13.1) (although it can be used beyond with considerable accuracy) • Valid in a temperature range of 15–35 °C within a difference of ± 0.2% (refer to Fig. 13.2) (although it can be used beyond with minute compromise on accuracy)	• All thermal-based desalination technologies • Seawater greenhouse • Electrodialysis (ED) • Reverse electrodialysis	Ghohua et al. (1994)

(continued)

Table 13.8 (continued)

Correlation -model	Advantages	Applications	Refs.
Fleming and Revelle [Eq. (13.9)]	• Valid in a salinity range of 10–35 g/kg within a difference of ± 0.4% (refer to Fig. 13.1); although, it can be used beyond 35 g/kg with considerable accuracy • Valid in a temperature range of 0–40 °C within a difference of ± 0.2% (refer to Fig. 13.2); although, it can be used beyond with a compromise on accuracy)	• All thermal-based desalination technologies • Seawater greenhouse • Electrodialysis (ED) • Reverse Electrodialysis	Fleming and Revelle (1939)
Sharqawy correlation (also reported in IAPWS-1992) [Eq. (13.5)]	• Valid in a temperature range of 0–40 °C within a difference of ± 2% (refer to Fig. 13.2); although, it can be used beyond with negligible compromise on accuracy	• All thermal-based desalination technologies • Seawater greenhouse • High-grade water recycling • Wave-powered desalination	IAPWS, (1992), Sharqawy et al. (2011)

13.5 EES Codes

The used EES codes for the above correlations are:

Nayar et al. Correlation

```
"Temperature"

T = 25 "C"

S = 35 "g/kg"

sigma = sigma_w * (1 + 3.766/10^4 * S + 2.347/10^6 * S * T)

Sigma_w = 235.8 * (1 – (T + 273.15)/(647.096))^1.256 * (1 – 0.625 * (1 – (T
+ 273.15)/(647.096)))
```

```
Sigma_SI = sigma/1000
```

Sharqawy Correlation

```
"Temperature"

T = 25 "C"
S = 35 "g/kg"

Sigma_w = 235.8 * (1 − (T + 273.15)/(647.096))^1.256 * (1 − 0.625 * (1 − (T
+ 273.15)/(647.096)))

Sigma_SI = sigma_w/1000

sigma = sigma_SI * (1 + (0.000226 * T + 0.00946) * ln(1 + 0.0331 * S))
```

Guohua et al. Correlation

```
T = 25 "C"
S = 35 "g/kg"

S = 1.00472 * S_p

Sigma = 75.64 − 0.1347 * T + 0.021352 * S_p − 0.00029529 * S_p * T

"SI units"
sigma_SI = sigma/1000
```

Krummel Correlation

```
T = 25 "C"
S = 35 "g/kg"

sigma = 77.09 − 0.17788 * T + 0.0221 * S

Sigma_SI = sigma/1000
```

Fleming and Revelle Correlation

```
T = 25 "C"
```

S = 35 “g/kg”

sigma = 75.64 − 0.144 * T + 0.0221 * S

Sigma_SI = sigma/1000

Dessouky and Ettouney Correlation for Pure Water (Saturated)

“Temperature in C”

T = 25

Sigma = 7.5798/10^2 − 1.469/10^4 * T − 2.2173/10^7 * T^2

Pure Water Correlation

“Temperature”

T = 25 “C”

Sigma_w = 235.8 * (1 − (T + 273.15)/(647.096))^1.256 * (1 − 0.625 * (1 − (T + 273.15)/(647.096)))

Sigma_SI = sigma_w/1000

References

Boniewicz-Szmyt K, Pogorzelski SJ (2016) Thermoelastic surface properties of seawater in coastal areas of the Baltic Sea. Oceanologia 58(1):25–38. https://doi.org/10.1016/j.oceano.2015.08.003

El-Dessouky HT, Ettouney HM (2002) Fundamentals of salt water desalination. Fundam Salt Water Desalin. https://doi.org/10.1016/b978-0-444-50810-2.x5000-3

Fleming RH, Revelle RR (1939) Physical process in the ocean, recent marine sediments. Am Assoc Petrol Geol Bull 48:48–141

Ghohua C, Jingzen S, Ling G, Lijun Z (1994) Study on the surface tension of sea water. Oceanologia Et Limnologia Sinica 25(3):306–311

IAPWS (1992) The international association for the properties of water and steam: revised (June 2014) supplementary release on saturation properties of ordinary water substance (vol 10, Issue August)

Kamali MJ, Kamali Z, Vatankhah G (2015) Thermodynamic modeling of surface tension of aqueous electrolyte solution by competitive adsorption model. J Thermodyn 2015:1–8. https://doi.org/10.1155/2015/319704

Krummel O (1907) Handbuch der ozeanographie (2. völlig). Stuttgart, J. Engelhorn. https://doi.org/10.5962/bhl.title.16991

Leonard C, Ferrasse JH, Boutin O, Lefevre S, Viand A (2018) Measurements and correlations for gas liquid surface tension at high pressure and high temperature. AIChE J 64(11):4110–4117. https://doi.org/10.1002/aic.16216

Leroy P, Lassin A, Azaroual M, André L (2010) Predicting the surface tension of aqueous 1:1 electrolyte solutions at high salinity. Geochim Cosmochim Acta 74(19):5427–5442. https://doi.org/10.1016/j.gca.2010.06.012

Li ZB, Li YG, Lu JF (1999) Surface tension model for concentrated electrolyte aqueous solutions by the pitzer equation. Ind Eng Chem Res 38(3):1133–1139. https://doi.org/10.1021/ie980465m

Lima ERA, De Melo BM, Baptista LT, Paredes MLL (2013) Specific ion effects on the interfacial tension of water/hydrocarbon systems. Braz J Chem Eng 30(1):55–62. https://doi.org/10.1590/S0104-66322013000100007

Nayar KG (2014) An experimental investigation of the surface tension of seawater (Issue December 2014). MIT

Nayar KG, Panchanathan D, McKinley GH, Lienhard JH (2014) Surface tension of seawater. J Phys Chem Ref Data 43(4). https://doi.org/10.1063/1.4899037

Nayar KG, Sharqawy MH, Banchik LD, Lienhard JH (2016) Thermophysical properties of seawater: a review and new correlations that include pressure dependence. Desalination 390:1–24. https://doi.org/10.1016/j.desal.2016.02.024

Okur HI, Chen Y, Wilkins DM, Roke S (2017) The Jones-Ray effect reinterpreted: Surface tension minima of low ionic strength electrolyte solutions are caused by electric field induced water-water correlations. Chem Phys Lett 684:433–442. https://doi.org/10.1016/j.cplett.2017.06.018

Peter KW, Robert JP (1996) Surface tension of aqueous solutions of electrolytes: relationship with ion hydration, oxygen solubility, and bubble coalescence. J Colloid Interface Sci 184:550–563. https://doi.org/10.1006/jcis.1996.0651

Qasem NAA, Generous MM, Qureshi BA, Zubair SM (2021) A comprehensive review of saline water correlations and data: Part II—thermophysical properties. Arab J Sci Eng 46(3):1941–1979. https://doi.org/10.1007/s13369-020-05020-5

Salameh A, Vorka F, Daskalakis V (2016) Correlation between surface tension and the bulk dynamics in salty atmospheric aquatic droplets. J Phys Chem C 120(21):11508–11518. https://doi.org/10.1021/acs.jpcc.6b01842

Sato A, Aoki M, Watanbe M (2010) Single rising bubble motion in aqueous solution of electrolyte. J Fluid Sci Technol 5(1):14–25. https://doi.org/10.1299/jfst.5.14

Sharqawy MH, Lienhard VJH, Zubair SM (2011) Thermophysical properties of seawater: a review of existing correlations and data. Desalination Water Treat 29(1–3):355–355. https://doi.org/10.5004/dwt.2011.2947

Součková M, Klomfar J, Pátek J (2008) Measurement and correlation of the surface tension-temperature relation for methanol. J Chem Eng Data 53(9):2233–2236. https://doi.org/10.1021/je8003468

Stairs RA (2006) Calculation of surface tension of salt solutions: effective polarizability of solvated ions. Can J Chem 73(6):781–787. https://doi.org/10.1139/v95-098

Wang X, Chen C, Binder K, Kuhn U, Pöschl U, Su H, Cheng Y (2018) Molecular dynamics simulation of the surface tension of aqueous sodium chloride: from dilute to highly supersaturated solutions and molten salt. Atmos Chem Phys 18(23):17077–17086. https://doi.org/10.5194/acp-18-17077-2018

Marchus Y (2015) Surface tension of aqueous electrolytes and ions. Encycl Surf Colloid Sci Third Edition 4:7111–7115. https://doi.org/10.1081/e-escs3-120000581

Chapter 14
Electrical Conductivity

Abstract Saltwater electrical conductivity is one of the essential physical properties in the desalination and water treatment industries, especially for electrically driven technologies such as electrodialysis. Saline water with higher salinity is considered a better conductor than pure water. This chapter discusses the multicomponent models that estimate the saline water electrical conductivity at different salinity and temperature values. Experimental data of seawater electrical conductivity are used to compare the estimated electrical conductivity values from the methods and assess their accuracy. Some methods are recommended based on their accuracy and operating conditions. Besides, the codes of methods used to calculate the electrical conductivity are provided.

Keywords Electrical conductivity · Multicomponent · Ionic strenghth · Electrodialysis · Saline water

14.1 Introduction

The electrical conductivity of saline water is one of the essential properties in the desalination industry, especially for electrically driven technologies such as electrodialysis. Although, water in its pure form is considered a very weak electrolyte such that its conductivity increases almost linearly with added salt concentration. Whenever salt is added to water, water molecules break up the salt crystals into their respective ionic species. Although it is the most prevailing thought that these ionic species accelerate electron transfer through an aqueous solution, a definite electric circuit is a mystery (Sauerheber and Heinz 2015). The electrical conductivity of saline water is proportional to the ionic concentration in the saline water. Therefore, salinity is a vital parameter in the determination of electrical conductivity (Szymczak et al. 1975). This section summarizes the various approaches used to determine the electrical conductivity of aqueous solutions with any given salinity under a specific accuracy.

N. A. A. Qasem et al., *Thermodynamic and Thermophysical Properties of Saline Water*, Springer Water, https://doi.org/10.1007/978-3-031-35193-8_14

The electrical conductivity of saline water is also sensitive to temperature, pressure, and even PH (Zhu et al. 2016). Temperature is the second most important parameter that can affect the electrical conductivity of saline water up to a large extent (Bradshaw and Schleicher 1980). Therefore, models dealing with temperature effects are also summarized in this chapter.

Different approaches are used to predict the electrical conductivity of saline water involving ionic strength, ionic mobility, mean spherical approximation (Turq et al. 1995), and electrical conductivity based on the ionic diffusion coefficient (Felderhof 2003).

14.2 Ionic Strength and Salinity

The ionic strength approach proved to be the best tool to predict the electrical conductivity of an aqueous ionic solution. The ionic strength (I) for multi-ionic species can be obtained by the following relation,

$$I = \frac{1}{2}\sum_{i}^{n} m_i Z_i^2 \tag{14.1}$$

where m_i and Z_i are the molality and ionic charge of the ith ion, respectively. The ionic strength is also related to the salinity scale in g/kg by the following correlation,

$$I = \frac{19.92732\ S}{1000 - 1.005109\ S} \tag{14.2}$$

Sometimes, aqueous solutions may have complexes instead of simple ionic species. Their limiting molar conductivities can be obtained by the following relation:

$$\lambda^{o}_{complex} = \frac{|Z_{complex}|}{\left(\sum_{i}^{n}\left(\frac{\lambda_i}{\lambda_i^{o}}\right)^{3}\right)^{\frac{1}{3}}} \tag{14.3}$$

where $|Z_{complex}|$ is the mode value of the charge of complex species, λ_i^o Limiting the equivalent conductivity of ions in the complex species and λ_i Equivalent conductivity of ions in the complex.

14.3 Multiple Components Models

Pawlowicz (2008) proposed a multi-component approach for the molal ionic equivalent conductivities at various temperatures and ionic strengths. Equation (14.4) summarizes the calculation procedure of electrical conductivity (K_e) of multicomponent solution.

$$K_e = \sum_i^n \lambda_i m_i \tag{14.4}$$

Molal ionic equivalent conductivities are calculated as,

$$\lambda_i = \frac{\lambda_i^o}{1 + a_i I^{0.5}} \tag{14.5}$$

where λ_i^o is the equivalent conductivity of ions (in $\mu S\ cm^{-1}\left(mol\ L^{-1}\right)^{-1}$) at a given temperature (in °C). It can be written as,

$$\lambda_i^o = \lambda_i^{o25}(1 + f_i \Delta T) \tag{14.6}$$

Temperature difference from reference temperature and parameter f_i can be calculated as

$$\Delta T = (T - T_o)$$

$$T_o = 25\,°C$$

$$f_i = F_1\Delta T + F_2\Delta T + F_3\Delta T$$

Data for factors as $F_1\left(C^{-1}\right)$, $F_2\left(C^{-2}\right)$, $F_3\left(C^{-3}\right)$ for different species are given in Table 14.1. The ionic strength (I) can be calculated using Eq. (14.1).

where m_i and Z_i are the molality and ionic charge of the ith ion, respectively. The factor a_i in $\left(mol\ L^{-1}\right)^{-\frac{1}{2}}$, at any given temperature, can be obtained by the following expression:

$$a_i = a_i^{25}(1 + \beta\Delta T)$$

The numerical values of a_i^{25} can be obtained from Table 14.1. It can also be calculated from the following expression for ionic species:

$$a_i^{25} Z_i = \frac{1}{\sqrt{I}}\left(\frac{\lambda_i^{o25}}{\lambda_i^{25}} - 1\right)$$

Table 14.1 Ion-dependent numerical parameter (Pawlowicz 2008)

Ion	Z_i	λ_i^{o25} mS/cm (mol/L)$^{-1}$	a_i (mol/L)$^{-1/2}$	F_1 (°C)$^{-1}$	F_2 (°C)$^{-2}$	F_3 (°C)$^{-3}$
Ca^{2+}	2	59.47	0.85	2.213×10^{-2}	9.854×10^{-5}	-1.027×10^{-6}
Mg^{2+}	2	53.00	0.90	1.901×10^{-2}	2.096×10^{-5}	–
Na^{+}	1	50.08	0.80	2.189×10^{-2}	9.159×10^{-5}	-1.127×10^{-6}
NH_4^{+}	1	73.50	0.62	1.931×10^{-2}	1.113×10^{-5}	2.546×10^{-6}
K^{+}	1	73.48	0.62	1.941×10^{-2}	5.786×10^{-5}	-1.593×10^{-7}
Cl^{-}	1	76.31	0.62	2.012×10^{-2}	6.217×10^{-5}	-4.921×10^{-8}
HCO_3^{-}	1	44.50	0.91	1.909×10^{-2}	1.573×10^{-5}	-1.320×10^{-6}
NO_3^{-}	1	71.42	0.74	1.874×10^{-2}	6.764×10^{-5}	8.948×10^{-7}
F^{-}	1	55.40	0.80	2.089×10^{-2}	–	–
CO_3^{2-}	2	69.30	0.74	1.772×10^{-2}	-6.001×10^{-5}	–
SO_4^{2-}	2	80.00	0.74	2.119×10^{-2}	6.746×10^{-5}	–
H^{+}	1	349.65	0.30	1.388×10^{-2}	-3.152×10^{-5}	-6.372×10^{-7}
OH^{-}	1	198.00	0.31	1.644×10^{-2}	1.403×10^{-5}	4.500×10^{-6}
Ba^{2+}	2	63.60	0.83	2.084×10^{-2}	8.885×10^{-5}	–
Sr^{2+}	2	59.40	0.84	2.096×10^{-2}	7.322×10^{-5}	–
Ag^{+}	1	61.90	0.62	1.969×10^{-2}	4.308×10^{-5}	–
Li^{+}	1	38.66	0.96	2.294×10^{-2}	1.313×10^{-4}	4.451×10^{-7}
Br^{-}	1	78.10	0.60	1.960×10^{-2}	6.678×10^{-5}	4.371×10^{-7}
I^{-}	1	76.80	0.61	1.950×10^{-2}	6.370×10^{-5}	7.755×10^{-7}
ClO_4^{-}	1	67.30	0.66	1.724×10^{-2}	-2.359×10^{-5}	–
Zn^{2+}	2	52.80	0.80	2.200×10^{-2}	1.287×10^{-4}	–
Cu^{2+}	2	53.60	0.80	2.329×10^{-2}	1.676×10^{-4}	–
La^{3+}	3	69.70	0.81	2.191×10^{-2}	8.119×10^{-5}	–
$Fe(CN_6)^{4+}$	4	110.4	0.62	–	–	–

β is constant for all species and can be taken as:

$$\beta = \frac{1}{300} (^{\circ}\mathrm{C})^{-1}$$

where λ_i and m_i molal ionic conductivities and molality of ions, respectively. This model is tested up to the salinity of 42 g/kg and temperature below boiling point with an accuracy of ± 4%.

Park (1964) method can be used to estimate electrical conductivity (K_e) of aqueous solutions. The main equation is provided as Eq. (14.7).

$$1000\, K_e = \sum_{i}^{n} \lambda_i C_i \tag{14.7}$$

where C_i is the concentration in equivalent per liter, and K_e is the specific conductance. The equivalent ionic conductance (λ_i) can be split into cations and anions.

$$1000 K_e = \sum_{i}^{n} \lambda_i^- C_i^- + \sum_{i}^{n} \lambda_i^+ C_i^+ \tag{14.8}$$

The data for partial equivalent ionic conductance of various constituents are listed in Table 14.2. The model is tested up to 35 g/kg at 0.1 MPa and 23 °C with an accuracy of ± 1%.

McCleskey et al. (2012a, b) proposed a most accurate model. The model can be summarized in Eq. (14.9).

$$K_e = \sum_{i}^{n} \lambda_i m_i \tag{14.9}$$

Molal ionic conductivities (in mS/cm) at various temperatures and ionic strength can be obtained by using Lattey's equation:

Table 14.2 Data for park method for seawater: partial equivalent ionic conductance of major ions in seawater at 23 °C, assuming the cation transference number of potassium chloride in seawater is 0.49 (cm^2/Ω) (Park 1964)

Cation	λ^+	Anion	λ^-
Na^+	31	Cl^-	59
K^+	57	SO_4^{2-}	21
Mg^{2+}	13	HCO_3^-	15
Ca^{2+}	19	Br^-	63
Sr^{2+}	21	CO_3^{2-}	− 8

$$\lambda_i = \lambda_i^o(T) - \frac{A_i(T)I^{0.5}}{1 + B_i I^{0.5}} \tag{14.10}$$

where B is an empirical constant, λ_i^o (refer to Table 14.3) is a function of temperature (°C), and A (refer to Table 14.3) is a function of temperature. The ionic strength (I) and its relationship with salinity can be used from Eq. (14.1) and Eq. (14.2), respectively. The limiting molar conductivities of the complex could be obtained from Eq. (14.3). The model generates good results up to 1 mol/kg of ionic strength and in the temperature range of 0–95 °C with an accuracy of $0.923 \leq R^2 \leq 1$.

Appelo method (McCleskey et al. 2012a, b) for electrical conductivity (K_e) in S/m as a function limiting conductivity (λ_i^o), activity coefficient (γ_i), and molality (m_i) of ions as follows,

$$K_e = \sum_i^n \lambda_i^o \gamma_i m_i \tag{14.11}$$

The limiting ionic conductivity (λ_i^o) is reported to be calculated by the following relations,

$$\lambda_i^o = \frac{Z^2 F^2}{RT} D_w \tag{14.12}$$

where D_w is the diffusion coefficient of water. The model generates good results up to 1 M of ionic strength and in the temperature range of 0–95 °C.

There is a less accurate method based on the ionic mobility approach (Noyes and Coolidge 1904; Pelizzetti 1999). The electrical conductivity (K_e) of an aqueous solution can be given by,

$$K_e = \left(\sum_i^n \lambda_i m_i \right) \rho \tag{14.13}$$

Ionic conductivities can be calculated from Eq. (14.14).

$$\lambda_i = Z\,u\,F \tag{14.14}$$

where Z is the ionic charge, u is the ionic mobility, and F is Faraday's constant. The ionic mobilities of various ions can be seen in Table 14.4.

The model based on ionic mobility can be used at 25 °C and 0.1 MPa for all salinity values. However, it is a less accurate method and is not recommended (accuracy within ± 0.02 S/cm).

Poisson and Papaud method (Poisson and Papaud 1983; Sato et al. 1996) can also be used to calculate electrical conductivities through aqueous solutions. The method is summarized in Eq. (14.15).

Table 14.3 McCleskey et al. (2012a, b) model for equations used to calculate individual ionic-molal conductivities (ms.kg/cm.mol) and for the ions applicable to natural waters (T is in °C)

Ion	λ°	A	B
K^+	0.003046 T^2 + 1.261 T + 40.70	0.00535 T^2 + 0.9316 T + 22.59	1.5
Na^+	0.003763 T^2 + 0.8770 T + 26.23	0.00027 T^2 + 1.141 T + 32.07	1.7
H^+	– 0.01414 T^2 + 5.355 T + 224.2	– 0.00918 T^2 + 1.842 T + 39.23	0.3
Li^+	0.002628 T^2 + 0.7079 T + 19.20	0.00412 T^2 + 0.4632 T + 13.71	0.2
Cs^+	0.003453 T^2 + 1.249 T + 43.94	0.00646 T^2 + 0.7023 T + 21.79	1.3
NH_4^+	0.003341 T^2 + 1.285 T + 39.04	0.00132 T^2 + 0.6070 T + 11.19	0.3
Ca^{2+}	0.009645 T^2 + 1.984 T + 62.28	0.03174 T^2 + 2.334 T + 132.3	2.8
Mg^{2+}	0.01068 T^2 + 1.695 T + 57.16	0.02453 T^2 + 1.915 T + 80.50	2.1
Ba^{2+}	0.01059 T^2 + 2.090 T + 68.10	0.03127 T^2 + 2.248 T + 93.91	1.9
Sr^{2+}	0.006649 T^2 + 2.069 T + 61.63	0.00702 T^2 + 0.9009 T + 33.41	0.1
SO_4^{2-}	0.01037 T^2 + 2.838 T + 82.37	0.03324 T^2 + 5.889 T + 193.5	2.6
Cl^-	0.003817 T^2 + 1.337 T + 40.99	0.00613 T^2 + 0.9469 T + 22.01	1.5
F^-	0.002764 T^2 + 1.087 T + 26.66	0.00178 T^2 + 0.6202 T + 19.34	0.5
Br^-	0.000709 T^2 + 1.477 T + 40.91	0.00251 T^2 + 0.5398 T + 12.01	0.1
CO_3^{2-}	– 0.000326 T^2 + 2.998 T + 64.03	– 0.00181 T^2 + 5.542 T + 120.2	2.3
HCO_3^-	0.000614 T^2 + 0.9048 T + 21.14	– 0.00503 T^2 + 0.8957 T + 10.97	0.1
NO_3^-	0.001925 T^2 + 1.214 T + 39.90	0.00118 T^2. 5045 T + 23.31	0.1
OH^-	0.003396 T^2 + 2.925 T + 121.3	0.00933 T^2 + 0.1086 T + 35.90	0.01
Al^{3+}	0.02376 T^2 + 3.227 T + 90.24	0.06484 T^2 + 5.149 T + 76.79	3.0
Cu^{2+}	0.00818 T^2 + 1.939 T + 53.26	0.0292 T^2 + 6.745 T + 151.5	8.0
Fe^{2+}	0.009939 T^2 + 1.878 T + 54.80	0.03997 T^2 + 3.217 T + 164.5	4.0
Fe^{3+}	0.02077 T^2 + 4.390 T + 82.42	– 0.09676 T^2 + 20.76 T – 22.18	4.0
Mn^{2+}	0.01275 T^2 + 2.109 T + 46.19	0.1071 T^2 + 9.023 T + 135.4	7.6
Zn^{2+}	0.01249 T^2 + 1.912 T + 48.20	0.08284 T^2 + 5.188 T + 75.73	7.0
KSO_4^-	– 0.00243 9T^2 + 4.253 T + 129.7	– 0.01576 T^2 + 6.210 T + 146.8	1.3
$NaSO_4^-$	0.002309 T^2 + 5.459 T + 219.2	0.01454 T^2 + 5.193 T + 253.6	0.5
HSO_4^-	0.000927 T^2 + 0.8337 T + 29.56	0.02887 T^2 + 0.8730 T + 36.25	7.0
$NaCO_3^-$	0.00336 T^2 + 3.845 T + 89.51	0.00061 T^2 + 6.387 T + 141.7	2.0

$$K_e = \frac{R^2}{FT} \times \left(\sum_{i=1}^{n} D_i C_i Z_i^2 \right) \tag{14.15}$$

where D is the diffusion coefficient ($\frac{m^2}{s}$), , R is molar gas constant (8.314 J/mol. K), C_i is ionic concentration, and Z_i is the ionic charge. The diffusion coefficient data is listed in Tables 14.5, 14.6 and 14.7. The model can be used at 25 °C and 0.1 MPa for all salinity values with an accuracy of ± 1%.

Table 14.4 Ionic mobilities

Ions	Mobility (u × 10^{-8}) ($m^2 S^{-1} V^{-1}$)
H^+	36.23
Na^+	5.19
Ca^{++}	6.17
Mg^{++}	5.50
K^+	7.62
NH_4^+	7.62
Cl^-	7.92
F^-	5.74
Br^-	8.09
I^-	7.96
SO_4^{--}	8.29
NO_3^-	7.41
CO_3^{--}	7.18

14.4 Case Study

The methods of McCleskey et al. (2012a, b) and Pawlowicz (2008) and the experimental data of Little (1962) are plotted at 25 °C and 0.1 MPa for various salinity values, as shown in Fig. 14.1. It is clear that McCleskey et al. (2012a, b) predict closer values to the experimental data (Little 1962), while the Pawlowicz (2008) model has lower values. The difference between these two methods from the experimental data (Little 1962) is exhibited in Fig. 14.1b. The electrical conductivity values estimated by McCleskey et al. (2012a, b) are the most accurate by about ± 10% error.

The effect of temperature on electrical conductivity is shown in Fig. 14.2. The increase in temperature values improves the electrical conductivity values. The values obtained by McCleskey et al. (2012a, b) are compared to the experimental data of Little (1962) for a temperature range of 0–24 °C at 35 g/kg and 0.1 MPa. The difference from the experimental data is within ± 1%.

Electrical conductivity is of great significance for electrically driven desalination technologies such as electrodialysis. The entire system of an ED plant is based on the calculation of the concept of charge transfer. Therefore, it may be disastrous to have wrong calculations for the electrical conductivity of saline water. Therefore, We have enlisted and compared various correlations and provided associated advantages and applications. However, we do suggest that more studies should be conducted and a comprehensive model should be developed that can be used for all possible compositions. However, it must be supported by experimental data. Based on the above discussion, some methods and correlations could be recommended to be used, as summarized in Table 14.8.

Table 14.5 Diffusion constants of inorganic cations (Sato et al. 1996)

Cation	|z|	$D_i.\ 10^{-9}\ [m^2/s]$	$\Lambda^o_{m,i}[S.cm^2/mol]$	$\Lambda^o_{eq,i}[S\ .cm^2/mol]$
Ag^+	1	1.648	61.9	61.9
Al^{3+}	3	0.559	188.9	63.0
Ba^{2+}	2	0.848	127.4	63.7
Be^{2+}	2	0.599	90.0	45.0
Ca^{2+}	2	0.793	119.1	59.6
$CaHCO_3^+$	1	0.506	19.0	19.0
Cd^{2+}	2	0.717	107.7	53.8
Co^{2+}	2	0.732	110.0	55.0
Cr^{3+}	3	0.595	201.1	67.0
Cu^{2+}	2	0.733	110.1	55.0
Fe^{2+}	2	0.719	108.0	54.0
Fe^{3+}	3	0.604	204.1	68.0
H^+	1	9.310	349.6	349.6
Hg^{2+}	2	0.913	137.2	68.6
K^+	1	1.960	73.6	73.6
Li^+	1	1.030	38.7	38.7
Mg^{2+}	2	0.705	105.1	53.0
$MgHCO_3^+$	1	0.478	18.0	18.0
Mn^{2+}	2	0.688	103.4	51.7
Na^+	1	1.330	50.0	50.0
NH_4^+	1	1.980	74.4	74.4
Pb^{2+}	2	0.945	142.0	71.0
Sr^{2+}	2	0.794	119.3	59.6
UO_2^{2+}	2	0.426	64.0	32.0
Zn^{2+}	2	0.715	107.4	53.7

Table 14.6 Diffusion constants of inorganic anions (Sato et al. 1996)

Anion	|z|	$D_i.10^{-9}[m^2/s]$	$\Lambda^o_{m,i}[S.cm^2/mol]$	$\Lambda^o_{eq,i}[S.cm^2/mol]$
Br^-	1	2.010	75.5	75.5
Cl^-	1	2.030	76.2	76.2
CO_3^{2-}	2	0.955	143.5	71.7
CN^-	1	2.077	78.0	78.0
CNO^-	1	1.720	64.6	64.6
CrO_4^{2-}	2	1.132	170.0	85.0
F^-	1	1.460	54.8	54.8

(continued)

Table 14.6 (continued)

Anion	|z|	$D_i.10^{-9}[m^2/s]$	$A^o_{m,i}[S.cm^2/mol]$	$A^o_{eq,i}[S.cm^2/mol]$
$H_2AsO_4^-$	1	0.905	34.0	34.0
$H_2PO_4^-$	1	0.846	31.8	31.8
HCO_3^-	1	1.180	44.3	44.3
HPO_4^{2-}	2	0.690	103.6	51.8
HS^-	1	1.730	65.0	65.0
HSO_4^-	1	1.330	50.0	50.0
I^-	1	2.045	76.8	76.8
KSO_4^-	1	0.746	28.0	28.0
MnO_4^-	1	1.632	61.3	61.3
MoO_4^{2-}	2	1.984	298.0	149.0
$NaCO_3^-$	1	0.585	22.0	22.0
$NaSO_4^-$	1	0.618	23.2	23.2
NO_2^-	1	1.910	71.7	71.7
NO_3^-	1	1.900	71.4	71.4
OH^-	1	5.270	197.9	197.9
PO_4^{-3}	3	0.612	206.8	68.9
S^{2-}	2	0.731	109.8	54.9
SeO_4^{2-}	2	1.008	151.4	76.7
SO_4^{2-}	2	1.070	160.7	80.4

Table 14.7 Diffusion constants of organic anions (Sato et al. 1996)

Organic anion	|Z|	$D_i.10^{-9}[m^2/s]$	$A^o_{m,i}[S\ cm^2/mol]$	$A^o_{eq,i}[S\ cm^2/mol]$
Acetate-	1	1.089	40.9	40.9
Benzoate-	1	0.863	32.4	32.4
Butyrate-	1	0.868	32.6	32.6
Citrate-3	3	0.623	210.6	70.2
Dihydrogencitrate-	1	0.799	30.0	30.0
Formate-	1	1.454	54.6	54.6
Hydrogenoxalate-	1	1.070	40.2	40.2
Isovalerate-	1	0.871	32.7	32.7
Lactate-	1	1.033	38.8	38.8
Malate-2	2	0.783	117.6	58.8
Maleate-2	2	0.824	123.8	61.9
Oxalate-2	2	0.987	148.3	74.1
Phenylacetate-	1	0.815	30.6	30.6

(continued)

Table 14.7 (continued)

Organic anion	|Z|	$D_i.10^{-9}[m^2/s]$	$A^o_{m,i}[S\ cm^2/mol]$	$A^o_{eq,i}[S\ cm^2/mol]$
o-Phtalate-2	2	0.696	104.6	52.3
m-Phtalate-2	2	0.728	109.4	54.7
Pivalate	1	0.849	31.9	31.9
Salicylate-	1	0.959	36.0	36.0
Suberate-2	2	0.479	72.0	36.0
Succinate-2	2	0.783	117.6	58.8
Tartarate	2	0.794	119.3	59.6

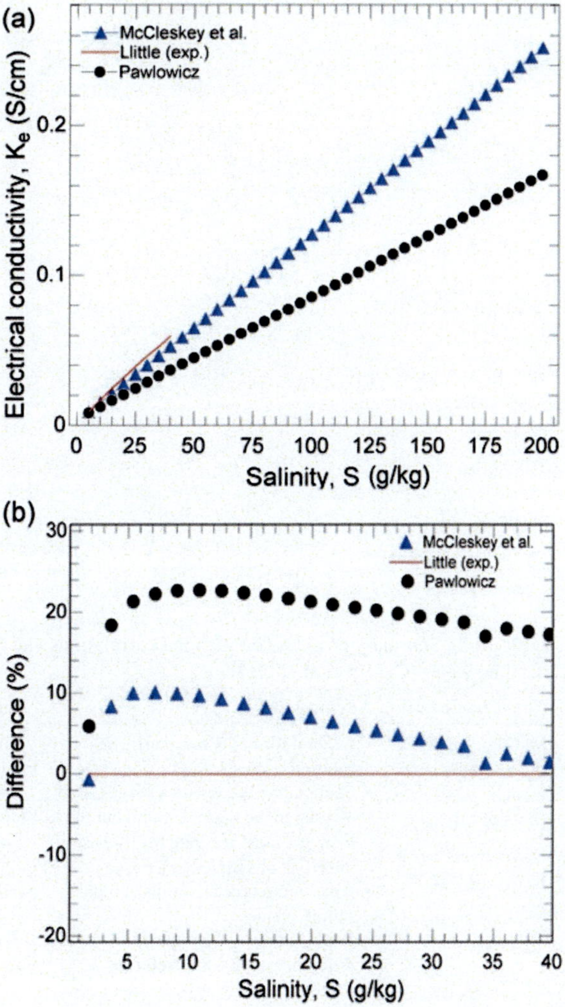

Fig. 14.1 **a** Salinity versus electrical conductivity and **b** the difference against the experimental data of Little (1962) for the methods of McCleskey et al. (2012a, b) and Pawlowicz (2008) at 25 °C and 0.1 MPa

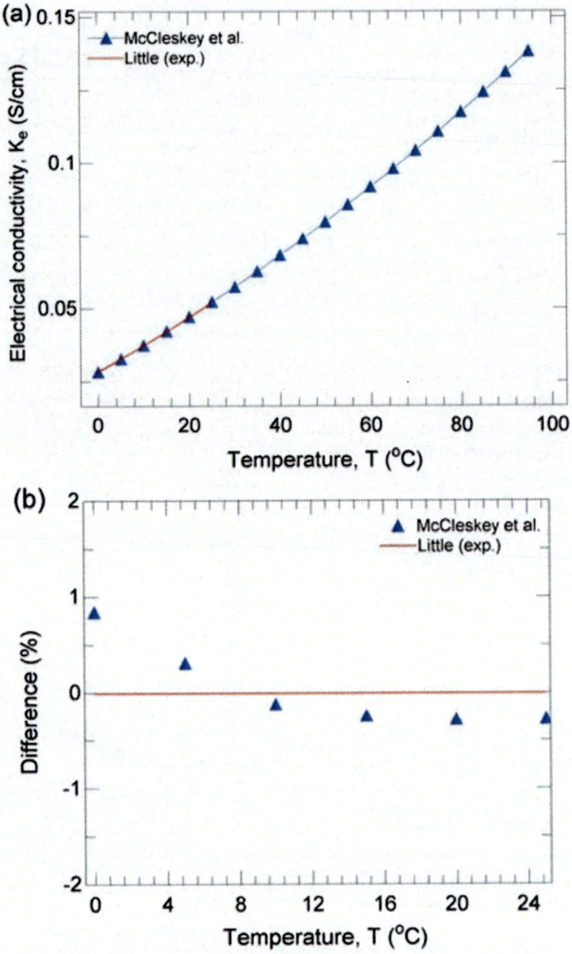

Fig. 14.2 **a** Temperature versus electrical conductivity and **b** the difference against the experimental data of Little (1962) for the method of McCleskey et al. (2012a, b) at MPa. 35 g/kg and 0.1 MPa

Table 14.8 Summary of the recommended correlations and models for the electrical conductivity of saline water (Qasem et al. 2021)

Correlation-model	Advantages	Applications	Ref
McClesky et al. (multi-component model) [Eq. (14.4)]	• General for multi-component saline solution • Studied in a limited data range (up to 40 g/kg in the present study) but found to be the most accurate for studying the effect of salinity (refer to Fig. 14.1) • Gives more precise results at higher salinity values • It is a function of temperature within a difference of ± 0.8% (refer to Fig. 14.2)	• All thermal-based desalination technologies • Membrane distillation (MD) • Electrodialysis (ED) • Reverse electrodialysis (RED)	(McCleskey et al. 2012a, b)

14.5 EES Codes

The used EES codes for the above correlations are:

Pawlowicz Multi-component Approach

```
"Pawlowicz Method"

T = 25
T_o = 25 "C"
beta = 1/300 "C^−1"

"Concentrations in mol/kg"
"Please note that 1: H2O, 2: Cl−, 3: Na+, 4: Mg++, 5: SO4−−, 6: Ca++, 7: K+"
n[1] = 53.6
n[2] = 0.546
n[3] = 0.469
n[4] = 0.0528
n[5] = 0.0282
n[6] = 0.0103
n[7] = 0.0102

"Charge of each species"
z[2] = 1
z[3] = 1
z[4] = 2
z[5] = 2
z[6] = 2
z[7] = 1

I_s = 0.5 * sum(n[i] * Z[i]^2, i = 2, 7)

"molar conductivities at 25C mS kg/cm mol"
lambda_oo[2] = 76.31
lambda_oo[3] = 50.08
lambda_oo[4] = 53
lambda_oo[5] = 80
lambda_oo[6] = 59.47
lambda_oo[7] = 73.48

F_i[2] = 2.012/10^2
F_i[3] = 2.189/10^2
F_i[4] = 1.901/10^2
```

```
F_i[5] = 2.119/10^2
F_i[6] = 2.213/10^2
F_i[7] = 1.941/10^2

F_j[2] = 6.217/10^5
F_j[3] = 9.159/10^5
F_j[4] = 2.096/10^5
F_j[5] = 6.746/10^5
F_j[6] = 9.854/10^5
F_j[7] = 5.786/10^5

F_k[2] = − 4.921/10^8
F_k[3] = − 1.127/10^6
F_k[4] = − 1.127/10^6
F_k[5] = 8.948/10^7
F_k[6] = − 1.027/10^6
F_k[7] = − 1.593/10^7

"a at 25 C"

a_o[2] = 0.62
a_o[3] = 0.80
a_o[4] = 0.90
a_o[5] = 0.74
a_o[6] = 0.85
a_o[7] = 0.62

"a at any given temperature."

Duplicate i = 2, 7
a[i] = a_o[i] * (1 + (beta * (T − T_o)))
End

"g_i at any given temperature."

Duplicate i = 2, 7
g[i] = (F_i[i] * (T − T_o)) + (F_j[i] * (T − T_o)) + (F_k[i] * (T − T_o))
End

"lambda_o at any given temperature."

Duplicate i = 2, 7
```

```
Lambda_o[i] = Lambda_oo[i] * (1 + (g[i] * (T − T_o)))
End

"Lambda_i can be found by the following relations."

Duplicate i = 2, 7
Lambda[i] = lambda_o[i]/(1 + (a[i] * z[i] * I_s^0.5))
End

"Specific Conductivity in mS/cm"

K = sum(lambda[i] * n[i], i = 2, 7)

"specific conductivity in S/cm"
K_s = K/1000
```

Park Method

```
"Concentrations in mol/kg"
"Please note that 1: H2O, 2: Cl−, 3: Na+, 4: Mg++, 5: SO4−−, 6: Ca++, 7: K+"
n[1] = 53.6
n[2] = 0.546 * 1.12
n[3] = 0.469 * 1.12
n[4] = 0.0528 * 1.12
n[5] = 0.0282 * 1.12
n[6] = 0.0103 * 1.12
n[7] = 0.0102 * 1.12

"Partial Equivalent Ionic Conductance at 25 C in S-Cm^2"

lambda[2] = 76.35
lambda[3] = 50.15
lambda[4] = 13 * 2
lambda[5] = 21 * 2
lambda[6] = 19 * 2
lambda[7] = 73.50

1000 * K = sum(lambda[i] * n[i], i = 2, 7)
```

McCleskey et al. Model (The Most Accurate)

```
T = 25 "C"

"Concentrations in mol/kg"
```

```
"Please note that 1: H2O, 2: Cl−, 3: Na+, 4: Mg++, 5: SO4−−, 6: Ca++, 7: K+"

n[1] = 53.6
n[2] = 0.546
n[3] = 0.469
n[4] = 0.0528
n[5] = 0.0282
n[6] = 0.0103
n[7] = 0.0102

"Charge of each species"
z[2] = − 1
z[3] = + 1
z[4] = + 2
z[5] = − 2
z[6] = + 2
z[7] = + 1

"Ionic Strength"

I_s = 0.5 * sum(n[i] * Z[i]^2, i = 2, 7)

Duplicate i = 2, 7
lambda[i] = lambda_o[i] − (A[i] * I_s^0.5)/(1+ (B[i] * I_s^0.5))
End

lambda_o[2] = (0.003817 * T^2) + (1.337 * T) + 40.99
lambda_o[3] = (0.003763 * T^2) + (0.8870 * T) + 26.23
lambda_o[4] = (0.01068 * T^2) + (1.695 * T) + 57.16
lambda_o[5] = (0.01037 * T^2) + (2.838 * T) + 82.37
lambda_o[6] = (0.009645 * T^2) + (1.984 * T) + 62.28
lambda_o[7] = (0.003046 * T^2) + (1.261 * T) + 40.70

A[2] = (0.00613 * T^2) + (0.9464 * T) + 22.01
A[3] = (0.00027 * T^2) + (1.141 * T) + 32.07
A[4] = (0.02453 * T^2) + (1.915 * T) + 80.50
A[5] = (0.03324 * T^2) + (5.889 * T) + 193.50
A[6] = (0.03174 * T^2) + (2.334 * T) + 132.3
A[7] = (0.00535 * T^2) + (0.9316 * T) + 22.59

B[2] = 1.5
B[3] = 1.7
```

```
B[4] = 2.1
B[5] = 2.6
B[6] = 2.8
B[7] = 1.5

"Specific Conductivity in milli-siemen/cm"

k = sum(lambda[i] * n[i], i = 2, 7)

"Specific Conductivity in S/cm"
k_siemen = k/1000
```

Ionic Mobility Approach (Less Accurate)

```
T=25 "C"

"Concentrations in mol/kg"
"Please note that 1: H2O, 2: Cl−, 3: Na+, 4: Mg++, 5: SO4−−, 6: Ca++, 7: K+"
n[1] = 53.6
n[2] = 0.5461
n[3] = 0.469
n[4] = 0.0528
n[5] = 0.0282
n[6] = 0.0103
n[7] = 0.0102

"Mode of charge"

z[2] = 1
z[3] = 1
z[4] = 2
z[5] = 2
z[6] = 2
z[7] = 1
F = 96485.3329 "S.A / mol"

"Ionic Mobility (m^2.S^−1.V^−1)"

mu[2] = 7.92/10^8
mu[3] = 5.19/10^8
mu[4] = 5.50/10^8
mu[5] = 8.29/10^8
```

```
mu[6] = 6.17/10^8
mu[7] = 7.62/10^8

"Molar Conductivity in S.m^2/mol"
Duplicate i = 2, 7
Lambda[i] = mu[i] * Z[i] * F
End

"Specific Conductivity in siemen/m"

k = sum(lambda[i] * n[i], i = 2, 7) * rho

"Conductivity in S/cm"

k_s = k/100

"Correlation Reported by F.J. Millerio"

"Temperature"

"Salinity in g/kg"

S = 35 "g/kg"

"Correlation Reported by F.J. Millero"

rho - rho_o = A * S_a + B * S_a^1.5 + C * S_a^2

"Absolute Salinity Correlation (S_a)"

S_a = (35.16504/35) * S

"Correlations for Constants A, B, C"

A = (0.824493) - (4.0899/10^3 * t) + (7.6438/10^5 * t^2) - (8.2467/10^7 *
t^3) + (5.3875/10^9 * t^4)
```

B = (−5.72466/10^3) + (1.0227/10^4 * t) − (1.6546/10^6 * t^2)

C = 4.8314/10^4

"Correlation for density of pure water as function of temperature"

rho_o = (999.842594) + (6.793952/10^2 * t) − (9.095290/10^3 * t^2) + (1.001685/10^4 * t^3) − (1.120083/10^6 * t^4) + (6.536332/10^9 * t^5)

"this correlation is valid for salinity range of 5–70 g/kg and temperature range of 273.15 K to 363.15 K"

References

Bradshaw AL, Schleicher KE (1980) Electrical conductivity of seawater. IEEE J Oceanic Eng 5(1):50–62. https://doi.org/10.1109/JOE.1980.1145449

Felderhof BU (2003) Comment on "transport equations for concentrated electrolyte solutions: reference frame, mutual diffusion" [J. Chem. Phys. 116, 2085 (2002)]. J Chem Phys 118(17):8114–8117. https://doi.org/10.1063/1.1563604

Little AD (1962) Defence documentation center for scientific and technical information, Cameron Station Virginia

McCleskey RB, Nordstrom DK, Ryan JN (2012a) Comparison of electrical conductivity calculation methods for natural waters. Limnol Oceanogr Methods 10(NOVEMBER):952–967. https://doi.org/10.4319/lom.2012.10.952

McCleskey RB, Nordstrom DK, Ryan JN, Ball JW (2012b) A new method of calculating electrical conductivity with applications to natural waters. Geochim Cosmochim Acta 77:369–382. https://doi.org/10.1016/j.gca.2011.10.031

Noyes AA, Coolidge WD (1904) The electrical conductivity of aqueous solutions at high temperatures. 1.—Description of the apparatus. Results with sodium and potassium chloride up to 306°. J American Chem Soc 26(2):134–170. https://doi.org/10.1021/ja01992a002

Park K (1964) Partial equivalent conductance of electrolytes in sea water. Deep-Sea Res Oceanographic Abs 11(5):729–736. https://doi.org/10.1016/0011-7471(64)90946-5

Pawlowicz R (2008) Limnology oceanography: method of calculating the conductivity of natural waters. Limnol Oceanogr 6:489–501

Pelizzetti AGE (1999) Chemistry of marine water sediments. In: Cold-adapted organisms: ecology, physiology, enzymology and molecular biology. https://doi.org/10.1007/978-3-662-06431-3

Poisson A, Papaud A (1983) Diffusion coefficients of major ions in seawater. Mar Chem 13(4):265–280. https://doi.org/10.1016/0304-4203(83)90002-6

Qasem NAA, Generous MM, Qureshi BA, Zubair SM (2021) A comprehensive review of saline water correlations and data: part ii—thermophysical properties. Arab J Sci Eng 46(3):1941–1979. https://doi.org/10.1007/s13369-020-05020-5

Sato H, Yui M, Yoshikawa H (1996) Ionic diffusion coefficients of Cs+, Pb2+, Sm3+, Ni2+, SeO2-4 and TcO−4 in free water determined from conductivity measurements. J Nucl Sci Technol 33(12):950–955. https://doi.org/10.1080/18811248.1996.9732037

Sauerheber R, Heinz B (2015) Temperature effects on conductivity of seawater and physiologic saline, mechanism and significance. Chem Sci J 6(4):4–8. https://doi.org/10.4172/2150-3494.1000109

Szymczak J, Holyk P, Ander P (1975) Electrical conductivity of aqueous solutions of monovalent salts of polystyrenesulfonate. J Phys Chem 79(3):269–272. https://doi.org/10.1021/j100570a015

Turq P, Blum L, Bernard O, Kunz W (1995) Conductance in associated electrolytes using the mean spherical approximation. J Phys Chem 99(2):822–827. https://doi.org/10.1021/j100002a053

Zhu X, Nordstrom DK, McCleskey RB, Wang R (2016) Ionic molal conductivities, activity coefficients, and dissociation constants of HAsO42—and H2AsO4− from 5 to 90 °C and ionic strengths from 0.001 up to 3 mol kg− 1 and applications in natural systems. Chem Geol 441:177–190. https://doi.org/10.1016/j.chemgeo.2016.08.006

Chapter 15
Thermal Expansivity

Abstract Saltwater thermal expansivity could be an essential thermophysical property when expansivity is important such as "gate height" for the precise design of multi-effect thermal desalination technologies. This chapter introduces the thermodynamic approach, multicomponent models, correlations, and data that estimate the saline water thermal expansivity at different salinity, temperature, and pressure values. Existing experimental data of seawater thermal expansivity are used to compare and assess the accuracy of the values obtained from the correlations and methods. Some correlations and methods are recommended based on their accuracy and operating conditions. In addition, the codes of methods used to calculate the saltwater thermal expansivity are provided.

Keywords Thermal expansivity · Thermodynamics · Multicomponent · Methods · Saline water

15.1 Introduction

The thermal expansivity of saline solutions could be an essential parameter in the design of thermal desalination technologies. For example, multi-effect thermal desalination technologies use a parameter called "gate height" that allows the brine to flow from one effect to the next. It is important to know the relation of the expansivity of water with temperature, pressure, and salinity to decide the "gate height" for the precise design of multi-effect thermal desalination technologies. Thermal expansivity is measured by estimating the thermal expansion coefficient of saline water at a given pressure for unit temperature variation (Meersman et al. 2008). It is important to note that the correlations of density can be used to derive thermal expansivity. The objective of this section is to summarize some correlations to conveniently calculate the (isobaric) thermal expansivity.

N. A. A. Qasem et al., *Thermodynamic and Thermophysical Properties of Saline Water*, Springer Water, https://doi.org/10.1007/978-3-031-35193-8_15

15.2 Thermodynamic Approach

The isobaric thermal expansion can be calculated using Eq. (15.1).

$$\alpha_P = \frac{1}{V}\left(\frac{\delta V}{\delta T}\right)_P \tag{15.1}$$

where V is the specific volume which is the reciprocal of density. Equation (15.1) can be written in terms of density as:

$$\alpha_P = \frac{1}{\left(\frac{1}{\rho}\right)}\left(\frac{\delta\left(\frac{1}{\rho}\right)}{\delta T}\right)_P \tag{15.2}$$

The final form of thermal expansivity (thermal expansion coefficient) is provided in Eq. (15.3).

$$\alpha_P = \frac{-1}{\rho}\left(\frac{\delta \rho}{\delta T}\right)_P \tag{15.3}$$

The equations above emphasize the fact that the isobaric thermal expansivity of saline water can be calculated by using density difference per unit temperature difference. It implies that density models and correlations in Chap. 1 can be used to calculate thermal expansivity too. Therefore, limited correlations are provided for this section because of their direct relevance to the density. Similarly, the isothermal compressibility can also be calculated from the following well-known thermodynamic relation,

$$\beta_T = \frac{1}{\rho}\left(\frac{\delta \rho}{\delta P}\right)_T \tag{15.4}$$

15.3 Multiple Components Models

Multicomponent models were proposed to be general for all types of saltwater under different salinity values. Major constituents of the aqueous ionic solutions are cations, anions, and water. The density of an ionic solution can be used to model thermal expansivity. Laliberté and Cooper's model (Laliberté and Cooper 2004; Meersman et al. 2008; Pandey et al. 2002) uses specific volume to calculate thermal expansivity. For temperature (T) in K, the thermal expansivity (α) in 1/K, Eq. (15.1) can be used to calculate thermal expansivity. The specific volume (v) can be calculated from the density model proposed by Laliberté and Cooper (2004), as described in Chap. 2 (Density) of this book. The model can be used with a standard deviation of $\pm 4.59\%$.

15.4 Correlations and Data

Correlations are the easiest way to calculate the saltwater density as they are fitted from the experimental data. Most of the available thermal expansivity correlations in the literature are for seawater. Bradshaw and Schleicher (1970) proposed a correlation for thermal expansivity of seawater in terms of temperature (T) in °C, pressure (P) in bar, and salinity in g/kg, the thermal expansivity (α) in 1/°C is,

$$\alpha = \frac{1}{V}\left(\frac{\delta V}{\delta T}\right)_p \tag{15.5}$$

where $\frac{\delta V}{\delta T}$ can be expressed as,

$$\frac{\delta V}{\delta T} = a_0 + a_1\,T + a_2\,T^2 + a_3 T^3 \tag{15.6}$$

The constants are expressed as,

$$\begin{aligned}10^6 \times a_0 = {} & 52.185 + 0.26803\,P - 0.82612 \times 10^{-4}P^2 + 0.6737 \times 10^{-8}P^3 \\ & + \left(2.8725 - 0.2724 \times 10^{-2}P + 0.7636 \times 10^{-6}P^2\right)(S - 35) \\ & - 0.125 \times 10^{-1}(S - 35)^2\end{aligned}$$

$$\begin{aligned}a_1 \times 10^6 = {} & 12.178 - 0.10502 \times 10^{-1}P + 0.438 \times 10^{-6}P^2 + 0.23596 \times 10^{-8}P^3 \\ & + \left(-0.13336 + 0.1437 \times 10^{-3}P - 0.556 \times 10^{-7}P^2\right)(S - 35)\end{aligned}$$

$$\begin{aligned}a_2 \times 10^6 = {} & -0.133445 + 0.4669 \times 10^{-4}P + 0.35098 \times 10^{-6}P^2 - 0.29588 \times 10^{-9}P^3 \\ & + \left(0.299 \times 10^{-2} - 0.28264 \times 10^{-5}P + 0.139 \times 10^{-8}P^2\right)(S - 35)\end{aligned}$$

$$\begin{aligned}a_3 \times 10^6 = {} & 0.10907 \times 10^{-2} + 0.211 \times 10^{-5}P - 0.10289 \times 10^{-7}P^2 + 0.7556 \times 10^{-11}P^3 \\ & - 0.3463 \times 10^{-4}(S - 35)\end{aligned}$$

The correlation is valid in the salinity range of 0–40 g/kg, pressure of 0.1–1.01 MPa, and temperature of – 2–30 °C with an accuracy of 1×10^6 cm^3/g °C.

Chen and Millero (1976) proposed a correlation for the thermal expansivity of seawater. For salinity (S) in g/kg, the thermal expansivity (α) in 1/K is,

$$\alpha = \alpha_W + AS + BS^{1.5} \tag{15.7}$$

where α_W is thermal expansivity of pure water. α_W and constants 'A' and 'B' were expressed as:

$$\begin{aligned}\alpha_W = &-6.80977\times 10^{-5} + 1.82945\times 10^{-5}T - 3.2519\times 10^{-7}T^2 + 6.4062\times 10^{-9}T^3 \\ &- 8.8997\times 10^{-11}T^4 + 5.846\times 10^{-13}T^5 \\ &+ \left(3.7409\times 10^{-7} - 1.5881\times 10^{-8}T + 1.741\times 10^{-14}T^4\right)P \\ &+ (-9.6658\times 10^{-11} + 4.8741\times 10^{-12}T - 1.10169\times 10^{-13}T^2 \\ &+ 9.941\times 10^{-16}T^3)P^2 + 4.75\times 10^{-16}P^3\end{aligned}$$

$$\begin{aligned}A = &\, 4.04385\times 10^{-6} - 1.3990\times 10^{-7}T + 1.727\times 10^{-9}T^2 - 1.944\times 10^{-11}T^3 \\ &+ \left(-3.2190\times 10^{-9} + 1.0353\times 10^{-10}T - 8.365\times 10^{-13}T^2\right)P \\ &+ \left(7.183\times 10^{-13} - 1.636\times 10^{-14}T\right)P^2\end{aligned}$$

$$\begin{aligned}B = &-1.058\times 10^{-7} + 1.308\times 10^{-9}T + 2.125\times 10^{-11}T^2 + 1.69\times 10^{-12}T^3 \\ &+ \left(8.446\times 10^{-11} - 1.88\times 10^{-12}T\right)P\end{aligned}$$

The correlation is valid in the salinity range of 0–40 g/kg, pressure of 0–100 MPa, and temperature of 0–40 °C within a standard deviation of $306 \times 10^{-6}(°C)^{-1}$.

El-Dessouky and Ettouney (2002) provided a correlation for the thermal expansivity of pure water. For temperature in °C, the thermal expansivity of pure water (α_w) in 1/K, Eq. (15.1) can be used with constants provided in Table 15.1. The specific volume can be calculated from Eq. (15.8).

$$V = \left(\frac{T_c}{T + 273.15} - 1\right)\sum_{i=1}^{6} f_i((T + 273.15))^{(i-1)} \tag{15.8}$$

$$T_c = 647.286\ K$$

$$V_c = 0.003172222\frac{m^3}{kg}$$

The correlation is valid in the temperature range of $5 - 200\,°C$, with an accuracy of $\pm$ 0.05%.

Table 15.1 Constants of the correlation of El-Dessouky and Ettouney (2002) for pure water

Constants	Values	Constants	Values
f_1	−2.781015567	f_4	3.636115 $\times 10^{-9}$
f_2	0.002543267	f_5	− 5.358938 $\times 10^{-11}$
f_3	9.845047 $\times 10^{-6}$	f_6	7.019341 $\times 10^{14}$

15.5 Case Study

The thermal expansivity of saline water is a strong function of salinity and temperature, while it is slightly affected by pressure. The methods and correlations of Chen and Millero (1976) and Bradshaw and Schleicher (1970) are compared to the experimental data of Nayar et al. (2014) as a function of the salinity at 20 °C and 0.1 MPa, as shown in Fig. 15.1. The values obtained from the Bradshaw and Schleicher (1970) correlation show a good agreement with those of Nayar et al. (2014) for a salinity up to 40 g/kg within a maximum difference of $\pm$ 3%, as shown in Fig. 15.1b.

The effect of temperature on the thermal expansivity is shown in Fig. 15.2 at 0.1 MPa and 30 g/kg for the correlations of Bradshaw and Schleicher (1970) and Chen and Millero (1976) against the experimental data of Chen and Millero (1976). It can be seen from the figure that both correlations show an increasing trend with the increase in the temperature of saline water. The accuracy of both correlations is also remarkable. Both methods have a difference of $\sim$ 5% when compared with the experimental data (Chen and Millero 1976) in a temperature range of $0 - 40\ ^\circ C$, as shown in Fig. 15.2b.

The pressure-dependent thermal expansivity is estimated by Bradshaw and Schleicher (1970) correlation at 35 g/kg and 25 °C, as shown in Fig. 15.3a. The correlations values differ from the experimental data of (1976) by $\pm$ 5.8%. This difference is around $\pm$ 3% when the pressure is below 50 MPa, as shown in Fig. 15.3b.

Thermal expansivity is the inherent property of a substance to expand upon heating. It is also prevalent in saline water. However, saline water is available in various compositions, and it is used under various operating conditions. Therefore, the development of a comprehensive model for thermal expansivity calculation is required. Another way to look at this problem can be the development and multi-component density model as specific volume is just reciprocal of density. Table 15.2 proposes the recommended correlations and methods to be used to calculate the thermal expansivity of saline water.

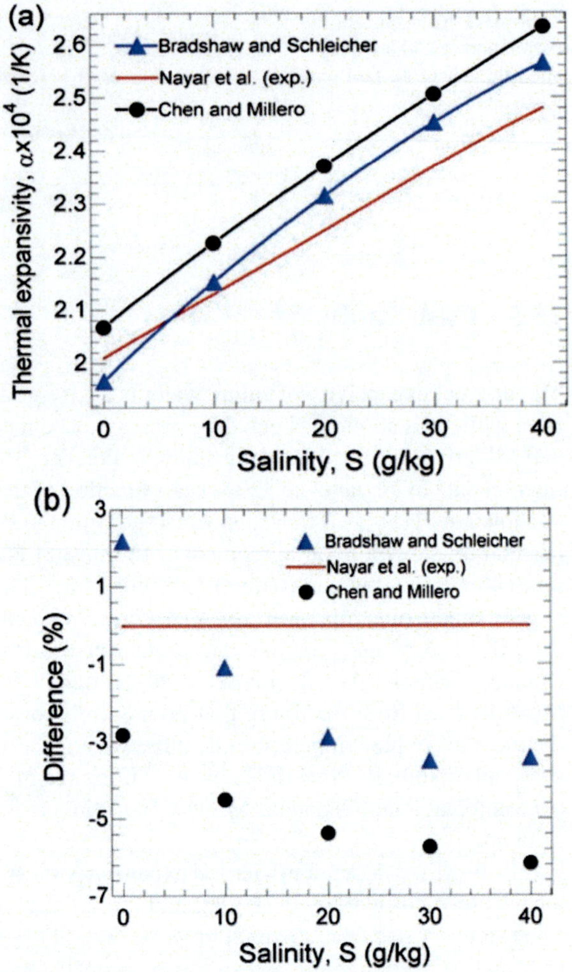

Fig. 15.1 Thermal expansivity coefficient of saline water as a function of salinity for **a** Bradshaw and Schleicher (1970), Chen and Millero (1976) and data of Nayar et al. (2014), **b** difference of Bradshaw and Schleicher (1970) and Chen and Millero (1976) from Nayar et al. (2014) at 20 °C and 0.1 MPa

15.6 EES Codes

The used EES codes to generate the above results are:

Laliberté and Cooper Model

```
T = 25 "C"

"moles"
n[1] = 0.424310
n[2] = 0.029245
```

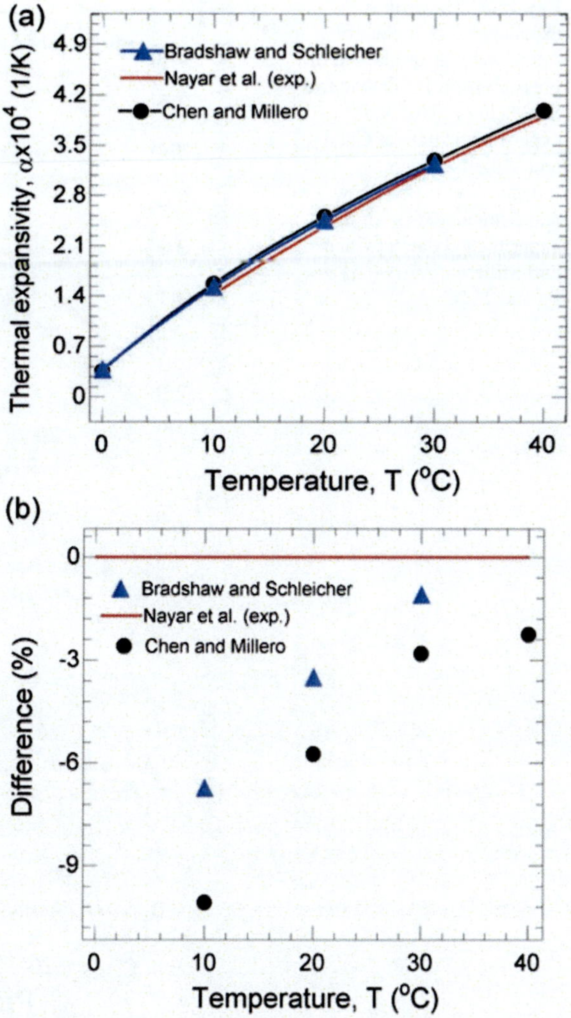

Fig. 15.2 Thermal expansivity coefficient of saline water as a function of temperature for **a** Bradshaw and Schleicher (1970), Chen and Millero (1976) and data of Nayar et al. (2014), **b** difference of Bradshaw and Schleicher (1970) and Chen and Millero (1976) from Nayar et al. (2014) at 0.1 MPa and 30 g/kg

n[3] = 0.002418
n[4] = 0.009412
n[5] = 0.055211
n[6] = 0.010707
n[7] = 53.6

"molar masses"
m[1] = 58.44
m[2] = 142.04

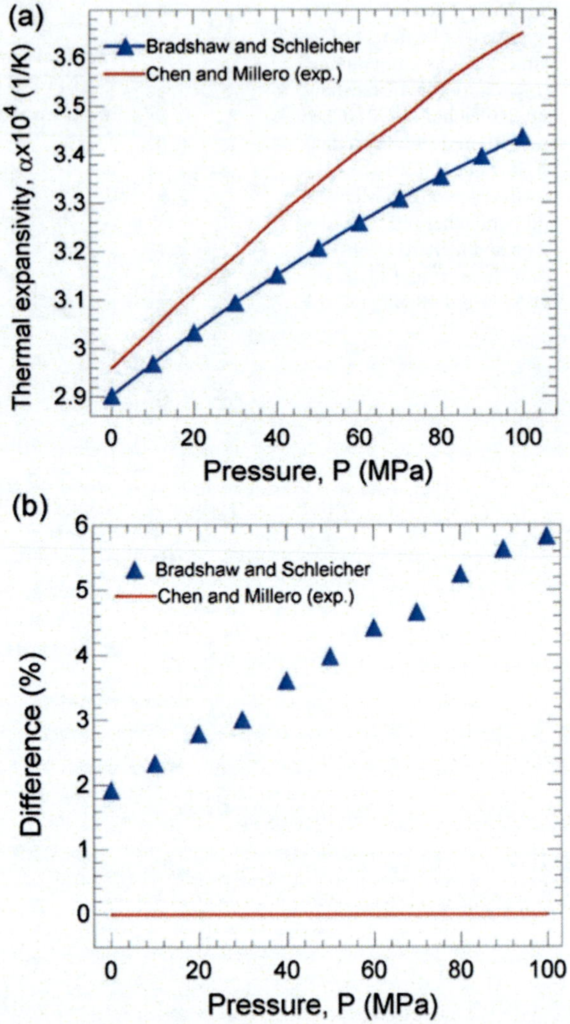

Fig. 15.3 Thermal expansivity coefficient of saline water as a function of pressure for **a** Bradshaw and Schleicher (1970) and experimental data of Chen and Millero (1976), **b** difference of Bradshaw and Schleicher (1970) from experimental data of Chen and Millero (1976) at 35 g/kg and 25 °C

```
m[3] = 84.007
m[4] = 74.55
m[5] = 95.211
m[6] = 110.98
m[7] = 18.05

"mass fraction"
m_t = sum(n[i] * m[i], i = 1, 7)
```

Table 15.2 Summary of the recommended correlations and models for the thermal expansivity of saline water (Qasem et al. 2021)

Correlations-model	Advantages	Applications	Refs.
Bradshaw and Schleicher [Eq. (15.5)]	• It can be used up to a salinity of 40 g/kg within a difference of ± 3% (refer to Fig. 15.1) • It can be used up to the temperature of 30 °C within a percent difference of less than ± 5% (refer to Fig. 15.2) • It is a function of pressure (can be used up to 100 MPa) within a difference of ± 5.8% when compared with experimental data. This difference reduces to ± 3% up to 50 MPa (refer to Fig. 15.3)	• All thermal- and membrane-based desalination technologies, including forward osmosis (FO) and reverse osmosis (RO)	Bradshaw and Schleicher (1970)
Chen and Millero [Eq. (15.7)]	• It can be used up to the temperature of 40 °C within a difference of less than ± 5% (refer to Fig. 15.2)	• Membrane distillation (MD) • Electrodialysis (ED) • Reverse electrodialysis (RED)	Chen and Millero (1976)

```
Duplicate i = 1, 7
X[i] = m[i] * n[i]/m_t
End

"1. NaCl, 2. Na2SO4, 3. NaHCO3, 4. KCl, 5. MgCl2, 6. CaCl2, 7. H2O"

C_o[1] = – 0.00433
C_1[1] = 0.09471
C_2[1] = 1.014624
C_3[1] = 0.014624
C_4[1] = 3315.6

C_o[2] = – 1.2095/10^7
C_1[2] = 4.3474/10^7
C_2[2] = 0.15364
C_3[2] = 0.0072514
C_4[2] = 4731.5
```

```
C_o[3] = – 9.4794/10^8
C_1[3] = 1.5657/10^7
C_2[3] = 0.9912
C_3[3] = 0.022644
C_4[3] = 4900.2

C_o[4] = – 0.46782
C_1[4] = 4.30800
C_2[4] = 2.3780
C_3[4] = 0.022044
C_4[4] = 2714

C_o[5] = – 0.00051500
C_1[5] = 0.001344
C_2[5] = 0.58358
C_3[5] = 0.0085832
C_4[5] = 3843.6

C_o[6] = – 0.63254
C_1[6] = 0.93995
C_2[6] = 4.2785
C_3[6] = 0.048319
C_4[6] = 3180.9

Duplicate i = 1, 6
V[i] = (X[i] + C_2[i] + C_3[i] * T)/((C_o[i] * X[i] + C_1[i]) * exp(0.000001 *
(T + C_4[i])^2))
End

"initial volume"
V_w = 1/1000

rho_1 = 1/((X[7] * V_w) + sum(x[i] * v[i], i = 1, 6))

V_i = 1/rho_1

"final volume"

T_j = T + 1
Duplicate i = 1, 6
```

```
V_j[i] = (X[i] + C_2[i] + C_3[i] * T_j)/((C_o[i] * X[i] + C_1[i]) * exp(0.000001
* (T_j + C_4[i])^2))
End

rho_2 = 1/((X[7] * V_w) + sum(x[i] * v_j[i], i = 1, 6))

V_f = 1/rho_2

"Change in volume"

dV = V_f − V_i

alpha = (rho_1 − rho_2)/rho_1
```

Bradshaw and Schleicher Correlation

```
T = 25 "C"
P = 1.01 "bars"
S = 35 "g/kg"

alpha = a_0 + a_1 * T + a_2 * T^2 + a_3 * T^3

10^6 * a_0 = 52.185 + 0.26803 * P − 0.82612/10^4 * P^2 + 0.6737/10^8 * P^3
+ (2.8725 − 0.2724/10^2 * P + 0.7636/10^6 * P^2) * (S − 35) − 0.125/10^1 *
(S − 35)^2

10^6 * a_1 = 12.178 − 0.10502/10^1 * P + 0.438/10^6 * P^2 + 0.23596/10^8
* P^3 + (−0.13336 + 0.1437/10^3 * P − 0.556/10^7 * P^2) * (S − 35)

10^6 * a_2 = − 0.133445 + 0.4669/10^4 * P + 0.35098/10^6 * P^2 − 0.29588/
10^9 * P^3 + (0.299/10^2 − 0.28264/10^5 * P + 0.139/10^8 * P^2) * (S − 35)

10^6 * a_3 = 0.10907/10^2 + 0.211/10^5 * P − 0.10289/10^7 * P^2 + 0.7556/
10^11 * P^3 − 0.3463/10^4 * (S − 35)
```

Chen and Millero (Chen and Millero 1976)

```
"Chen and Millero For Salinity in g/kg"
T = 25
S = 35
P_mpa = 1
P = P_mpa * 10 "bar"
```

```
alpha = alpha_W + A * S + B * S^1.5

alpha_W = – 6.80977/10^5 + 1.82945/10^5 * T – 3.2519/10^7 * T^2 + 6.4062/
10^9 * T^3 – 8.8997/10^11 * T^4 + 5.846/10^13 * T^5 + (3.7409/10^7 –
1.5881/10^8 * T + 1.741/10^14 * T^4) * P + (–9.6658/10^11 + 4.8741/10^12
* T – 1.10169/10^13 * T^2 + 9.941/10^16 * T^3 ) * P^2 + 4.75/10^16 * P^3

A = 4.04385/10^6 – 1.3990/10^7 * T + 1.727/10^9 * T^2 – 1.944/10^11 *
T^3 + (–3.2190/10^9 + 1.0353/10^10 * T – 8.365/10^13 * T^2 ) * P + (7.183/
10^13 – 1.636/10^14 * T) * P^2
B = – 1.058/10^7 + 1.308/10^9 * T + 2.125/10^11 * T^2 + 1.69/10^12 * T^3
+ (8.446/10^11 – 1.88/10^12 * T) * P
```

Pure Water Correlation (El-Dessouky and Ettouney)

```
T = 25 "C"

"Constansts"
f[1] = – 2.781015567
f[2] = 0.002543267
f[3] = 9.845047/10^6
f[4] = 3.636115/10^9
f[5] = –5.358938/10^11
f[6] = 7.019341/10^14

T_c = 647.286 "K"
V_c = 0.003172222 "kPa"

V = V_c * (T_c/(T + 273.15) – 1) * exp(sum(f[i] * (T + 273.15)^(i – 1), i =
1, 6))

T2 = T+ 1

V_2 = V_c * (T_c/(T2 + 273.15) – 1) * exp(sum(f[i] * (T2 + 273.15)^(i – 1),
i = 1, 6))

dV = V_2 – V

Alpha = dV/V
```

References

Bradshaw A, Schleicher KE (1970) Direct measurement of thermal expansion of sea water under pressure. Deep-Sea Res Oceanogr Abs 17(4). https://doi.org/10.1016/0011-7471(70)90035-5

Chen CT, Millero FJ (1976) The specific volume of seawater at high pressures. Deep-Sea Res Oceanogr Abs 23(7):595–612. https://doi.org/10.1016/0011-7471(76)90003-6

El-Dessouky HT, Ettouney HM (2002) Fundamentals of salt water desalination. Fundam Salt Water Desalin. https://doi.org/10.1016/b978-0-444-50810-2.x5000-3

Laliberté M, Cooper WE (2004) Model for calculating the density of aqueous electrolyte solutinos. J Chem Eng Data 49(5):1141–1151. https://doi.org/10.1021/je0498659

Meersman F, Nies E, Heremans K (2008) On the thermal expansion of water and the phase behavior of macromolecules in aqueous solution. Zeitschrift Fur Naturforschung—Sect B J Chem Sci 63(6):785–790

Nayar KG, Panchanathan D, McKinley GH, Lienhard VJH (2014) Surface tension of seawater. J Phys Chem Ref Data 43(4). https://doi.org/10.1063/1.4899037

Pandey JD, Tripathi SB, Sanguri V (2002) Thermal expansivity of multicomponent liquid systems. J Mol Liq 100(2):153–161. https://doi.org/10.1016/S0167-7322(02)00016-8

Qasem NAA, Generous MM, Qureshi BA, Zubair SM (2021) A comprehensive review of saline water correlations and data: part ii—thermophysical properties. Arab J Sci Eng 46(3):1941–1979. https://doi.org/10.1007/s13369-020-05020-5

Chapter 16
Freezing Point

Abstract The freezing point is a thermodynamic property representing the temperature at which the substance changes from liquid to solid. It is equal to the melting point that happens with the change of the solid phase to the liquid phase at the same temperature. This chapter discusses the correlation and data used to estimate the freezing point at different salinity and pressure values. In addition, the calculated freezing point values from the correlations are compared and assessed against the experimental data of seawater. This chapter suggests some correlations based on their accuracy and operating conditions. Moreover, the codes used for the estimation of the investigated correlations are supplied.

Keywords Freezing point · Boiling point · Thermodynamics · Correlations · Cooling · Saline water

16.1 Introduction

The freezing point is the temperature at which the substance phase changes from liquid to solid. Oppositely, the melting point happens with the change of the solid phase to the liquid phase at the same temperature. For example, the freezing point of pure water is at 0.002519 °C at atmospheric pressure (101,325 Pa) (Liangjun et al. 2019). However, the situation is different for the saline water. During the process of cooling, the water's density increases and initiates vertical convection that leads to mixing. As the process proceeds further, the ice begins to precipitate at the top surface of the water, reducing the quantity of brine solution. This phenomenon further increases the salinity of the available saline solution and, therefore, further decreases the freezing point of saline water (Liangjun et al. 2019; Millero 2010).

Strictly speaking, the freezing point of the saline water decreases with an increase in pressure and salinity. More broadly, the freezing point is a nonlinear function of salinity and a linear function of pressure (Millero 1978). The objective of this chapter is to provide available correlations to estimate the freezing point of saline water.

N. A. A. Qasem et al., *Thermodynamic and Thermophysical Properties of Saline Water*, Springer Water, https://doi.org/10.1007/978-3-031-35193-8_16

16.2 Correlations for Seawater

Most probably the best equation is the UNESCO-approved equation of the estimation of the freezing point provided by Millero (1978) and Millero and Leung (1976),

$$T_f = a_o S + a_1 S^{1.5} + a_2 S^2 + bP \tag{16.1}$$

where S is the salinity of the seawater (in g/kg) and P is the gauge pressure (in MPa), $a_o = -0.0575$ °C psu^{-1}, $a_1 = 1.710523 \times 10^{-3}$ °C psu$^{-1.5}$, $a_2 = -2.154996 \times 10^{-4}$ °C psu^{-2}, $b = -7.53 \times 10^{-2}$ °C MPa^{-1} where psu is the practical salinity unit approximately equal to ppt. This correlation is valid in the pressure range of 0–5 MPa (gauge pressure).

Kundsen (Miyake 1939) provided a formula for the freezing point of seawater based on the density of seawater:

$$T_f = -0.00866 - 0.064633\sigma_r - 0.0001055\sigma_r^2 \tag{16.2}$$

where $\sigma_r = (\rho_r - 1) \times 1000$ and ρ_r is the ratio of the density of seawater at 0 °C to the density of seawater at 4 °C. Thomas and Thompson (Liangjun et al. 2019) provided another correlation of the estimation of the freezing point based on the chlorinity of the seawater,

$$T_f = -0.0966\,Cl - 0.00000520\,Cl^3 \tag{16.3}$$

where *Cl* is the chlorinity of seawater in grams of chlorine per kg of seawater; similarly, Miyake et al. (1939) found a linear relation between freezing point depression and chlorine in seawater as:

$$\Delta T_f = 0.102710\,Cl \tag{16.4}$$

Fujino et al. (1974) provided a correlation for the freezing point depression that is valid for a salinity range of up to 40 g/kg:

$$T_f = -0.036 - 0.0499\,S - 0.000112\,S^2 \tag{16.5}$$

Doherty and Kester (1974) proposed a correlation based on supercooling, equilibrium, and flow, which is valid at 3.8 to 40.2 g/kg and atmospheric pressure as:

$$T_f = -0.0137 - 0.051990\,S - 0.00007725\,S^2 \tag{16.6}$$

Jackett et al. (2006) proposed a correlation based on rational functions. They defined the freezing point as:

Table 16.1 Coefficients for polynomials in Eq. (16.7)

Terms	$P_n^T(S, P)$ coefficients	$P_d^T(S, P)$ coefficients
Constant	$2.5180516744541290 \times 10^{-3}$	1.0
S	$-5.8946669548576310 \times 10^{-2}$	
$S^{1.5}$	$2.4811422319110776 \times 10^{-3}$	
S^2	$-3.1930091631496098 \times 10^{-4}$	
$S^{2.5}$		$-4.3301568126998630 \times 10^{-7}$
S^4	$1.5637174143955485 \times 10^{-8}$	
P	$-7.4276961814810053 \times 10^{-4}$	$-1.9625518786831890 \times 10^{-6}$
P^2	$-1.4312216596227918 \times 10^{-8}$	$7.0588565064816584 \times 10^{-11}$

$$T_f(S, p) = \frac{P_n^T(S, P)}{P_d^T(S, P)} \tag{16.7}$$

$P_n^T(S, P)$ and $P_d^T(S, P)$ are the polynomials those are defined in Table 16.1.

Similarly, Feistel (2008) used the concept of gibs function and chemical potential to provide a relation for depression in freezing point due to salt in air-saturated salt-water, as:

$$T_f = 2.519 + t_{90}\left(t_f^{DK74} - \frac{S_A}{70}\right) \tag{16.8}$$

where T_f is in mK, S_A is the salinity in g/kg, and t_{90} is the temperature according to the ITS-90. The value of t_f^{DK74} are those reported by Doherty and Kester (1974).

To facilitate the model simulations for high pressures, two methods were suggested: MITgcm (Adcroft et al. 2008) and ROMs (Galton-Fenzi 2009), as follows:

$$T_f = a_o S + a_1 + +bP \tag{16.9}$$

where S is the salinity of the seawater (in g/kg) and P is the gauge pressure (in MPa). For MITgcm, $a_o = -0.0573\,°\text{C psu}^{-1}$, $a_1 = 0.0901\,°\text{C}$, $b = -0.0761\,°\text{C MPa}^{-1}$. For ROMs, $a_o = -0.0573\,°\text{C psu}^{-1}$, $a_1 = 0.0832\,°\text{C}$, $b = -0.0761\,°\text{C MPa}^{-1}$.

16.3 Case Study

There are various correlations reported in the literature for the estimation of the freezing point of seawater in various salinity ranges. However, the Millero method (1978), was approved by the United Nation Scientific and Cultural Organization (UNESCO) for a pressure range of 0 to 5 MPa. The results generated by this correlation are compared with those of Doherty and Kester (1974), Fujino et al. (1974),

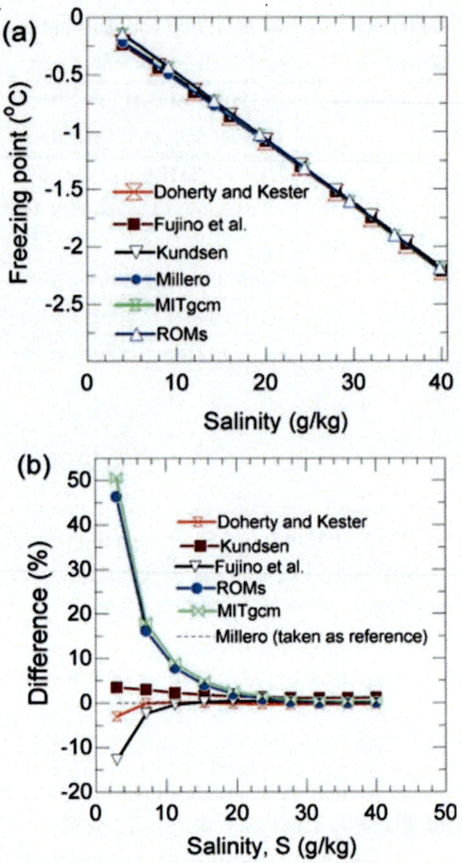

Fig. 16.1 **a** Salinity versus freezing point for Millero (1978), Doherty and Kester (1974), Fujino et al. (1974), Kundsen (Miyake 1939), MITgcm (Adcroft et al. 2008), and ROMs (Galton-Fenzi 2009); and **b** the percentage difference between Millero method and others for low pressure, 0.1 MPa(abs)

Kundsen (Miyake 1939), MITgcm (Adcroft et al. 2008), and ROMs (Galton-Fenzi 2009). The results of all correlations are almost identical, with higher values at low salinities for MITgcm (Adcroft et al. 2008) and ROMs (Galton-Fenzi 2009), as shown in Fig. 16.1a. It is an excellent indication that all correlations could be used for high salinity, while ROMs, MITgcm, and Fujino methods are not recommended for low salinities at atmospheric pressure (Fig. 16.1b). However, due to the need for high pressure in some situations, such as Amery Ice Shelf (25 MPa) and Lake Vostok (not seawater, 35 MPa) in the Antarctic (Siegert et al. 2001), MITgcm, ROMs, and Millero methods are recommended, as shown in Fig. 16.2. More specifically, at this range of pressure (< 5 MPa), Millero method is strongly recommended, while MITgcm and ROMs methods are recommended for higher pressure (5–35 MPa) (Liangjun et al. 2019).

Based on the above discussion, some methods and correlations could be recommended to be used, as summarized in Table 16.1.

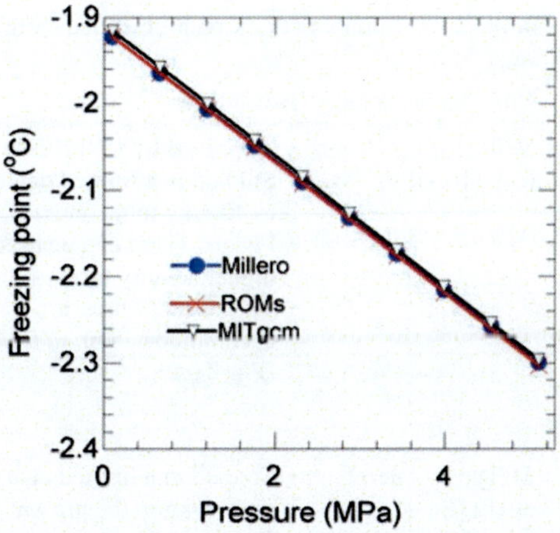

Fig. 16.2 Effect of pressure on the freezing point using Millero (1978), MITgcm (Adcroft et al. 2008), and ROMs (Galton-Fenzi 2009) at a salinity of 35 g/kg

Some criteria should be considered to select a proper correlation to estimate the freezing point, such as the type of saline water, operating limits (salinity and pressure), and fitting accuracy. Some correlations are valid for a particular application, whereas others are not. Table 16.2 proposes the recommended correlations and methods to be used to calculate the freezing point of saline water.

16.4 EES Codes

The used EES codes for the above correlations are:

Millero (1978)

```
"Millero and Leung (1978) correlation for freezing point estimation"

"Given"
S = 35 [g/kg]
P = 0.1 [MPa] // (absolute)

a_o = – 0.0575
a_1 = 0.001710523
a_2 = – 0.0002154996
b = – 0.0753

T_f_Millero = a_o * S + a_1 * (S)^(3/2) + a_2 * (S)^2 + b * (P – 0.1)
```

Table. 16.2 Summary of the recommended correlations and models for the freezing point of saline water

Correlation-model	Advantages	Applications	Refs.
Millero in [Eq. (16.1)]	• Approved by UNESCO • Studied in a limited data range (up to 40 g/kg in the present study, Fig. 16.1) but can be used for higher salinity too • The effect of pressure is accommodated up to 5 MPa (Fig. 16.2)	• Research of Arctic and Antarctic Sea Ice • Thermal energy storage • Other solidification and melting applications	Millero (1978)
MITgcm (Adcroft et al.) [Eq. (16.9)]	• Studied in a limited data range (up to 40 g/kg in the present study, Fig. 16.1) but can be used for higher salinity too • It is recommended when the pressure is between 5 and 35 MPa (Liangjun et al. 2019)	• Research of Arctic and Antarctic Sea Ice • Thermal energy storage • Other solidification and melting applications	Adcroft et al. (2008)
ROMs (Galton-Fenzi) [Eq. (16.9)]	• Studied in a limited data range (up to 40 g/kg in the present study, Fig. 16.1) but can be used for higher salinity too • It is recommended when the pressure is between 5 and 35 MPa (Liangjun et al. 2019)	• Research of Arctic and Antarctic Sea Ice • Thermal energy storage • Other solidification and melting applications	Galton-Fenzi (2009)

Doherty and Kester (1974)

```
"Doherty and Kester (1974) correlation for freezing point estimation"

"Given"

S = 35 [g/kg]

T_f = -0 .0137 - 0.051990 * S - 0.00007725 * S^2
```

Kundsen Correlation

```
"Kundsen correlation for freezing point estimation"

"Given"
S = 35 [g/kg]
P = 0.1 [MPa]

T_f = = −0.00866 – 0.064633 * (sw_density(0, S, P)/1000 − 1) * 1000 –
0.0001055 * ((sw_density(0, S, P)/1000 − 1) * 1000)^2
```

Fujino Correlation

```
"Fujino correlation for freezing point estimation"

"Given"
S = 35 [g/kg]

T_f = = − 0. 036 − 0.0499 * S − 00.000112 * S^2
```

ROMs (Galton-Fenzi) Correlation

```
"ROMs correlation for freezing point estimation"

"Given"
S = 35 [g/kg]
P = 0.1 [MPa]//(absolute)

a0 = − 0.0573
b0 = 0.0832
c0 = − 0.0761

T_f_ROMS = a0 * S + b0 + c0 * (P − 0.1)
```

MITgcm (Adcroft et al.) Correlation

```
"MITgcm correlation for freezing point estimation"

"Given"
S = 35 [g/kg]
```

P = 0.1 [MPa]//(absolute)

a1 = − 0.0573
b1 = 0.0901
c1 = − 0.0761

T_f_MITgcm = a1 * S + b1 + c1 * (P − 0.1)

References

Adcroft A, Campin JM, Dutkiewicz S (2008) MITgcm user manual. Massachusetts Institute of Technology

Doherty BT, Kester DR (1974) Freezing point of seawater. J Mar Res 32(2):285–300

Feistel R (2008) A Gibbs function for seawater thermodynamics for − 6 to 80 °C and salinity up to 120 g kg-1. Deep Sea Res Part 1 Oceanogr Res Pap 55(12):1639–1671. https://doi.org/10.1016/j.dsr.2008.07.004

Fujino K, Lewis EL, Perkin RG (1974) Freezing point of seawater at pressures up to 100 bars. J Geophys Res 79(12):1792–1797. https://doi.org/10.1029/JC079i012p01792

Galton-Fenzi BK (2009) Modelling ice-shelf/ocean interaction. University of Tasmania, Australia, Hobart

Jackett DR, McDougall TJ, Feistel R, Wright DG, Griffies SM (2006) Algorithms for density, potential temperature, conservative temperature, and the freezing temperature of seawater. J Atmos Oceanic Tech 23(12):1709–1728. https://doi.org/10.1175/JTECH1946.1

Liangjun YAN, Zhaomin W, Chengyan LIU, Yang WU (2019) On the determination and simulation of seawater freezing point temperature under high pressure. December. https://doi.org/10.13679/j.advps.2019.0018

Millero FJ (1978) Freezing point of seawater. Annex 6, Eight Rep Joint Panel Oceanogr Tables Stan (JPOTS). UNESCO Technical Papers in Marine Sciences 28:29–35

Millero FJ (2010) History of the equation of state of seawater. Oceanography 23(3):18–33. https://doi.org/10.5670/oceanog.2010.21

Millero FJ, Leung WH (1976) Thermodynamics of seawater at one atmosphere. Am J Sci 276(9):1035–1077. https://doi.org/10.2475/ajs.276.9.1035

Miyake Y (1939) Chemical studies of the western pacific ocean. III. freezing point, osmotic pressure, boiling point, and vapour pressure of sea water. Bull Chem Soc Japan, 14(3):58–62. https://doi.org/10.1246/bcsj.14.58

Siegert MJ, Ellis-Evans JC, Tranter M, Mayer C, Petit JR, Salamatin A, Priscu JC (2001) Physical, chemical and biological processes in Lake Vostok and other Antarctic subglacial lakes. Nature 414(6864):603–609. https://doi.org/10.1038/414603a

Chapter 17
Conclusion

Keywords Conclusion · Recommendations · Saline water properties

17.1 Concluding Remarks

The thermodynamic properties of saline water, such as its boiling point, vapor pressure, and enthalpy of vaporization, determine how much energy is required to separate the salt from the water. The higher the thermodynamic properties of the saline water, the more energy is needed to desalinate it. This means that understanding and controlling these thermodynamic properties can help reduce the cost and energy requirements for desalination processes. The thermophysical properties of saline water are important for desalination and relevant applications because they determine the efficiency of the process. For example, the thermal conductivity of saline water affects the rate at which heat can be transferred to the water, which in turn affects how quickly it can be heated and evaporated. The viscosity of saline water also affects how easily it can be pumped through a desalination system. Knowing the thermodynamic and thermophysical properties helps engineers design more efficient desalination systems.

This book compiles and discusses the correlations and data of thermodynamic and thermophysical properties of saline water. The thermodynamic properties involve solution concentration, density, specific heat capacity, latent heat of vaporization, boiling point elevation, vapor pressure, specific enthalpy, specific entropy, and freezing point, while the thermophysical properties include viscosity, surface tension, electrical conductivity, thermal conductivity, osmotic coefficient, activity coefficient, and thermal expansivity. The vital key parameters such as multicomponent (ions) concentration, salinity, temperature, and pressure are considered to target all saline water types such as brackish water, seawater, high saline water from basins, lakes, and produced water from oil and gas hydraulic fracturing.

N. A. A. Qasem et al., *Thermodynamic and Thermophysical Properties of Saline Water*, Springer Water, https://doi.org/10.1007/978-3-031-35193-8_17

To get a property value of saline water, the approaches suggested are thermodynamic relations, multicomponent models, and correlations. The correlations are the easiest ways to obtain properties with relatively low required inputs, such as salinity, temperature, and pressure. Most of the correlations were proposed for seawater. Multicomponent models are general for all types of saltwater under different salinity values but with more detailed inputs. Notably, the multicomponent models are valid to obtain solution properties under any number of constituents, while the binary models mainly account for solutions that have Na and Cl as the major compositions (e.g., seawater).

The accuracy of models and correlations for saline water properties is also limited by the availability of data. Many properties lack detailed measurements of salinity, temperature, pressure, and other factors that affect saline water properties, making it difficult to develop accurate models or correlations for these properties.

The presented correlations and methods are discussed based on the operating limitations and accuracy. A comparison between the properties data obtained from the correlations with the available experimental data (which are mostly for seawater) is exhibited in plots and discussed for each property. In addition, a guideline for selecting some correlations is provided and tabulated.

17.2 Recommendations

Based on the material provided by this book, it is also important to point to some recommendations as follows:

- Most of the experimental data are available for NaCl solution; thus, many correlations suggested are valid only for seawater. Massive experimental work is still needed to evaluate the properties of different multicomponent solutions and to obtain general correlations for all saline water types (brackish water, seawater, and sources of high salinity).
- Most of the correlations are functions of salinity and, secondly, temperature; therefore, the pressure-dependent properties are essential to be suggested, especially for high-pressure-driven applications, such as reverse osmosis desalination.
- The suggested correlations for the latent heat of vaporization are constrained to a salinity of 120 g/kg. Further studies are required to propose more correlations for high salinities up to 200 g/kg.
- The activity coefficient and osmotic coefficient are presented by limited correlations; they can be further investigated under different temperatures and pressure values; thus, the corresponding correlations can be obtained.
- The Pitzer model is the most accurate multicomponent approach for some properties (such as specific enthalpy, specific entropy, activity coefficient, and osmotic coefficient) but with a cumbersome calculation procedure. More straightforward, accurate correlations as functions of salinity and temperature are highly required to simplify the Pitzer model.

- Saline water also contains many dissolved substances that can affect its properties, such as nutrients, metals, and organic compounds. These substances can vary widely in concentration and composition depending on the source of the water, making it difficult to accurately predict properties with a single model or correlation. Thus, models consider these dissolved substances are needed.